Innovations in Cybersecurity Education

Kevin Daimi • Guillermo Francia III
Editors

Innovations in Cybersecurity Education

Springer

Editors
Kevin Daimi
University of Detroit Mercy
Detroit, MI, USA

Guillermo Francia III
University of West Florida
Pensacola, FL, USA

ISBN 978-3-030-50243-0 ISBN 978-3-030-50244-7 (eBook)
https://doi.org/10.1007/978-3-030-50244-7

This Springer imprint is published by the registered company Springer Nature Switzerland AG.
The registered company address is: Gewerbestrasse 11, 6330 Cham, Switzerland

Preface

Cyberinfrastructures across the globe continue to evolve in increasing complexity. Coupled with this progress are the threats that exploit their vulnerabilities. It is therefore incumbent on every public or private entity, who depend on these systems, to secure and protect them. Recognizing this need, the National Initiative for Cybersecurity Education (NICE) embarked on a mission to build on existing programs to increase the number of skilled cybersecurity professionals through change and innovation in cybersecurity education and workforce development. Indeed, cybersecurity education is at a crossroads.

Recent and multiple media reports have documented the fact that hundreds of thousands of cybersecurity job positions remain unfilled. Demand simply overwhelms supply in the cybersecurity job market. The pressure to produce as much qualified and skilled workforce as possible in cybersecurity has never been great. In times of great need, the ability to innovate arises out of human nature. The cliché that states "in desperate times, ordinary people produce extraordinary results" appears to never fail to emerge. Members of academia across the globe answered the challenge by introducing advancements in cybersecurity education.

Innovations in Cybersecurity Education offers a compendium of works in diverse topics including curriculum development, professional and faculty development, community outreach, laboratory improvement, and student learning. The reader should expect to find engaging pedagogical tools and methods through puzzles and games, a realistic simulation framework for anomaly detection, secure software development, Blockchain technology, information visualization, scenario-based learning, E-governance systems, Internet of Things, and various laboratory enhancements for the cybersecurity curriculum. In general, it illustrates both novel and proven concepts, techniques, methods, approaches, and trends in cybersecurity education that can be adopted by cybersecurity specialists and educators for the benefit of the future workforce. Furthermore, it provides a glimpse of future directions where cybersecurity techniques, policies, applications, and theories are headed. The book is a rich collection of carefully selected and reviewed manuscripts written by multinational cybersecurity educators and professionals and edited by prominent cybersecurity researchers and specialists.

Detroit, MI, USA Kevin Daimi
Pensacola, FL, USA Guillermo Francia III

Acknowledgments

The Innovations in Cybersecurity Education book could not have been published without the cooperation and support of many people. We would like first to recognize all authors of the chapters of this book who contributed their knowledge and expertise in Cybersecurity Education to make this book a reality. We are also indebted to the hard work of all chapters' reviewers listed below, who invested extensive time in the review process. Finally, our gratitude to Mary James, Zoe Kennedy, Brian Halm, and Nandhakumar Sundar at Springer for their kindness, courtesy, and professionalism.

Mohammed Akour, Yarmouk University, Jordan
Abeer Alsadoon, Charles Sturt University, Australia
Allen Ashourian, ZRD Technology, USA
Souvik Bhattacharyya, University of Burdwan, India
Ajay Biswas, University of Burdwan, India
Mitchell Buchman, ManTech International, USA
Violeta Bulbenkiene, Klaipeda University, Lithuania
María Calle, Universidad del Norte, Barranquilla, Colombia
Kevin Daimi, University of Detroit Mercy, USA
Ioanna Dionysiou, University of Nicosia, Cyprus
Luis Hernandez Encinas, Institute of Physical and Information Technologies, Spain
Reinaldo Padilha França, University of Campinas, Brazil
Guillermo A. Francia III, University of West Florida, USA
Tirthankar Ghosh, University of West Florida, Pensacola, USA
Samuel Ndueso John, Nigerian Defence Academy, Nigeria
Irene Kopaliani, Princeton University, USA
Malathi Letchumanan, University of Putra Malaysia, Malaysia
Víctor Gayoso Martínez, Institute of Physical and Information Technologies, Spain
Wojciech Mazurczyk, Warsaw University of Technology, Poland
Vickie McLain, Lake Superior College, USA
Esmiralda Moradian, Stockholm University, Sweden
Agustín Martín Muñoz, Institute of Physical and Information Technologies, Spain

Mais Nijim, Texas A&M University-Kingsville, USA
Saibal K Pal, Defense R&D Organization, India
Cathryn Peoples, The Open University, United Kingdom
Mustafa Saed, Hyundai-Kia America Technical Center, USA
Karpoor Shashidhar, Sam Houston State University, USA
Nicolas Sklavos, University of Patras, Greece
Cristina Soviany, Features Analytics SA, Belgium
Suzanne Mello-Stark, Rhode Island College, USA
Narasimha Rao Vajjhala, American University of Nigeria, Nigeria
Zheng Yan, Aalto University, Finland
Jianhua Yang, Columbus State University, USA
Andrew D. Wolfe, Jr., Loyola University New Orleans, USA
David Zeichick, California State University Chico, USA

Contents

About the Editors

Kevin Daimi received his PhD from the University of Cranfield, England. He has a long academic and industry experience. His research interests include Computer and Network Security with emphasis on vehicle network security, Software Engineering, Data Science, and Computer Science and Software Engineering Education. He has published a number of papers on vehicle security. He is the editor of Computer and Network Security Essentials book, which was published by Springer. He has been chairing the annual International Conference on Security and Management (SAM) since 2012. Kevin is a Senior Member of the Association for Computing Machinery (ACM), a Senior Member of the Institute of Electrical and Electronic Engineers (IEEE), and a Fellow of the British Computer Society (BCS). He is the recipient of the Outstanding Achievement Award from the 2010 World Congress in Computer Science, Computer Engineering, and Applied Computing (WORLDCOMP'10) in Recognition and Appreciation of his Leadership, Service and Research Contributions to the Field of Network Security. He is currently Professor Emeritus of Computer Science and Software Engineering at the University of Detroit Mercy.

Guillermo A. Francia III received his BS in Mechanical Engineering degree from Mapua Tech. His PhD in Computer Science is from New Mexico Tech. Before joining Jacksonville State University (JSU), he was the chairman of the Computer Science department at Kansas Wesleyan University. Dr. Francia is a recipient of numerous grants. His projects have been funded by prestigious institutions such as the National Science Foundation, Eisenhower Foundation, Department of Education, Department of Defense, and Microsoft Corporation. In 1996, Dr. Francia received one of the five national awards for Innovators in Higher Education from Microsoft Corporation. Dr. Francia served as a Fulbright scholar to Malta in 2007 and a Fulbright Cybersecurity research scholar to the United Kingdom in 2017. Before joining the Center for Cybersecurity at the University of West Florida (UWF), Dr. Francia served as the Director of the Center for Information Security and Assurance and held a Distinguished Professor position at JSU. Currently, he is Professor and Faculty Scholar at UWF.

Part I
Student Learning

Dynamic Difficulty Adjustment in Cybersecurity Awareness Games

Analysis of Player Behavior and the Potential of Electroencephalography

David Thornton and Falynn Turley

1 Introduction

This chapter describes a study of player behavior and electroencephalography (EEG) headset readings while playing a cybersecurity educational video game. While difficulty was progressively increased, player actions and EEG readings were recorded, along with a pretest and posttest of student knowledge and opinions regarding information security awareness and perceived immersion. This study employed Brute Force, a tower defense game that teaches players to choose strong, unique, and memorable passwords. Participants reported significantly more responsible attitudes regarding the importance of strong, unique passwords. More successful players who played the full 15 min tended to improve at identifying strong passwords from a list. After playing the game, participants were most likely to add password length and uniqueness as important password strategies.

This study was the first in a series, intended to examine differences in terms of learning outcomes and related metrics between various dynamic difficulty adjustment (DDA) approaches using EEG biofeedback with inexpensive headsets.

2 Dynamic Difficulty Adjustment

Dynamic difficulty adjustment, sometimes called dynamic game balancing, is an attempt to ensure that a game's difficulty is customized to every player's ability. One might attempt to adjust the difficulty based on a player's reported competence,

D. Thornton (✉) · F. Turley
Jacksonville State University, Jacksonville, AL, USA
e-mail: thornton@jsu.edu; fturley@jsu.edu

K. Daimi, G. Francia III (eds.), *Innovations in Cybersecurity Education*,
https://doi.org/10.1007/978-3-030-50244-7_1

but this often has little correlation to true ability [1]. Further, a player's ability is constantly changing, especially at the novice level. One might also diagnose that a game is too hard or easy for the player by simply monitoring in-game performance; however, newer approaches aimed at sensing the player's mental state have the potential for improved accuracy and responsiveness. Further, the recent technological advances in low-cost EEG headsets enable such applications outside of the classic laboratory setting.

The use of serious (i.e., nonentertainment) games offer great promise for education. According to Francia [2]:

> Digital games can develop cognitive, spatial, and motor skills. Teachers can use games to emphasize facts, principles, and complex problem solving. Games can also be used to increase creativity or to provide practical examples of concepts and rules that may be difficult to illustrate in the real world. Teachers can make use of games to perform experiments that could be dangerous when performed in real life, such as experiments that use hazardous materials and equipment. While games are often not explicitly educational, they possess intrinsic qualities that challenge learners' cognitive abilities. Playing well-designed games has the potential to increase the time students spend learning, increase difficulty along with their ability, and allow them to fail without fear.

Game-based learning, when well designed, can appeal to a broad audience and provide engaging experiences. Merrilea J. Mayo, former Director of Government—University—Industry Research Roundtable at the National Academies, went so far as to say: "While there will never be a silver bullet for science and engineering education, video games have the potential to be, perhaps, a bronze bullet" [3].

Without thoughtful pedagogical components, however, they may serve as little more than entertainment. Additionally, the best educational games will continually match the challenge to the growing competency of the learner [4].

Such qualities are especially important when the education need is great, as it is in cybersecurity information awareness. A recent Pew Research report [5] found that 25% of Americans could not correctly identify the most secure password from a list, and nearly half could not distinguish the phishing scam message among a list of legitimate messages.

To help address this need for greater cybersecurity awareness, the two educational cybersecurity games shown in Figs. 1 and 2 were developed by the primary author as part of NSF grant #H98230-12-1-0427 and were used in recent NSF/DoD-funded GenCyber teacher camps. Figure 1 illustrates the Brute Force game, a tower defense game that teaches players to choose strong, unique, and memorable passwords. Figure 2 depicts the Space Scams game, a wave shooting game that teaches players to identify phishing scams.

In the current version of each of these games, the difficulty steadily increases until a player eventually loses. Dynamic difficulty adjustment, in contrast, attempts to keep the player in a continual state of what linguistics and psychology of education researcher James Paul Gee [6] would call "pleasant frustration," or professor of psychology Mihaly Csikszentmihalyi's concept of "Flow" (depicted in Fig. 3), a state of complete absorption with the current activity. To achieve this, a player must be involved with a challenge that is "just about manageable" [7].

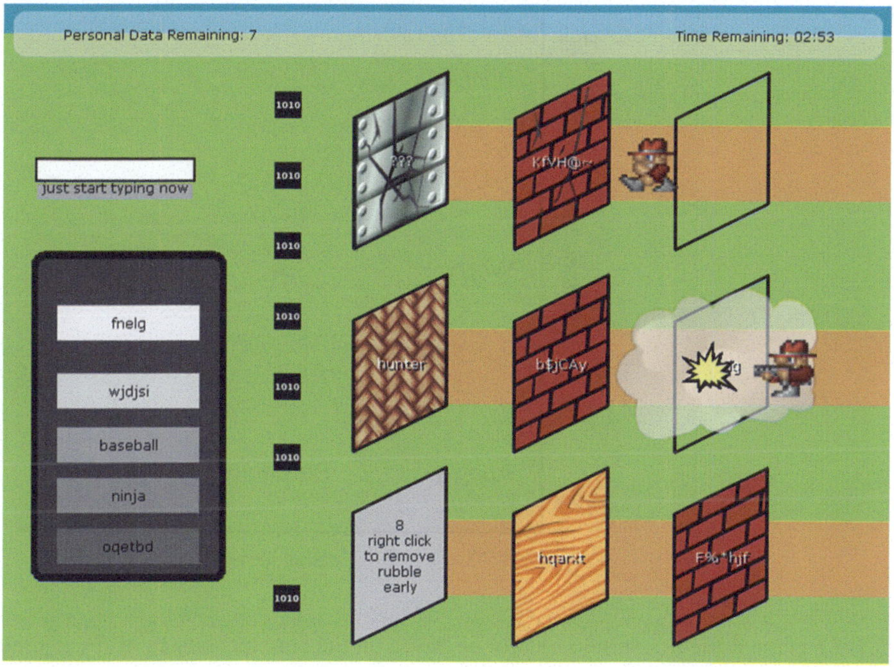

Fig. 1 Brute Force game

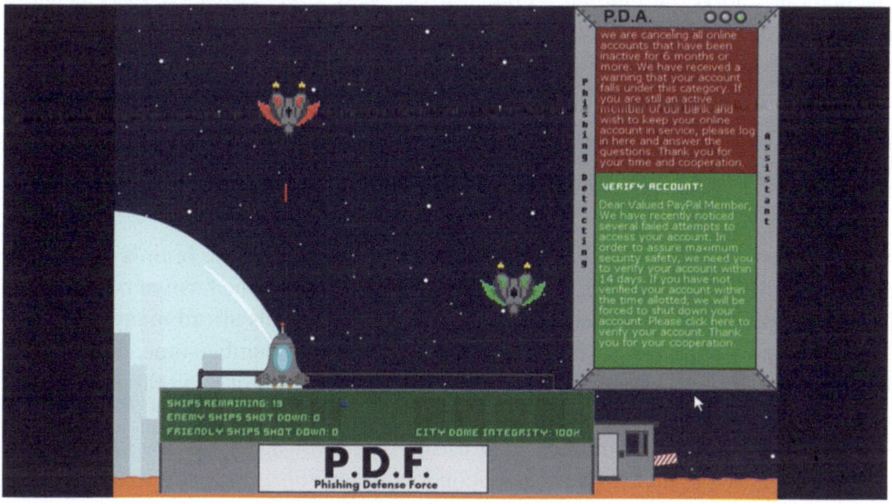

Fig. 2 Space Scams

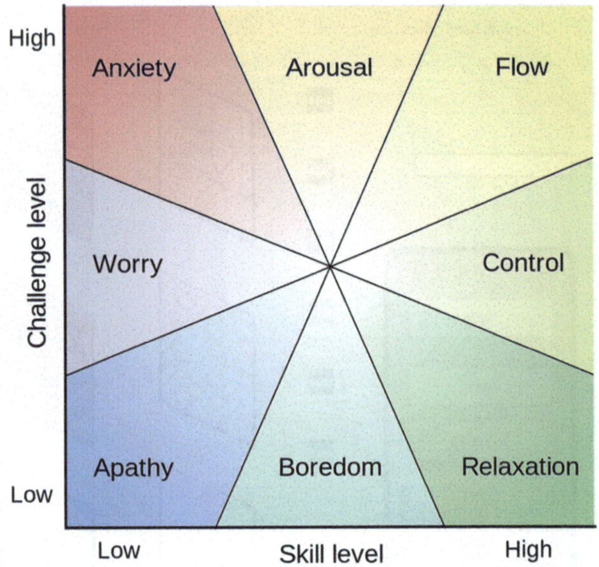

Fig. 3 An illustration of player experience (source: https://en.wikipedia.org/wiki/File:Challenge_vs_skill.svg#filehistory)

Scaling game difficulty should ideally be based on the core mechanics of the game, while also being invisible, continuous, and responsive [8]. That is, it ought to occur without the player's awareness, at all times, and without delay.

Determining a player's perception of difficulty is often achieved by a set of calculations called the *challenge function* [9]. This study collected EEG headset readings in order to help identify which readings would serve as the most indicative of a player's perception of difficulty.

3 Electroencephalography

Electroencephalography, or EEG for short, is the reading of the brain's electronic signals via electronic instrumentation [10]. Mostly this takes the form of noninvasive sensors worn as a headset. While the more advanced models afford many sampling points and greater accuracy, they are often heavier, more cumbersome, and require longer preparation before use. Moreover, a price tag upward of $25,000 makes such devices unfeasible for non-research environments.

In contrast, lower-end models such as the Emotiv Insight [11] are lightweight, cost less than $400, and deliver sub-second readings concerning the wearer's state of mind, including focus, engagement, stress, excitement, interest, and relaxation. As such, they make possible applications for EEG which have been heretofore out of reach, including personalized learning based not merely on the learner's *performance*, which may be laggy and coarse, but also on their affective factors. Figure 4 shows raw readings from a playthrough of Brute Force.

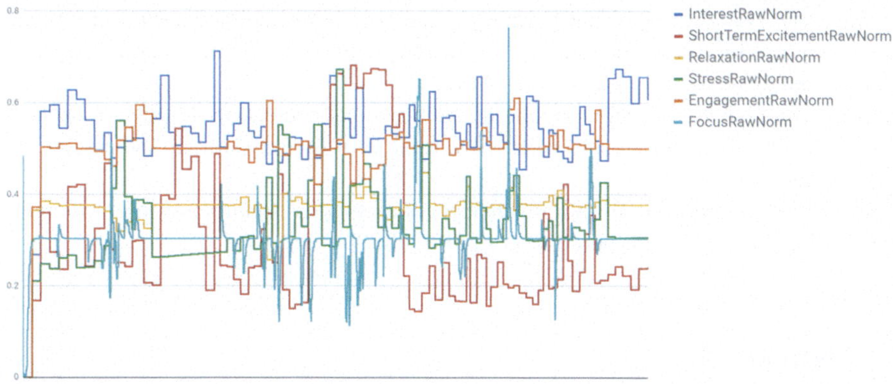

Fig. 4 EEG readings from an experiment participant

Fig. 5 Lightweight five-sensor EEG headset fitted on wearer (Permission to use graphics granted by Emotiv, Inc.; source: https://www.emotiv.com)

There are some practical considerations when using such a headset. Firstly, the sensors must make direct contact with the scalp. Secondly, results are better when a conductive solution is used to moisten the sensors. These two factors might make some experimental subjects unwilling or unable to participate. Otherwise, fitting and optimizing connection with a new wearer takes only about 2 min. Figure 5 shows the correct fit of a five-sensor EEG headset.

4 Research Questions

This study is the first in a series, intended to examine differences between various DDA approaches using EEG biofeedback. As such, it provides a baseline of learning outcomes and player experience which will be compared to the EEG-driven DDA in the next phase of the study.

Research Question 1: What learning outcomes (both knowledge and opinions) are produced by linear progressive difficulty adjustment in the context of the educational cybersecurity game Brute Force?

Research Question 2: How much engagement and immersion do players experience with the linear progressive difficulty version of the educational cybersecurity game Brute Force?

Research Question 3: Which electroencephalography signals are the most reliably indicative of players' perception of difficulty?

5 Methods

At the beginning of a session, participants read a description of the experiment and provided basic demographic information. Next, they took a brief pretest regarding their knowledge and attitudes toward password practices. After seating the EEG headset, players were led through a brief game tutorial, then they were made to play the Brute Force game with linear progressive difficulty adjustment for up to 15 min. This was followed by a posttest to measure any changes from the pretest scores, along with a questionnaire regarding player experience. Figure 6 depicts the user interface of the experiment dashboard.

The instruments used to collect measures are illustrated in Table 1.

Throughout the game, players defended themselves from "hackers" by selecting passwords from a randomized list, and by periodically creating their own custom passwords. The strength of these passwords was evaluated using the open-source library, zxcvbn [12] developed by Dropbox.

In order to gauge players' level of engagement, the authors employed the Game Engagement Questionnaire developed by Fox and Brockmeyer [13], while learning outcomes were measured by the relevant section of the Human Aspects of Information Security Questionnaire developed by McCormac et al. [14].

As discussed, this study employed the Emotiv Insight, a low-cost, multichannel, wireless, consumer-grade EEG headset [15]. While higher-end headsets would provide more accurate readings, the authors wanted to investigate the feasibility of using such hardware in the secondary and post-secondary educational setting, where budgets are often quite limited.

In order to test participants' knowledge about passwords, the pre-survey asked the following questions:

- What are some qualities of good passwords?
- What are some strategies for creating good passwords?
- Write three good passwords below.

The post-survey asked the same first two questions, and asked participants to try to remember the passwords supplied in the pre-survey.

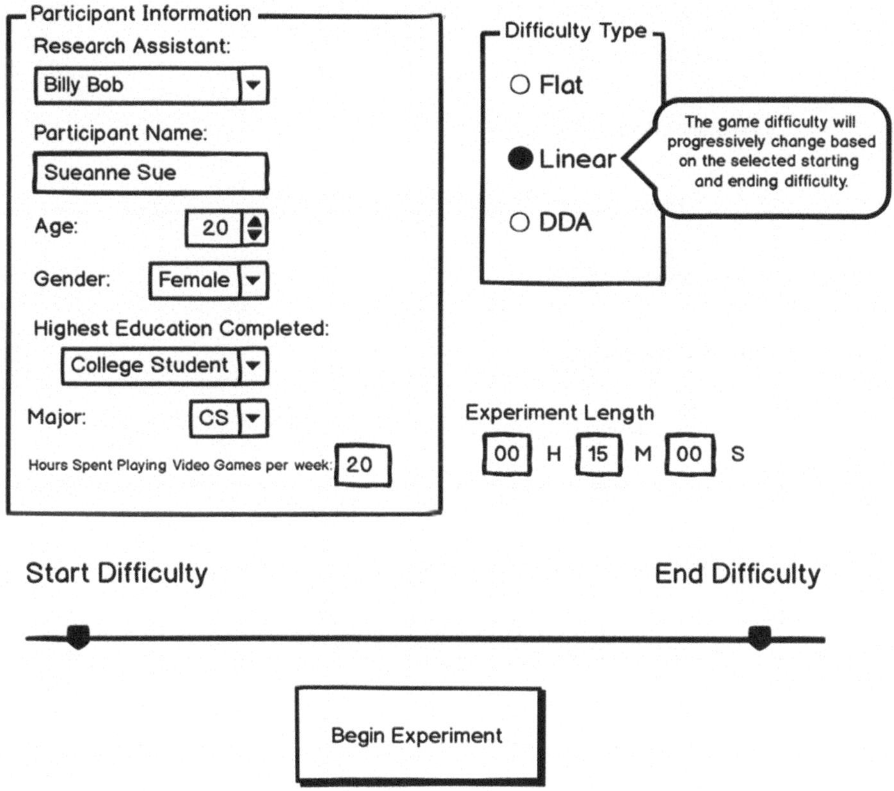

Fig. 6 DDA experiment dashboard screenshot

Table 1 Metrics and their corresponding instruments

Metric	Instrument
Created and selected passwords, play time	In-game analytics
Engagement score	Game Engagement Questionnaire
Password security attitudes	Human Aspects of Information Security Questionnaire (HAIS-Q)
Electroencephalography readings	Emotiv Insight headset
Password Strategy Knowledge	Pretest and posttest
Basic demographics	Survey

6 Population

Participants included 28 college students, with an average age of 23.68 (SD = 6.06) and a female-to-male ratio of 16:12. On average, participants reported spending approximately 2.04 h per week playing video games on their own time. There was a wide variety of majors represented with most participants from the School of

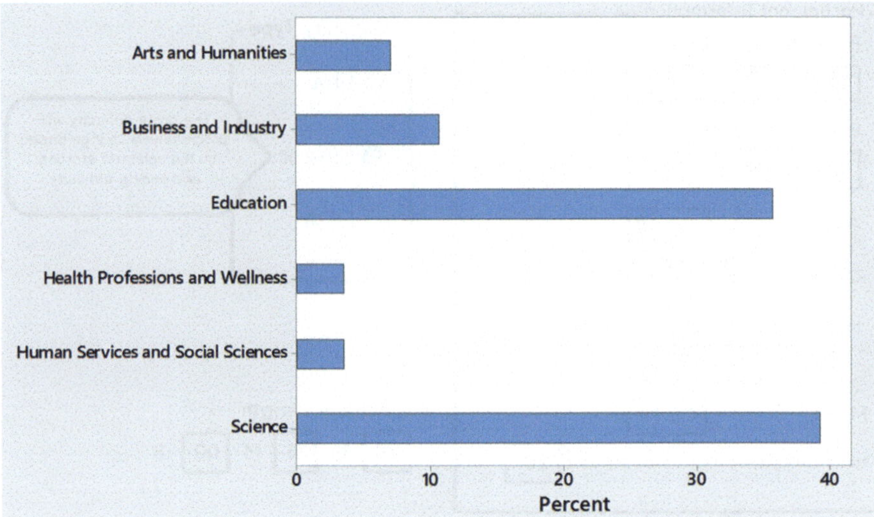

Fig. 7 Participant population by educational school

Education (35.7%) and the School of Science (39.3%). The breakdown by school is shown in Fig. 7.

7 Difficulty Scaling

As aforementioned, difficulty scaling is highly game dependent. Below, the game mechanics for Brute Force and the corresponding difficulty factor formulas are discussed.

In the lane-defense strategy game, Brute Force, players must defend their personal data from hackers by erecting password "walls" whose material is based on the password's strength. For instance, the weakest passwords are composed of straw or wood, while stronger passwords are bricks or steel. Walls can be created by selecting a password from a randomly generated list, or by creating custom passwords. An early concept sketch and prototype of Brute Force are shown in Fig. 8.

Players are encouraged to choose or create unique passwords, because a hacker who cracks one password automatically destroys all matching passwords instantly. Lastly, while custom passwords are designed to be the strongest type (and thus encouraged), players must remember the password to remove "rubble" from cracked passwords (discouraging random keyboard strokes). Together, these mechanics are designed to encourage the identification and synthesis of strong, unique, and memorable passwords.

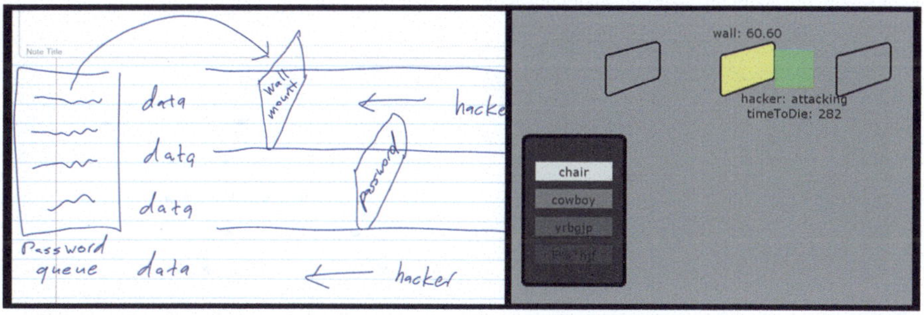

Fig. 8 Concept and playable prototype of Brute Force cybersecurity game

During gameplay, selecting a poor password prompted an instructional pop-up, such as "a strong password is at least 8 characters long" or "strong passwords contain combinations of letters, numbers, and symbols."

The following game features are dependent on the difficulty factor, which ranges from 0.1 (easiest) to 1.0 (most difficult):

- The hackers' movement speed increases with higher difficulty ($0.5 \times$ [difficulty] $+ 1$ pixel per frame)
- The hackers' cracking speed increases with higher difficulty ($0.6 \times$ [difficulty] $+ 0.4 \times$ (minutes elapsed)-0.3 health per frame)

A frame in this context corresponds to 1/30th of a second.

In this study, the game's difficulty began at 0.1, and climbed linearly to 1.0 over a period of 13 min.

8 Results

During the game, the amount of time spent by each participant playing was collected, along with the number and strength of custom passwords created by each participant. On average, students played the game for approximately 12.46 min and generated 13.2 custom passwords. Players were given an opportunity to create a new custom password every 45 s. This means that players spent an average of less than 15 s creating each custom password. The scatterplot in Fig. 9 shows the strong correlation that exists between the number of custom passwords and how long the participant is able to continue playing the game (Pearson's $r = 0.802$, $p < 0.001$).

While this finding is not unexpected, a closer look at password strength revealed that player password selection (i.e., choosing a password from a list) sometimes improves with longer play time. A summary of the Pearson value for selected and custom-created password strength is shown in Table 2.

These values would seem to indicate that over time, players were slightly more likely to choose worse passwords as the game became more difficult but create

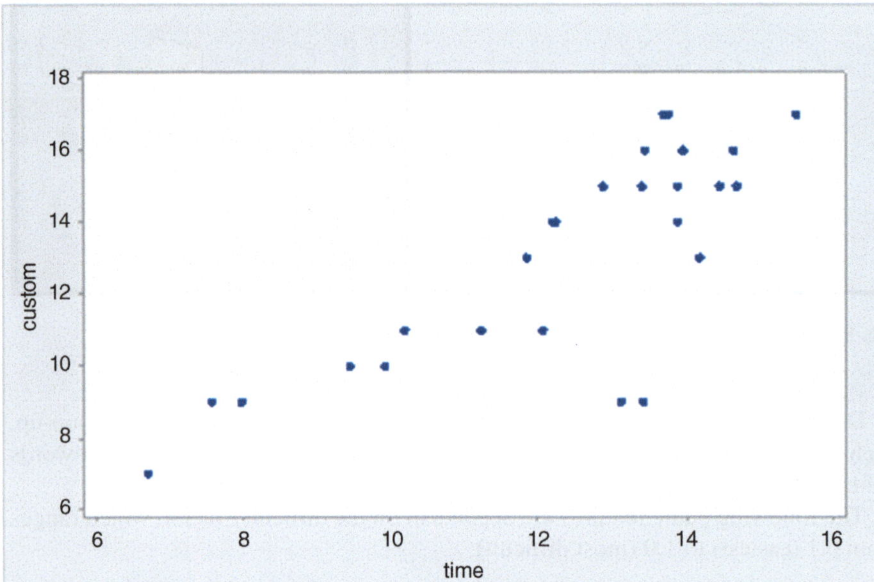

Fig. 9 Number of custom passwords versus time spent playing the game

Table 2 Pearson values for password strength for both selected and created passwords

Pearson value of password strength	Selected passwords	Created passwords
Positive correlation	11 participants	14 participants
Neutral	2 participants	1 participant
Negative correlation	15 participants	13 participants

better custom passwords. However, Fig. 10 shows that longer play time tends to yield a more positive Pearson correlation. This chart maps the Pearson correlation of selected password strength, measured by the aforementioned zxcvbn library, to the number of selected passwords. It indicates that more successful players who played the full 15 min tended to improve with password selection.

In contrast, when players were given periodic opportunities to *create* custom passwords, their performance did not improve as the game went on, despite the custom password prompt pausing the game's action. In other words, even though players were not rushed, and had the opportunity to create the strongest possible password, the strength of their custom-created passwords did not tend to improve as play time increased. Moreover, the variation in the Pearson correlation tightened toward the end of the graph, as shown in Fig. 11.

Two possible explanations for these results are that players who were most proficient had the least to learn about creating strong custom passwords, or that they leaned more heavily on the game mechanics of password selection for success.

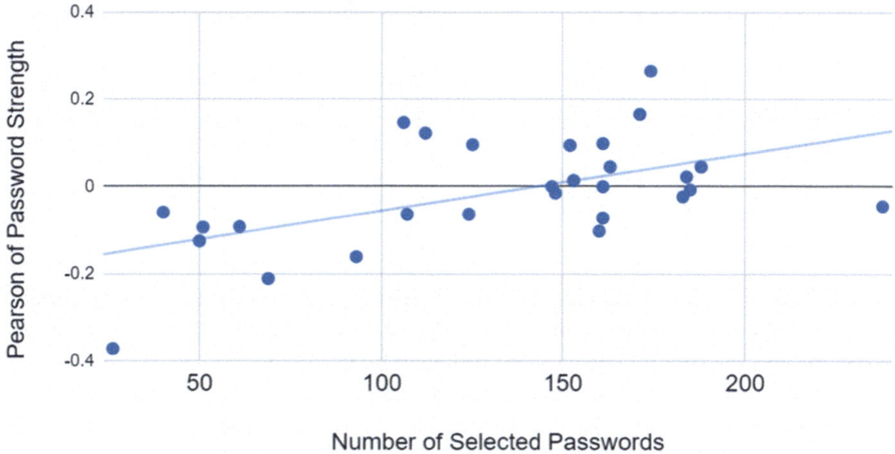

Fig. 10 Pearson r of password strength versus number of selected passwords

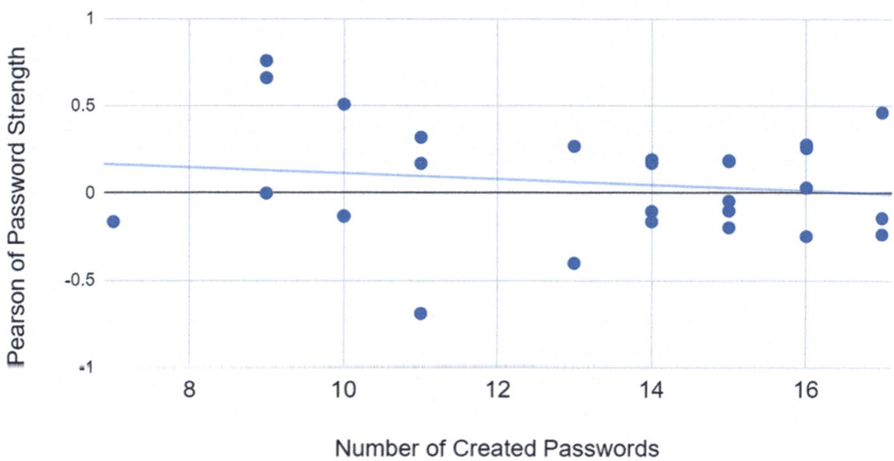

Fig. 11 Pearson r of password strength versus number of created passwords

A text analysis was conducted for the free response questions asked on the survey. Before completing the game, all 28 participants were asked to list three passwords he or she considered good. More than half of the respondents (57%) included one or more passwords that included a symbol (@, #, or!). After finishing the game, participants were asked to recall the passwords they had supplied about 15 min beforehand. Over half of participants (16 out of 28) were not able to recall any passwords.

Additionally, participants were asked what factors constituted a good password and to list password strategies before and after participating in the study. Before playing the game, 19 out of 28 (68%) indicated that good passwords are those which

Table 3 Mean scores for HAIS-Q excerpted questions regarding the importance of password strength and uniqueness

Question	Pre-score	Post-score
It is safe to use the same password	1.8 (1.1)	1.3 (0.7)
I use/will use different passwords	3.7 (1.5)	4.6 (0.7)
A safe password contains just letters	1.7 (0.8)	1.4 (0.7)
I use/will use combination passwords	4.5 (0.7)	4.7 (0.5)

include "letters, numbers, symbols." Only 20 participants were able to list additional qualities in the posttest. Out of those 20, the most common response to ensure a good password was password length (35%) followed by uniqueness.

Similarly, participants were also asked to list password strategies they relied on for creating good passwords. Before playing, most participants (54%) indicated creating a memorable/hard to guess password as a password strategy. However, after playing the game, this sentiment changed, as the most popular strategy (40% of respondents) involved choosing a unique/creative password.

The excerpted questions from the HAIS-Q measured participants' attitudes regarding the importance of strong, unique passwords. Responses values employed a 5-value Likert scale, from Strongly Disagree to Strongly Agree. Their responses were analyzed using paired t-test to detect if any scores significantly changed after playing the game. A significance level of $\alpha = 0.10$ was selected. Table 3 lists the mean score (standard deviation) for each question.

There was a significant reduction in the average score for the statement: *It is safe to use the same password.* This suggests that the participants have a much stronger feeling of disagreement with this statement after playing the game ($t = 3.10$, $p = 0.001$).

There was a significant increase in the average score for the statement: *I use/will use different passwords.* This suggests that participants tended to have a stronger feeling of agreement with this statement after playing the game ($t = -3.95$, $p = 0.001$).

There was a significant reduction in the average score for the statement: *A safe password contains just letters.* This suggests that the participants have a much stronger feeling of disagreement with this statement after playing the game ($t = 2.55, p = 0.017$).

There was a significant increase in the average score for the statement: *I use/will use combination passwords.* This suggests that participants tended to have a stronger feeling of agreement with this statement after playing the game ($t = -1.76$, $p = 0.090$).

While all of the above factors moved significantly in the desired direction, it would be naive to suppose that participants' habits will be greatly changed in the long run. Still, they provide positive results and a baseline with which to compare subsequent experiments.

After playing the game, participants were asked about their experience using questions from the Game Engagement Questionnaire. Responses values employed

Table 4 Mean scores and standard deviation for Game Engagement Questionnaire responses

Question	Mean	SD
My thoughts go fast	4.286	0.81
I really get into the game	4.036	0.793
I lose track of time	4.0357	0.5079
Things seem to happen automatically	3.75	0.967
Playing seems automatic	3.75	1.076
I play without thinking about how to play	3.667	1.074
I feel different	3.357	1.062
I get wound up	3.357	1.339
I play longer than I meant to	3.214	1.166
Time seems to kind of standstill or stop	3.107	1.257
I feel like I just can't stop playing	3.036	1.401
I feel spaced out	2.964	1.401
Playing makes me feel calm	2.893	1.197
I can't tell that I'm getting tired	2.893	1.227
I lose track of where I am	2.821	1.188
I don't answer when someone talks to me	2.821	1.219
The game feels real	2.714	1.243
If someone talks to me, I don't hear them	2.607	1.166
I feel scared	1.821	1.09

a 5-value Likert scale, from Strongly Disagree to Strongly Agree. Participant responses are sorted by arithmetic mean in Table 4.

These baseline readings will help to determine the relation between learning outcomes and engagement with other DDA approaches with the same game.

Another goal of this study was to help identify which readings are most salient to player's perception of difficulty. Other authors, such as Afergan et al. [16] and Mikami et al. [17] successfully used attention biosensor data from a similar headset to effectuate DDA. However, when inspecting the readings from participants, stress appears to be a more promising attribute. While the amplitude seems less informative, the frequency of spikes appears to be fairly well correlated to perception of difficulty. Figure 12 shows a typical reading profile for participants playing the linear progressive difficulty version of Brute Force. As such, large spikes in stress may indicate opportune moments to subtly lower difficulty.

9 Future Work

The next phase of the study will investigate the efficacy of the stress EEG reading as the sole input to the challenge function, effectuating dynamic difficulty adjustment based on the player's biofeedback. With all other methods and instruments alike, will there be a difference in engagement, play time, and most importantly, learning outcomes? Provided the results are promising, the third phase of the study will

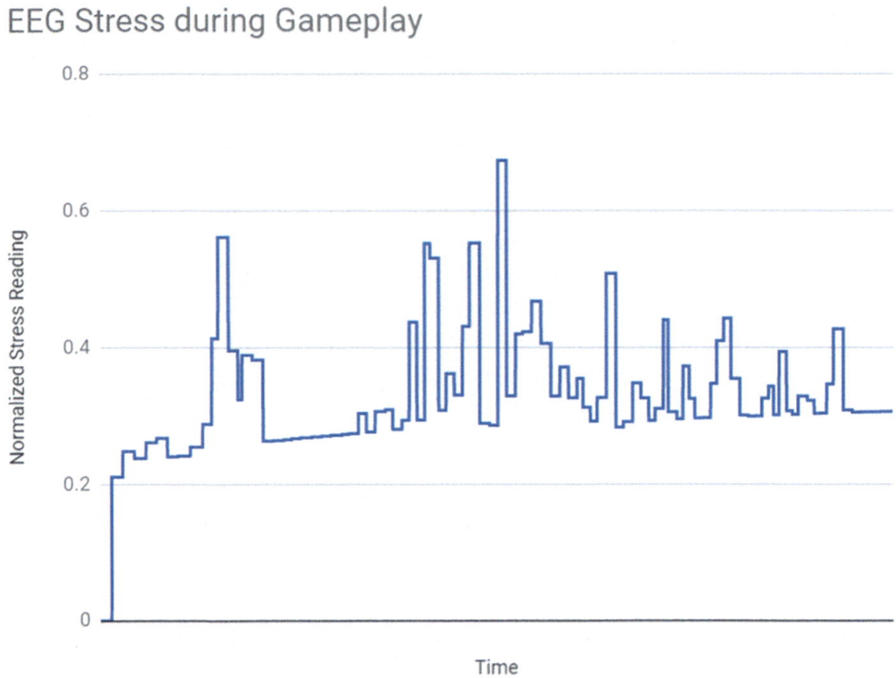

Fig. 12 Normalized stress readings from the Emotiv Insight while playing Brute Force

broaden to other cybersecurity games, including Space Scams, to test generalizability of the results.

References

1. R. Hunicke. The case for dynamic difficulty adjustment in games, in *Proceedings of the 2005 ACM SIGCHI International Conference on Advances in Computer Entertainment Technology*, ACM, pp. 429–433, 2005
2. G. Francia III, D. Thornton, M. Trifas, T. Bowden, Gamification of information security awareness training, in *Emerging Trends in ICT Security*, (Morgan Kaufmann, Boston, 2014), pp. 85–97
3. M. Mayo, Games for science and engineering education. Commun. ACM **50**(7), 30–35 (2007)
4. P. Felicia, *Developments in Current Game-Based Learning Design and Deployment* (IGI Global, Hershey, PA, 2012)
5. A. Smith, *What the Public Knows About Cybersecurity* (Pew Research Center on Internet and American Life, Washington, DC, 2017)
6. J.P. Gee, *Good Video Games + Good Learning: Collected Essays on Video Games, Learning, and Literacy*, vol 27 (Peter Lang, New York, 2007)
7. M. Csíkszentmihályi, *Finding Flow: The Psychology of Engagement with Everyday Life* (Basic Books, New York, 1997)

8. G. Andrade, G. Ramalho, H. Santana, V. Corruble, Extending reinforcement learning to provide dynamic game balancing, in *Proceedings of the Workshop on Reasoning, Representation, and Learning in Computer Games, 19th International Joint Conference on Artificial Intelligence (IJCAI)*, (2005), pp. 7–12
9. P. Demasi, J. de O Adriano, On-line coevolution for action games. Int. J. Intellig. Games Simul. **2**(2) (2003)
10. K. Blinowska, P. Durka, Electroencephalography (EEG), in *Wiley Encyclopedia of Biomedical Engineering*, (Wiley, Hoboken, NJ, 2006)
11. Emotiv, What Is Emotiv Insight? (December 2018). https://www.emotiv.com/knowledge-base/what-is-emotiv-insight/. Accessed 1 Apr 2020
12. D. Wheeler, zxcvbn: Realistic password strength estimation. *Dropbox Tech Blog*, 2012
13. C. Fox, J.H. Brockmyer, The development of the game engagement questionnaire: A measure of engagement in video game playing: Response to reviews. Interact. Comput. **25**(4), 290–293 (2013)
14. A. McCormac, D. Calic, M. Butavicius, K. Parsons, T. Zwaans, M. Pattinson, A reliable measure of information security awareness and the identification of bias in responses. Australas. J. Inf. Syst. **21** (2017)
15. C. Heunis, *Export and Analysis of Emotiv Insight EEG Data via EEGLab* (2016)
16. D. Afergan, E. Peck, E. Solovey, A. Jenkins, S. Hincks, E. Brown, R. Chang, R. Jacob. Dynamic difficulty using brain metrics of workload, in *Proceedings of the 32nd Annual ACM Conference on Human Factors in Computing Systems*, ACM, pp. 3797—3806, 2014
17. K. Mikami, K. Kondo, et al. Adaptable game experience based on player's performance and EEG, in *Nicograph International (NicoInt), 2017*, IEEE, pp. 1–8, 2017

SMAD: A Configurable and Extensible Low-Level System Monitoring and Anomaly Detection Framework

Basel Sababa, Karlen Avogian, Ioanna Dionysiou, and Harald Gjermundrod

1 Introduction

The global proliferation of technology has introduced new security challenges that span the devices themselves, their communication channels, and the systems that are connected to. Preventing security breaches in such a heterogeneous and diverse environment is nontrivial, despite the abundance of security products and technologies in the market, as the attack surface is simply too broad. The frequency of cyber attacks is increasing dramatically and organizations from both public and private sectors are struggling to identify and respond to those security breaches. Over the last few years, several instances of security breaches were brought to light with at least one thing in common: the response time was too long. According to the 2019 IBM Cost of a Data Breach report [1], the mean-time-to-identify (MTTI) a breach in 2019 was 206 days and the mean-time-to-contain (MTTC) was 73 days, with a notable 4.9% increase over the 2018 breach lifecycle. Taking into account these facts, one should expect that the security parameter of a system will be penetrated by an unauthorized party at some point, and the goal must be to identify the incident as quickly as possible and respond effectively. It is therefore of paramount importance to provide adequate practical training to the next-generation security experts, preferably using real data from past security incidents.

The identification of the security attacks relies on technological and human factors; the former one being the security tools that are integrated in the organization's network infrastructure and the latter one being the in-house security and system administrators who have the ultimate responsibility of all decision-making. Defense

B. Sababa · K. Avogian · I. Dionysiou (✉) · H. Gjermundrod
Department of Computer Science, School of Sciences and Engineering, University of Nicosia, Nicosia, Cyprus
e-mail: sababa.b@live.unic.ac.cy; avogian.k@live.unic.ac.cy; dionysiou.i@unic.ac.cy; gjermundrod.h@unic.ac.cy

K. Daimi, G. Francia III (eds.), *Innovations in Cybersecurity Education*,
https://doi.org/10.1007/978-3-030-50244-7_2

in depth, a multilayered approach that supports defensive mechanisms on various layers, is a popular approach to protect the network. However, the vast majority of the deployed security technologies focus on external attempts of bypassing the front lines of the system defense plan, without considering internal attacks orchestrated by individuals who are authorized users abusing their system privileges. As far as the human factor is concerned, it is generally argued that people are the *weakest link in the information security chain*. Rather than scanning and exploiting vulnerabilities in the deployed technology, it is comparatively easier to surpass the defenses of the human endpoints in the security chain using low tech but still sophisticated approaches such as phishing and social engineering.

This chapter proposes SMAD, a configurable and extensible **S**ystem **M**onitoring and **A**nomaly **D**etection framework based on Sysdig, which monitors kernel and system resources data (e.g., system calls, network connections, process info) based on user-defined configurations that initiate nonintrusive actions when alerts are triggered. SMAD is envisioned to be used not only by security-enthusiastic students with some technical skills to track their local Linux server health but also by university instructors who want to leverage the practical dimension of their security courses with the introduction of novel low-level system security monitoring tools. Monitoring raw kernel and system resources data is a tedious and error-prone task, if done manually. SMAD eases this overwhelming task by allowing users launching several system monitors via SMAD in a user-intuitive manner, alerting him/her on any unexpected behavior based on user-defined metrics, including CPU usage, directories visited, system errors, commands executed, HTTP requests, IP addresses connected to the system, and files opened. Furthermore, once an alert is triggered, raw data capturing could be initiated for a specified time period, giving the student the opportunity to correlate the alert triggering to the actual activities taking place in the kernel.

The rest of the chapter is organized as follows: Sect. 2 describes related research efforts on system monitoring. Section 3 presents the SMAD framework, followed by its experimental evaluation in Sect. 4. Section 5 discusses the value of SMAD in cybersecurity education. Section 6 concludes with future directions.

2 Related Work

System monitoring typically includes installing a surveillance software on a system, usually running as a background process monitoring the system's resources and performance, on the lookout for deviations from normal behavior. It runs concurrently with and independently from other types of system monitoring, such as operating system monitoring. In the event of unexpected behavior, alerts or notifications are sent to the system administrator. It is crucial to support system monitoring in real time while intrusive activities are in progress to minimize and/or contain the damage [2] as it leverages the user ability to control and maintain the monitored system [3].

There exist several academic and open source security-related system monitoring approaches as well as commercial solutions. Starting with the noncommercial approaches, Swatchdog [4] (formally known as Swatch [5]) is a log file monitoring system designed to address the challenges faced by system administrator when monitoring numerous servers continuously and simultaneously. Linux is configured to log security information to a central logging host system. In order to keep the system administrator from being overwhelmed by the size of logged data, Swatchdog monitors the log files and filters out unwanted and redundant data and supports an action system where an action is executed after filtering, specified by the user from a list of possible actions.

Haystack [6], a system designed to detect intrusions in air force computer systems, analyzes audit trail files daily, searching for user activity and comparing it against predefined security constraints and normal behavior models. The application generates a report that summarizes the activities analyzed. These reports can be analyzed by system administrators in order to locate possible intrusions.

Finally, Graph-based Intrusion Detection System (GrIDS) [7] is a system designed to monitor and analyze network activity on TCP/IP networks with thousands of hosts and possibly detect large-scale attacks. The application collects activities of individual computers and the networks among them, which are then aggregated into activity graphs. By analyzing various characteristics of these activity graphs, the application is able to automatically detect and report any network attacks that are happening in near real time.

There are other system monitoring tools that are not security oriented. For example, Benini et al. [8] have designed a system monitoring tool used for supporting dynamic power management in computers with tight power constraints. This is accomplished by monitoring and analyzing power consumption and dissipation on a device. The tool collects data on the use of system resources such as disks, CPU, keyboard, and mouse and analyzes the power consumption.

Commercial solutions also exist that monitor system activity either locally or on a cloud infrastructure. Nagios XI log monitoring system [9] integrates Swatchdog in a commercial enterprise IT infrastructure monitoring solution. Similarly, IBM [10] provides cloud monitoring using Sysdig [11] to collect monitoring data.

3 SMAD: System Monitoring and Anomaly Detection Framework

This section presents the design and functionality of SMAD,[1] an extensible framework for monitoring the state and various activities of a Linux server in an intuitive way. The SMAD framework could be perceived as a wrapper to Sysdig, therefore a brief overview of Sysdig is also given.

[1]The source code of the SMAD prototype can be found in Github repository https://github.com/kosnet2/sad

Sysdig Overview

Sysdig [11] is an open source, command-line utility for capturing all system calls residing within the Linux kernel. It could be perceived as the Wireshark for the end system. Every time an installed application performs a privileged operation (e.g., open/read file, open network port, read/write to any device), it invokes a system call that executes the operation on the behalf of the user's process. Capturing all invoked system calls could be viewed as passive sniffing of all the operations performed within the server. A subset of the Linux monitoring and debugging tasks that are bundled by Sysdig is {*strace, tcpdump, netstat, htop, iftop, lsof*}. Additional features of Sysdig include provision of chisels (lightweight Lua scripts) for processing captured system events, provision of simple filtering of output, support of system and application tracing, and support of Linux server attack analysis features for ethical hackers.

The large number of the executed system calls would quickly overwhelm the system, with respect to both processing time and storage. As a countermeasure, Sysdig supports the configuration of filters in order to capture specific system calls or a subset of them. In this way, the filtered events are those of interest to the user. However, on the downside, it is nontrivial to formulate the appropriate filters tailored to the deployment, usage, and threat landscape of the specific server. There are commercial solutions that provide intuitive user interfaces for monitoring large deployments of servers (as well as container deployment) using Sysdig. An example is the Sysdig Monitor Dashboard [12] developed by Sysdig, a commercial product targeted for enterprises deploying applications in cloud infrastructures. Similarly, IBM offers a front end to Sysdig [10] as part of its BlueMadator product [13].

SMAD in a Nutshell

The System Monitoring and Anomaly Detection (SMAD) framework is a modular framework that acts as a front end for Sysdig, utilized by a user possessing the required technical skills to monitor a personal server. SMAD's user-centric approach provides an intuitive environment to configure and operate a set of monitors, including start/stop operations, adding alerts to monitors whenever user-defined conditions are met, and capturing all system events for a specified time duration upon an alert trigger (optional action). The inspection and analysis of the captured log files are nontrivial and could overwhelm the novice system administrator. One of the goals of SMAD is to introduce the amateur user into low-level system events, acquiring and/or expanding their knowledge and skill set on the low-level operation of a system. To this end, a new SMAD component is currently under development for visualizing and/or graphically representing captured system activity. More details on SMAD functionality can be found in [14].

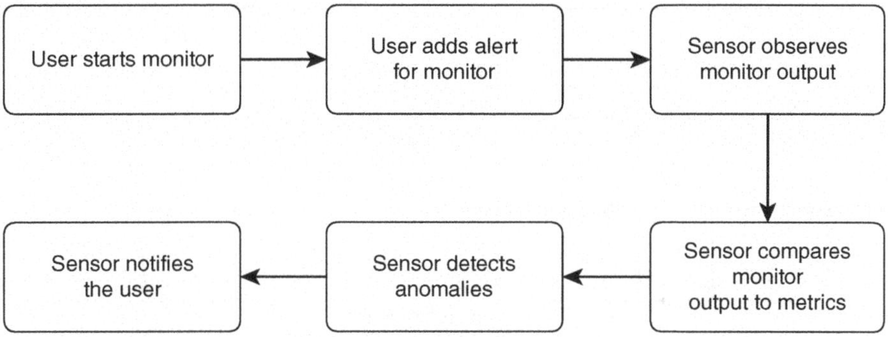

Fig. 1 SMAD components' workflow

SMAD Components

The SMAD baseline components along with the workflow of the system are illustrated in Fig. 1. The User interacts with the User Interface to issue commands (the command set is discussed in detail in the next subsections). The system executes the actions needed to fulfill the user's request, rendering the graphical interface with the execution output. If the stop/start monitor commands are issued, the User Interface interacts with the Monitoring Sensor that carries out the necessary actions. In the case where an event capture is attached to an alert trigger, the Monitoring Sensor starts storing events to Capture Files, based on the user-defined settings. The logged data is available for further analysis.

SMAD is extensible, supporting the integration of new components to provide additional functionality to the user. For example, one could develop a module that transforms the nonintrusive nature of SMAD into an active one by reacting to the alert trigger using actions other than event capturing, such as kill a process or close a port. The two baseline components, Monitoring Sensor and User Interface, are now presented in more detail.

Monitoring Sensor Component

The Monitoring Sensor component is a wrapper for Sysdig, executing Sysdig commands with the appropriate arguments that are automatically generated based on the user's preferences and selections set using the User Interface.

As mentioned earlier, SMAD supports a notification mechanism via alerts. An alert is assigned to a particular monitor and its configuration profile includes the metrics to be monitored, the user notification method, and whether or not event capturing upon triggering is required. The cumbersome task of setting the desired metrics and their parameters is alleviated by the user-centric approach of SMAD that allows the specification of metrics to be done in a user-friendly and intuitive manner

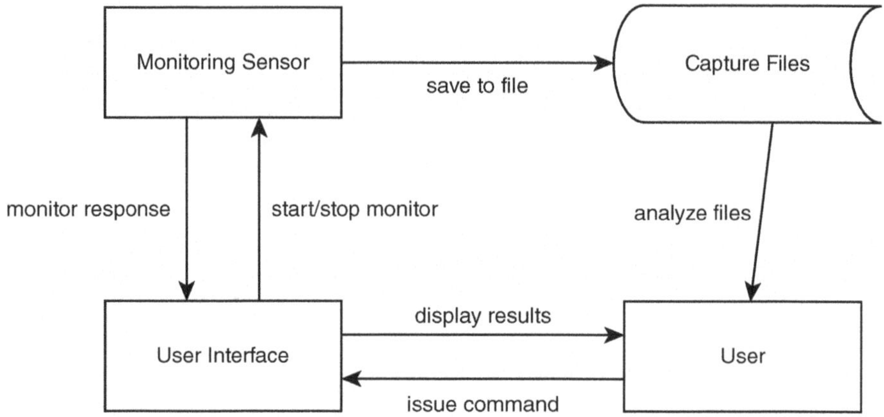

Fig. 2 Monitoring Sensor component

while converting them to the appropriate Sysdig commands. The Monitoring Sensor maintains a list of running monitors and automatically detects alerts for its running monitors and starts observing the monitor output according to the metrics in the alert configuration profile. Figure 2 illustrates the Monitoring Sensor process.

User Interface Component

The User Interface module was created using PyQt5, a Python-binding of the cross-platform GUI toolkit Qt [15]. It consists of three primary pages, namely, Monitors, Alerts, and Notifications, as shown in Fig. 3. These pages are now described in more detail.

Monitors Page

Monitors is the main SMAD page, responsible for managing monitors. A taxonomy of monitors is inherent in SMAD as it supports five monitor categories (CPU & Processes, Performance & Errors, Network, Security, and Application), and each category is represented in its own tab, as shown in Fig. 3. The monitor taxonomy was devised by taking into consideration usage scenarios for the target SMAD user. An easy-to-use technique to specify arguments is supported that allows monitor customization to meet the user's needs. Additional monitors could be added to each category as this taxonomy serves as a starting point to further extend the system.

Within each category, the user selects the monitors he/she would like to start by clicking on a checkbox. There are two types of monitors, one that requires further user input and another one that is self-contained. In the former type, no alerts are required to be configured and attached to the monitor. Clicking the "Start Monitors"

Fig. 3 Monitors: CPU & Processes tab

button launches Sysdig instances running as background processes. At the same time, the Running Monitors list is updated to include the newly launched monitors. The "Stop Monitor" button kills the process running the selected monitor and gets removed from the list.

As mentioned earlier, there are five monitor categories. Starting with the CPU & Process monitor category, as shown in Fig. 3, it consists of two monitors. The first monitor observes the output of a specific process, where the user specifies the process to be monitored. Once the monitor starts, the specified process is monitored.

The second one keeps track of the top CPU-consuming processes, requiring no further user input. However, unlike the previous monitor, this monitor will not start immediately the monitoring, but it rather waits for an alert to be triggered to do so. The user must specify a metric for the allowed CPU percentage consumption that a process must exceed in order to trigger the alert (see Fig. 8).

Figure 4 shows the Performance & Errors monitor category, supporting six monitors, as listed below:

1. Monitor to show all failed open file operations by given processes (this monitor requires user input but not an alert specification)
2. Monitor to observe system calls returning the most errors
3. Monitor to show files with the most input/output errors
4. Monitor to observe processes with the most input/output errors

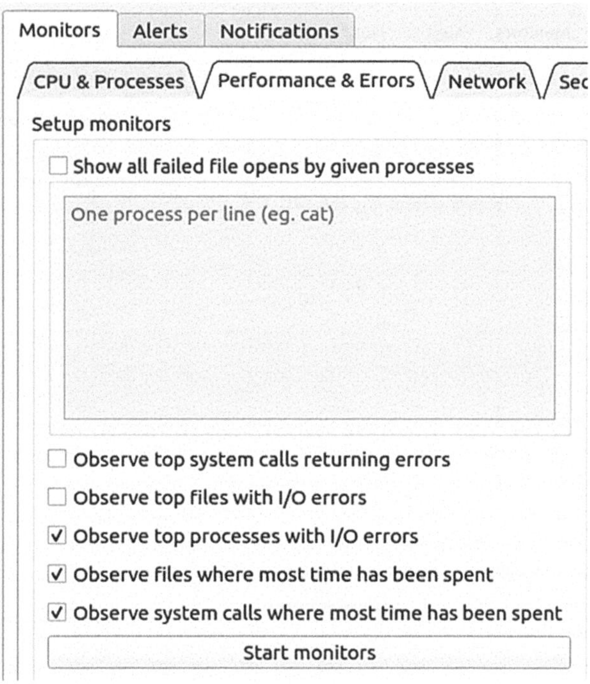

Fig. 4 Monitors: Performance & Errors tab

5. Monitor to observe files where the system has spent the most time
6. Monitor to observe system calls where the system has spent the most time

These monitors could be used to detect system misbehavior or sluggish performance, allowing to narrow down the root of the problem. Additionally, they could be used to assess whether or not current applications running on the system generate an abnormally high number of faults, probing the user to investigate alternative applications to execute the specific task.

Proceeding with the Network category, shown in Fig. 5, the monitors comprising it are the following:

1. Monitor for network data exchanged for given IP addresses
2. Monitor for the network connections that consume the most bandwidth
3. Monitor for the processes that consume the most bandwidth

The last two monitors require an alert configuration with regards to bandwidth (see Fig. 10), triggering an alert when a threshold is reached.

The Security category is arguably the most important one (shown in Fig. 6). It contains three monitors that do not require alerts:

1. Monitor to show all the commands that are being executed by a user logged into the system
2. Monitor to show all the directories that a user is visiting
3. Monitor to display all file open operations that occur inside the mentioned directories

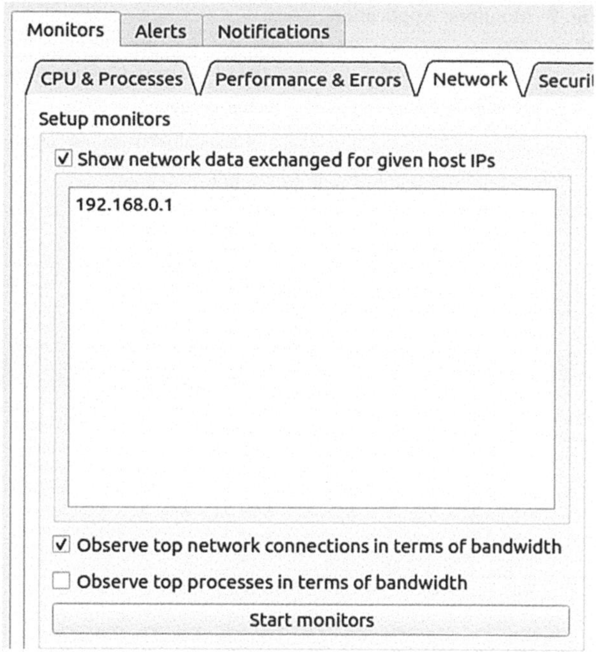

Fig. 5 Monitors: Network tab

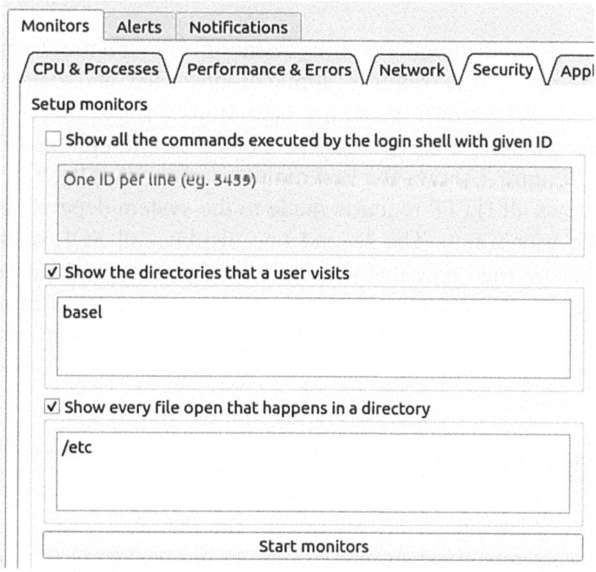

Fig. 6 Monitors: Security tab

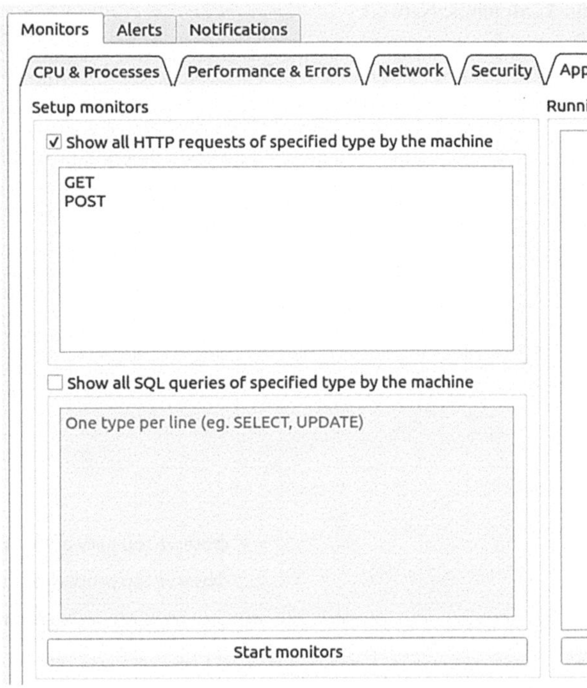

Fig. 7 Monitors: Application tab

The launch of the security monitors allows a user to monitor the actual operations executed by a running application/service and detect misuse of resources, for example perform read/copy operations on files that it shouldn't need to access or open the file that will access a camera or microphone (if present).

Figure 7 shows the two monitors of the Application category. The first monitor shows all HTTP requests made to the system depending on the type of request that the user wants. The second one displays all SQL queries made to the system of the specified type that the user chooses. The Application monitors complement the Security monitors as they also monitor the activity of an application, but they can also be used to debug and evaluate what and how a specific application is performing its functionality.

Alerts Page

The Alerts page is responsible for the alert configuration. An alert is a set of parameters attached to a specific monitor already running. Some monitors will not start monitoring events until at least one alert is attached to it since they require the metric from the alert specification profile. A metric violation yields an automatic notification posted on the Notifications page as well as initiating the capturing of events, if capturing is enabled for this alert. The alert parameters are as follows:

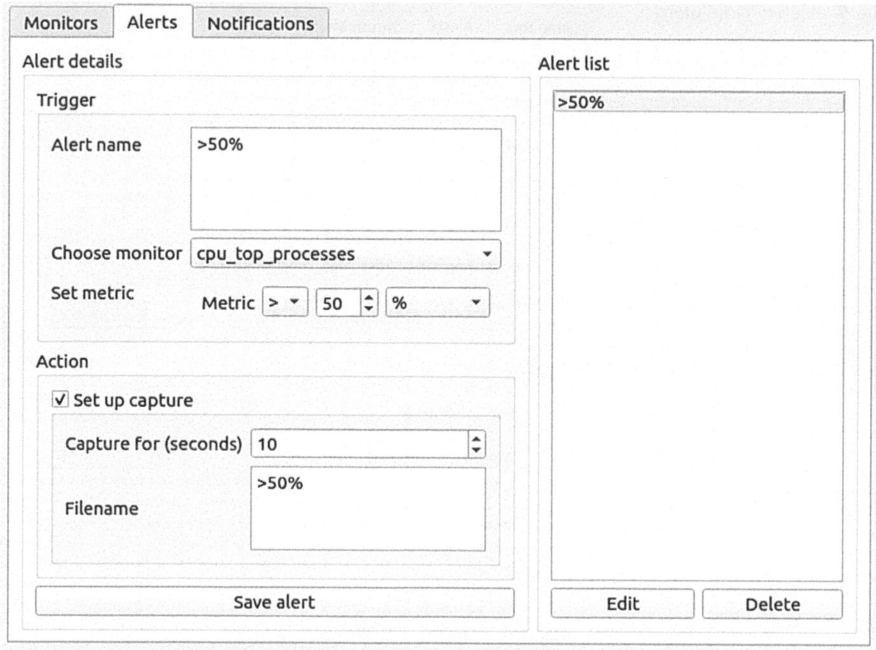

Fig. 8 Alerts: Number metric

1. Monitor it is assigned to
2. Metric to be monitored (number, time, size)

Figure 8 shows the alert configuration to be attached to the *cpu_top_processes* monitor. This alert utilizes a number metric representing a percentage. It triggers the specific monitor to start observing processes that exceed 50% CPU usage. Additionally, once triggered, it initiates capturing of all events for the duration of 10 s (user-customized setting) and save them to the user-specified file. The file contents could be analyzed at a later time to derive useful information. SMAD only allows the configuration of error-prone alerts as it supplies the metrics along with the permitted operations (e.g. <, >, =) on them as well as checking that the value is within a valid range.

Figure 9 presents an alert that uses the time metric, where time could be specified in four time units. The user specifies the desired time using his/her preferred time unit and the application handles any unit conversion, if needed. The alert is attached to the *errors_files_most_time_spent* monitor and gets triggered when a file has been used for more than 4 μs. The alert also triggers a 20-s capture session saved to the file *time_spent.*

The third and final type of metric is size, in terms of data. There are four units available, namely, byte, kilobyte, megabyte, and gigabyte. Similar to the time metric, the user is free to use any size unit and the application will handle

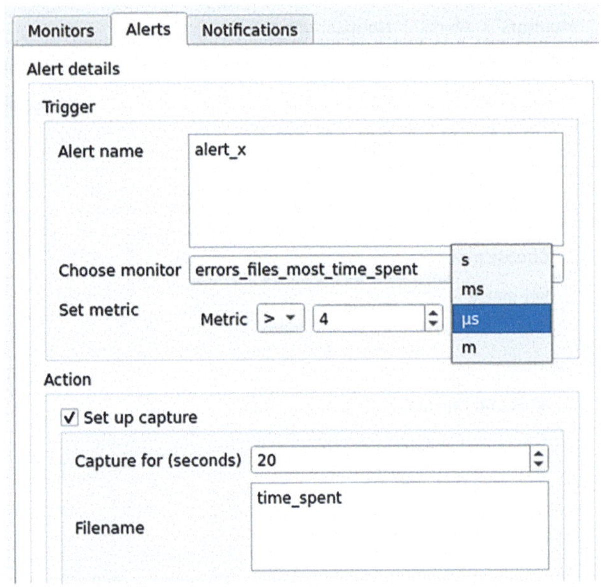

Fig. 9 Alerts: Time metric

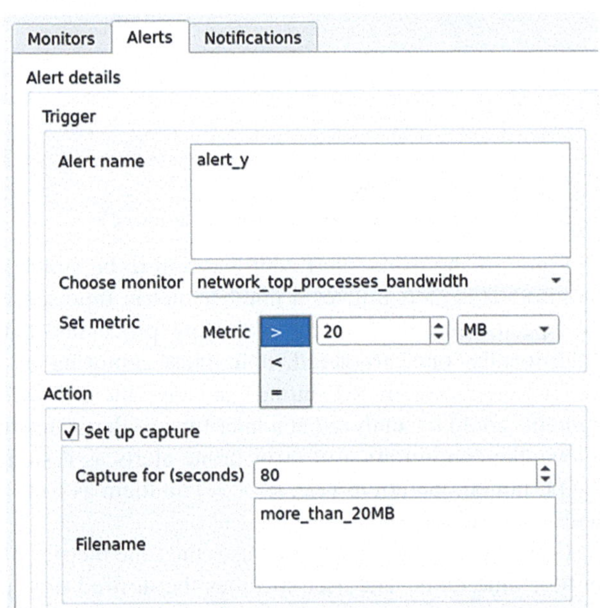

Fig. 10 Alerts: Size metric

the conversions. In the example shown in Fig. 10 an alert is attached to the *network_top_processes_bandwidth* monitor and gets triggered when the bandwidth exceeds 20 MB. It also starts capturing events for 80 s after the metric violation.

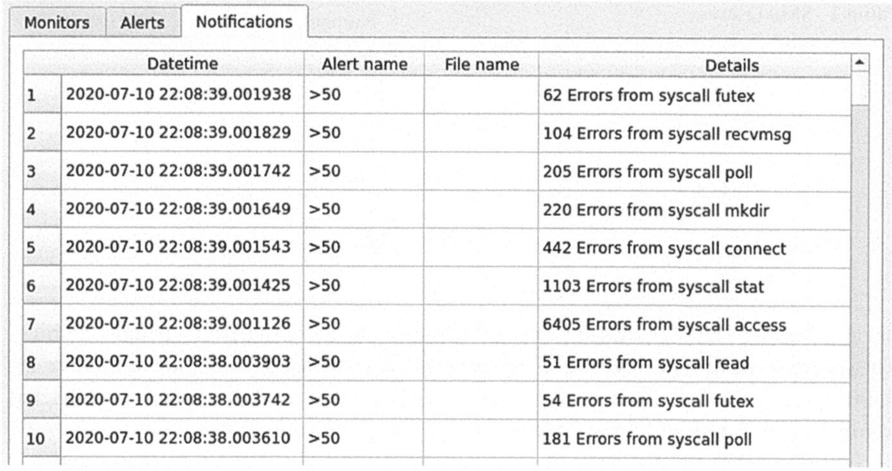

	Datetime	Alert name	File name	Details
1	2020-07-10 22:08:39.001938	>50		62 Errors from syscall futex
2	2020-07-10 22:08:39.001829	>50		104 Errors from syscall recvmsg
3	2020-07-10 22:08:39.001742	>50		205 Errors from syscall poll
4	2020-07-10 22:08:39.001649	>50		220 Errors from syscall mkdir
5	2020-07-10 22:08:39.001543	>50		442 Errors from syscall connect
6	2020-07-10 22:08:39.001425	>50		1103 Errors from syscall stat
7	2020-07-10 22:08:39.001126	>50		6405 Errors from syscall access
8	2020-07-10 22:08:38.003903	>50		51 Errors from syscall read
9	2020-07-10 22:08:38.003742	>50		54 Errors from syscall futex
10	2020-07-10 22:08:38.003610	>50		181 Errors from syscall poll

Fig. 11 Notifications tab

Notifications Page

The Notifications page displays events that have violated an alert's metric. The information shown is the alert responsible for the anomaly event, the date and time of the alert triggering, other information relevant to the event, and the name of the capture file if capturing was enabled for that alert.

In Fig. 11, one monitor is currently running that observes errors resulting from system calls. This monitor has only one alert attached with a metric to alert the user when the number of errors exceeds 50. In this case, the details provided are the number of errors and the system call where the errors originated from. The File name field is empty because capturing was not enabled by the user.

4 SMAD Experimental Evaluation

Three different evaluation tests were conducted to assess the SMAD system: stress testing, functionality testing, and vulnerability assessment.

Stress Testing

The stress testing was performed to assess the overhead of SMAD on the machine it runs. SMAD was tested under heavy workload with multiple monitors running, each with multiple configured alerts. The stress testing experiment was performed

Table 1 SMAD stress testing results

Number of monitors	CPU usage (%)
1	1
2	5
3	12
5	25
10	59
15	91
20	100

on an Ubuntu 18.04 operating system running on a VirtualBox virtual machine. The machine running the virtual machine had an x64-based processor Intel Core i5-9300H with 2.40 GHz frequency and 8 GB RAM. The virtual machine was restricted to only 2 GB RAM and 4 out of 8 cores.

Table 1 shows the findings of the experiment and the impact on the machine's CPU usage. Two alerts were attached per running monitor. Based on the findings, at 15 monitors, the application was heavily slowed down but was still functional. At 20 monitors, the application ultimately consumed the full capacity of the CPU. Therefore, it is not recommended to exceed 10 monitors running simultaneously.

Functionality Testing

A test was designed to detect server misuse by an authorized user with limited privileges. The default settings on a Linux server assigns the user group read access (but not write access) to various system files like the *etc/passwd* file in order to perform their assigned tasks. However, if a user manually browses contents of specific files, it may be considered a suspicious activity and could be part of an insider attack.

Monitors were configured to observe high-risk locations and resources for potential misuse by insider attackers, who had already authorized accounts on the server. The following monitors were launched:

1. Observe files where most time has been spent
2. Show network data exchanged with server with a 10-s capturing
3. Show all directories visited by users with a 30-s capturing

At the beginning of the experiment, there were no notable captured events that required further analysis. After some time, an alert was triggered and a post notification shown on the Notifications list indicated that a user accessed the */etc* directory that contains system configuration files that a normal user should not need to access directly. This incident on its own does not indicate malicious activity but prompted a further investigation of the captured events after the alert was triggered. It was discovered that the user read the contents of the *etc/passwd* file using the *cat* command and this activity could be part of the reconnaissance phase of an attack. Figure 12 shows a snapshot of the captured file that logged the user activity.

Vulnerability Assessment

The goal of the vulnerability assessment was to uncover ways that SMAD could be exploited. It was discovered that SMAD was vulnerable to command injection, giving a malicious user access to the system. In particular, a number of text fields used for monitor configuration are subject to this attack. However, a user must be permitted to enter data into these fields, otherwise the capabilities of SMAD would be limited. For example, a user would not be able to specify a specific process, user, or IP address to monitor.

The countermeasure against command injection is the filtering of special characters from the user input, whose presence could indicate a command injection. If those characters are present, the input is discarded, as shown in the regex below:

```
if re.search('[&|;#$]', line):
    return False
```

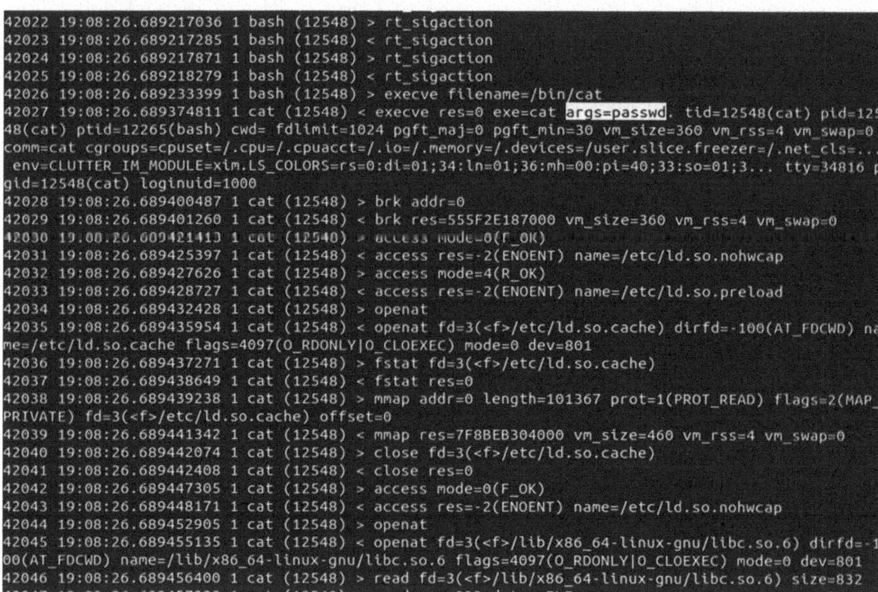

```
42022 19:08:26.689217036 1 bash (12548) > rt_sigaction
42023 19:08:26.689217285 1 bash (12548) < rt_sigaction
42024 19:08:26.689217871 1 bash (12548) > rt_sigaction
42025 19:08:26.689218279 1 bash (12548) < rt_sigaction
42026 19:08:26.689233399 1 bash (12548) > execve filename=/bin/cat
42027 19:08:26.689374811 1 cat (12548) < execve res=0 exe=cat args=passwd. tid=12548(cat) pid=125
48(cat) ptid=12265(bash) cwd= fdlimit=1024 pgft_maj=0 pgft_min=30 vm_size=360 vm_rss=4 vm_swap=0
comm=cat cgroups=cpuset=/.cpu=/.cpuacct=/.io=/.memory=/.devices=/user.slice.freezer=/.net_cls=...
 env=CLUTTER_IM_MODULE=xim.LS_COLORS=rs=0:di=01;34:ln=01;36:mh=00:pi=40;33:so=01;3... tty=34816 p
gid=12548(cat) loginuid=1000
42028 19:08:26.689400487 1 cat (12548) > brk addr=0
42029 19:08:26.689401260 1 cat (12548) < brk res=555F2E187000 vm_size=360 vm_rss=4 vm_swap=0
42030 19:08:26.689421413 1 cat (12548) > access mode=0(F_OK)
42031 19:08:26.689425397 1 cat (12548) < access res=-2(ENOENT) name=/etc/ld.so.nohwcap
42032 19:08:26.689427626 1 cat (12548) > access mode=4(R_OK)
42033 19:08:26.689428727 1 cat (12548) < access res=-2(ENOENT) name=/etc/ld.so.preload
42034 19:08:26.689432428 1 cat (12548) > openat
42035 19:08:26.689435954 1 cat (12548) < openat fd=3(<f>/etc/ld.so.cache) dirfd=-100(AT_FDCWD) na
me=/etc/ld.so.cache flags=4097(O_RDONLY|O_CLOEXEC) mode=0 dev=801
42036 19:08:26.689437271 1 cat (12548) > fstat fd=3(<f>/etc/ld.so.cache)
42037 19:08:26.689438649 1 cat (12548) < fstat res=0
42038 19:08:26.689439238 1 cat (12548) > mmap addr=0 length=101367 prot=1(PROT_READ) flags=2(MAP_
PRIVATE) fd=3(<f>/etc/ld.so.cache) offset=0
42039 19:08:26.689441342 1 cat (12548) < mmap res=7F8BEB304000 vm_size=460 vm_rss=4 vm_swap=0
42040 19:08:26.689442074 1 cat (12548) > close fd=3(<f>/etc/ld.so.cache)
42041 19:08:26.689442408 1 cat (12548) < close res=0
42042 19:08:26.689447305 1 cat (12548) > access mode=0(F_OK)
42043 19:08:26.689448171 1 cat (12548) < access res=-2(ENOENT) name=/etc/ld.so.nohwcap
42044 19:08:26.689452905 1 cat (12548) > openat
42045 19:08:26.689455135 1 cat (12548) < openat fd=3(<f>/lib/x86_64-linux-gnu/libc.so.6) dirfd=-1
00(AT_FDCWD) name=/lib/x86_64-linux-gnu/libc.so.6 flags=4097(O_RDONLY|O_CLOEXEC) mode=0 dev=801
42046 19:08:26.689456400 1 cat (12548) > read fd=3(<f>/lib/x86_64-linux-gnu/libc.so.6) size=832
42047 19:08:26.689457922 1 cat (12548) < read res=832 data=.ELF..............>.............@.....
```

Fig. 12 Snapshot of captured data

5 SMAD Role in Cybersecurity Education

SMAD could play a vital role in formal education with its subject-oriented structured curricula, while at the same time SMAD educational benefits could also be perceived in a nonformal education setting via flexible adult self-learning.

SMAD in Formal Education

Technology integration in course curricula could extend learning in powerful ways, demonstrating the application of theoretical concepts in practice. There is a plethora of cybersecurity tools and technologies that could be embedded in university courses, envisioning to develop multiskilled competent security practitioners. The SMAD framework could be an integral part of any security-oriented undergraduate/graduate course that aims in providing students with the sought-after technical knowledge and skills in cybersecurity system monitoring.

If one of the learning objectives of a security program were to also train the students to defend against zero-day vulnerabilities and novel attack techniques, then it would be imperative to comprehend the internal functionality of the security tools, moving aside the black-box practice of using tools and instead delve into the technical specifications of the tool modules. Educators often find it challenging to demonstrate low-level attack vectors and interactions that allow an attacker to circumvent the defense lines of a system to accomplish his/her mission (password cracking, file exfiltration, privilege escalation, backdoor installation, process manipulation, to just name a few). The current practice usually involves demonstration of point-and-click tools (usually GUI-based tools), low-level command-line commands/scripts (potential shell scripts in hex), rules files (for tools that support this), and log files. SMAD bridges the gap of black-box and white-box paradigms by using a point-and-click approach while at the same time uncovering what takes place underneath the hood (in the SMAD case, the kernel). Educators who use SMAD allow their students to gradually move from the point-and-click approach toward script environments, grasping the full technical profile of numerous attack scenarios.

Four SMAD-based scenarios, appropriate for the current laboratory part of a security course, are presented next. Each scenario is assigned a tentative level of difficulty based on the authors' experience teaching security courses, ranging from introductory courses (e.g., computer security, network security) to advanced graduate courses (ethical hacking, cyber warfare). It is strongly recommended that the course instructor prepares virtual machines (depending on the scenario) with known vulnerabilities to allow the students experiment in a secure testing environment, providing protection of operating within a sandbox environment to avoid accidental security incidents originating from the testing environment to an outside host.

Scenario 1: Capture the Intruder (Level of Difficulty—Easy)

One of the learning objectives of security courses covering topics related to system defenses and countermeasures against internal and external attacks is to expose the student in various attack methodologies. Ideally, the theoretical aspects of these methodologies should be demonstrated in practice in order for the student to acquire the practical dimension of the aforementioned methodologies.

The course instructor is taking an active role in this scenario and prepares a set of exercises following the capture-the-intruder style, where each exercise clearly states the service/resource/account the student is supposed to monitor using SMAD. The student is responsible for configuring the appropriate SMAD monitors, setting the event capture option on whenever an alert is triggered, and to monitor the assigned service/resource/account. The instructor decides when to launch the attack and executes the attack. Unlike other intrusion detection tools, SMAD logs all system calls once an intrusion is detected, giving the opportunity to the student to investigate the various system interactions once a threat is realized into an attack. The analysis of the captured system call set could yield an attack profile that the student could map into the various phases of an attack methodology.

Scenario 2: Red/Blue Team Exercise (Level of Difficulty—Medium)

This scenario is similar to the first one, with the main difference being that the course instructor has an observer/monitor role, who clearly sets the boundaries before the attack exercises commence. Students form red and blue teams, with the red team preparing an attack using penetration testing tools and/or attack frameworks whereas the blue team prepares the lines of defense using SMAD. It is recommended to hold two different exercises, allowing students to assume both red and blue team roles.

Scenario 3: Dissection of Malware (Level of Difficulty—High)

The postmortem analysis of an attack offers a useful insight into the attack pathway, allowing the formulation of an attack profile that could be utilized to detect (and perhaps prevent) future attacks based on the same or similar profile. The dissection of malicious code is challenging and nontrivial, going beyond the scope of most security courses. SMAD could be used to introduce the concept of the anatomy of an attack by running a malware in the sandbox test environment and studying the sequence of system calls (including their arguments) that were executed while the malware was running. The student should be able not only to form a timeline of the malware-related system calls but also determine what vulnerability was exploited, and how and when was it exploited. Depending on the level of the course and its learning objectives, the instructor could provide a set of malware, spanning relatively benign ones to more sophisticated ones.

Scenario 4: Extending SMAD (Level of Difficulty—High)

In general, graduate-level security courses focus on current trends and new research developments. Students interested in low-level security monitoring could extend SMAD and contribute their modules to the SMAD community. It is up to the instructor's discretion to decide what module could be developed. The instructor could also contact the authors, who could provide a list of potential modules.

SMAD in Nonformal Education

A nonformal education learner discovers and acquires skills and knowledge from nonformal activities, outside the educational institution, usually while being part of the workforce. It is not uncommon for nonstudents with technical skills to experiment at home on their personal Linux servers, trying to synthesize theory and practice on their own. There are several sources of learning cybersecurity in an environment that diminishes the contact of instructor and student, including the massive open online courses offered by online learning platforms such as Udemy, Coursera, and Lynda and the tutorial-style video clips posted on online platforms such as YouTube. Users pursuing nonformal education are interested in obtaining knowledge or enhancing their current knowledge on a specific topic, without being concerned about accreditation and/or certificates of any sort. Below are ways that SMAD could contribute toward nonformal learning in cybersecurity.

SMAD-Based Security Monitoring Tutorials

SMAD is an ideal tool to use for developing short tutorials that demonstrate the attack methodology followed to exploit a known vulnerability, while at the same time present the actual technical details throughout the lifetime of the attack. It is challenging to deliver interesting and user-intuitive tutorials using the command-line interface. SMAD's graphical-based configuration of the command-line interface commands creates a learning environment that is more intuitive to the average learner. It is recommended in the tutorial to follow the sequence of steps as shown below:

1. Configure SMAD monitors that are triggered accordingly once the exploit is launched on the SMAD host, making sure to activate the event-capturing option.
2. Run the exploit and observe how/when SMAD alerts are triggered.
3. Use the terminal-style interaction to view the captured log files and determine from the system call sequence when and how the vulnerability was exploited.

SMAD Interest Communities

Open source software greatly benefits through the community, an ad hoc group of contributors that inspect, modify, and enhance the software. It is environed that SMAD will attract the creation of two different communities to

1. Enhance its source code: Contributors belonging to this community advance the SMAD framework via submission of new modules or improved ones. This is a well-established activity in the open source community, with the source code hosted in a repository and contributions get accepted as long as they comply with the project's guidelines.
2. Enhance its use: Contributors in this community develop/configure VMs with different vulnerabilities or interesting misconfigurations along with guidelines of how an instructor or self-taught learner could use the specific VM in an educational/training session. This community is as equally important as the first one, but unfortunately it is underrepresented and not streamlined as the source-code contribution community. Lecturers/trainers/tutors spend a significant amount of time developing training material, including getting the systems configured/customized in order to demonstrate a particular topic. This work is mostly not shared back with the community. This practice could change, especially now with the rapid uptake of using virtualized environments (including droplets). Freemium models may also be appropriate alternatives for this sharing in order to motivate the contributions.

6 Conclusion

This chapter presented SMAD, a novel framework that monitors kernel and system resources data (e.g., system calls, network connections, process info) based on user-defined configurations that initiate nonintrusive actions when alerts are triggered. The user-centric SMAD environment allows the specifications of monitors and alerts to be done in a free-of-errors manner. A prototype system based on the framework was evaluated and its performance was assessed, yielding promising results. The only drawback of the system is that the amount of information that is captured might be overwhelming, making it tedious to browse and analyze the captured data. For example, the average capture file size generated after running a 60-s capture is 9 MB. A new SMAD component is currently under development for visualizing and/or graphically representing captured system activity and integrate this output with the Falco security system [16].

SMAD is also a security educational tool and its intended usage is by educators who want to leverage the practical aspect of their security courses as well as by students who wish to monitor the health of Linux servers. As technology integration in course curricula could extend learning in powerful ways, demonstrating the application of theoretical concepts in practice, SMAD could be part of any security-related undergraduate and graduate course. The easy-to-configure nature of SMAD makes it an ideal introductory tool to low-level security monitoring, allowing students to experiment with alert configuration based on low-level system commands and properties, view and analyze system activity upon alert triggering, and add new functionality by extending SMAD with new modules.

References

1. IBM Security and Ponemon Institute LLC, Cost of a Data Breach Report 2019. https://www.ibm.com/. Accessed Jan 2020
2. N. Ye, S. Vilbert, Q. Chen, Computer intrusion detection through EWMA for autocorrelated and uncorrelated data. IEEE Trans. Reliab. **52**(1), 75–82 (2003)
3. J.R. Harrow, F.P. Messinger, *System Monitoring Method and Device Including a Graphical User Interface to View and Manipulate System Information*. US Patent 5,375,199, 20 Dec 1994
4. Swatchdog, Simple Log Watcher. https://sourceforge.net/projects/swatch/les/swatchdog/. Accessed Jan 2020
5. S.E. Hansen, E.T. Atkins, Automated system monitoring and notification with swatch, in *Proceedings of the 7th USENIX Conference on System Administration*, USENIX Association, Monterey, CA, USA, 1993, pp. 145–152
6. S.E. Smaha, Haystack: an intrusion detection system, in *Proceeding of Fourth Aerospace Computer Security Applications*, Orlando, FL, USA, 1988, pp. 37–44
7. S. Staniford-Chen, S. Cheung, R. Crawford, M. Dilger, J. Frank, J. Hoagland, K. Levitt, C. Wee, R. Yip, D. Zerkle, Grids—a graph based intrusion detection system for large networks, in *Proceedings of the 19th National Information Systems Security Conference*, Baltimore, MD, USA, 1996, pp. 361–370
8. L. Benini, A. Bogliolo, S. Cavallucci, B. Ricco, Monitoring system activity for OS-directed dynamic power management, in *Proceedings of the 1998 International Symposium on Low Power Electronics and Design (IEEE Cat. No. 98TH8379)*, Monterey, CA, USA, 1998, pp. 185–190
9. Nagios, Nagios Enterprises Log Monitoring with Swatchdog. https://assets.nagios.com/downloads/nagiosxi/docs/Log-Monitoring-With-Swatch.pdf. Accessed Jan 2020
10. IBM, IBM Cloud Monitoring with Sysdig. https://www.ibm.com/cloud/sysdig. Accessed Jan 2020
11. Sysdig, Sysdig Open Source. https://github.com/draios/sysdig. Accessed Jan 2020
12. Sysdig, Sysdig Monitor Dashboards. https://sysdig.com/products/monitor/dashboarding/. Accessed Jan 2020
13. BlueMatador, Alert Automation for your Cloud Infrastructure. https://www.bluematador.com. Accessed Jan 2020
14. B. Sababa, System monitoring and anomaly detection application. Final Year Project Report, Department of Computer Science, University of Nicosia, 2020
15. Qt, Qt Open Source Widget Toolkit for GUI and Cross-platform Applications. https://www.qt.io. Accessed Jan 2020
16. Sysdig, Sysdig Falco. https://sysdig.com/opensource/falco/. Accessed Jan 2020

Thinking Outside the Box: Using Escape Room Games to Increase Interest in Cyber Security

Suzanne Mello-Stark, MaryAnn VanValkenburg, and Emily Hao

1 Background

For the past 5 years, we have directed a GenCyber [1] cyber security summer camp for rising high school juniors and seniors. The camp was located at Worcester Polytechnic Institute (WPI) for the first four summers and at Rhode Island College (RIC) last summer. Sponsored jointly by the National Security Agency's National Cryptologic School and the National Science Foundation, a critical goal of the camp is to supply much-needed cyber security training, at no cost, to K–12 teachers and students. More importantly, the camps are meant to pique the camper's interest and inspire them to further explore careers in cyber security.

Although the GenCyber camp has specific goals, each camp director creates the curricula and designs the format for the camp. Our camp currently runs for five consecutive weekdays from 8 am to 4:30 pm. The lab used is a traditional classroom that consists of ~30 personal computers. During camp week, there are 30 h of instruction in cybersecurity and 10 h in other related activities, such as how to write a resume, listening to an invited speaker, or learning an outside game.

Our cyber security program consists of mostly hands-on activities with very little lecture. Many games were invented for the camp. We have scavenger hunts, puzzles to solve, and ice breaker games. In the "Amazing Crypto Race" game, teams are given a set of encrypted clues to decipher. Each clue leads them to another part of campus to find the next clue. As teams return to home base, they win prizes.

S. Mello-Stark (✉)
Rhode Island College, Providence, RI, USA
e-mail: smellostark@ric.edu

M. VanValkenburg · E. Hao
Worcester Polytechnic Institute, Worcester, MA, USA
e-mail: mevanvalkenburg@wpi.edu; eyhao@wpi.edu

K. Daimi, G. Francia III (eds.), *Innovations in Cybersecurity Education*,
https://doi.org/10.1007/978-3-030-50244-7_3

We used Raspberry PIs [2] to illustrate building a computer from scratch. We utilized Virtual Box [3] to create a virtual environment that consisted of three cloned Kali Linux [4] machines where students could practice with networking tools such as NMAP, Ping, and Wireshark, and observe hacking techniques safely. Students used John the Ripper to learn password cracking (and therefore creating safe password) techniques. Students also learned enough Python to create a fun Mad Lib program to share with each other.

One of the favorite games we developed was the Escape Room. The Escape Room is a fun social activity where the participants have to problem solve their way out of a situation. It holds their attention because they are trying to solve a mystery. It can be used to assess whether or not students learned the material as well as they mastered other soft skills, such as teamwork and patience. Escape Rooms are an innovative way to assess whether or not students learn material and also to teach cybersecurity principles.

2 Escape Room Mutations

In our first Escape Room iteration, we turned a conference room into a physical escape room.

We hid letters around the room that when pieced together, spelled a common cyber security idiom (originally from Sun Tzu's *Art of War*), "Know your Enemy." Some of the letters were hidden in plain sight in graphics on the walls. We had a checkerboard with glued down pieces, when tipped over spelled out a letter. We had letters highlighted in books that were scattered around the room. We also had a few puzzles that earned you a letter. We ran teams of five through the room, timed them, and then interviewed them about their experiences. It was fun, but hard to scale or move. Although it exercised teamwork, it didn't teach much of anything else.

In our next iteration, we decided to create something on a smaller scale that we could easily set up and move around. We wanted to teach cyber security concepts and create something uncomplicated, that teachers with no cyber security experience would be comfortable using with their students. At the same time, we wanted a challenge for students that would be fun and exciting and would also really show them that they learned something without being intimidating. We settled on the Escape the Briefcase game that uses simple ciphers from classic cryptography and the cyber security concept of least privilege. We will describe these cyber security concepts in detail in the next section.

We were then invited to hold a workshop on using escape rooms as a teaching method at Women in Cybersecurity (WiCyS) 2018 [5]. We again ran into the issue of scalability. How could we bring enough briefcases for everyone to play? We created a card game that simulated the Escape the Briefcase game. The card game has the same puzzles but can be played using a set of cards. Players enter their answers into Python scripts to see if they are correct. Both versions of this game were featured at the 2018 USENIX Workshop on Advances in Security Education [6].

Students and professors from other universities have taken an interest in both versions of the game. The card game has been turned into an IOS application as part of a graphic design project by Cole Weinhauer at The George Washington University under the direction of Professor Emeritus Shelly Heller [7]. The physical Escape the Suitcase game has morphed into Escape the Luggage and Escape the Purse at other universities. It also has been used at many GenCyber cybersecurity camps around the country. We were the proud winner of the National Cyber Watch Center's Innovation in Cybersecurity Education for 2019 [8].

3 Skills Learned

Throughout history, concerns for privacy and secrecy have prompted the need for concealment, ciphers, and codes. Encryption uses a cipher to take an original message called a plain text and obscures it to an unrecognizable message called a "cipher text." According to Simon Singh, author of *The Code Book* [9], codes are constantly under attack by code breakers and have been for thousands of years. Studying classic cryptography enforces one of the basic tenets of cyber security which is to assume the code breaker has the code. In this game we chose the following ciphers and cyber security principles:

Caesar Cipher: It is documented in Suetonius' Lives of the Caesars LVI (AD 75–150) [10] that Caesar used a basic substitution cipher to send important military messages. The cipher was a simple shift of three characters as follows:

Plain text:	A	B	C	D	E	F	G	H	I	J	K	L	M	N	O	P	Q	R	S	T	U	V	W	X	Y	Z
Cipher text:	D	E	F	G	H	I	J	K	L	M	N	O	P	Q	R	S	T	U	V	W	X	Y	Z	A	B	C

The encryption of the plain text message, "CYBERSECURITY" results in the cipher text, "FBEHUVHFXULWB."

Rail Fence Cipher: The rail fence cipher is an example of a transposition cipher, where the plaintext characters are permuted in some way previously agreed upon by the sender and receiver [10]. For the rail fence cipher, your message is written in a zigzag pattern, and then the cipher is given by reading across the columns. The plaintext is written upon a table of an agreed upon number of rows and columns as follows:

R		I		F		N		E
	A		L		E		C	

The ciphertext is RIFNERALEC for this example.

Mixed Alphabet Cipher: The mixed alphabet cipher is a monoalphabetic cipher where one letter substitutes for another letter and the substitution is agreed upon by the senders and receivers. In our puzzle, we provide the substitution (Figs. 13 and 14), but this is a popular jumble game that you can find in many newspapers.

Steganography: Steganography is hiding a message inside a message. On the surface, you don't know that there is a message being concealed. We provide a letter that looks like an ordinary letter to a friend, but when you discover that there is a set of code words, a new message is uncovered (Figs. 8 and 9).

Least Privilege: A person should only be given the privileges they need to complete the task [11]. In our game, once the first puzzle is solved, all the other puzzles are exposed.

Strong Passwords: This game demonstrates several examples of weak passwords. Passwords that are easy to guess and passwords that are left in plain site are used.

Soft Skills: Teamwork and patience are among the soft skills that are needed to successful solve the mystery. We set a strict time limit on our game and obfuscated the clues to ensure that it was infeasible for one person to complete the puzzles alone.

4 Teaching Strategies

Puzzle-based learning is an accepted training method in cyber security where it is important to learn to think outside the box. The escape room game supplies a noncompetitive alternative to the capture the flag game. Our version is easy for teachers to implement and can be a challenge for students to master. The game can be used to apply problem-based learning techniques for learning by doing, multimodal learning, or assessment approaches.

Learning by doing or trial and error are fundamental methods of problem-solving. Since we supply all the necessary clues to solve the puzzles in the game itself, it can easily be used as a standalone activity in the classroom. Multimodal learning is when students are introduced to a topic in several different ways. Escape room games can be used in addition to short explanatory lectures or other activities to reinforce the material.

It has been shown in similar science-based disciplines that people learn best when they are engaged in an activity [12] and they learn best when they are given the same material in multiple ways [13, 14]. Finally, the escape room can be used as an assessment in place of a traditional test, although more work needs to be done in this area.

5 Related Research

The US Naval Postgraduate School and University of Washington each independently developed informal security-themed tabletop games in which players collaborate as white-hat hackers [15].

The researchers share our goals in trying to expose broader audiences to computer security through nondigital games.

In [16], the researchers take a nondigital game, and make it digital for another GenCyber cyber security camp. This is interesting because their approach is opposite to ours. They also conclude that their approach is more effective in male high school students than in female students, which also adds to our point.

In [17], there is an interesting perspective on CTFs and their perception. We would like to compare CTFs to nondigital game play at a later date. Nondigital games could complement CTFs and provide an easier gateway in the future.

6 The Puzzles and Their Secrets

It is a who-done-it. In this section we describe the puzzles we created and in the interest of academics, we give away the secrets to solving each puzzle. Before beginning, the materials needed are as follows:

- Briefcase with combination lock (6 spinning numbers)
- Computer, preferably without major functionality (to prevent distracting players)
- Phone that allows for text passwords on lock screen
- Wallet with combination lock (3 spinning numbers)
- Wristwatch
- Printed Clues to hide (Figs. 1, 2, 7, 8, 12, 13, 14, and 15)
- E-Clues (Figs. 6, 9, and 10)

There are seven puzzles to piece together and solve in the briefcase. The puzzles can be deciphered in a number of orders, but all the puzzles must be worked out before the entire mystery is revealed.

Puzzle 1: Briefcase Trespass

In the first puzzle, the players work together to break into a locked briefcase. They are given just two items, the briefcase (Fig. 3) and a birthday card (Fig. 1).

To solve the puzzle, players must realize that birthdays are a common password and are easy to crack. Using the date on the card, and the fact that the card's recipient turned 30 on that date, the combination on the suitcase, her birthday, 01–30–88 is derived. This puzzle demonstrates the need for strong passwords. It also illustrates the cyber security principle of least privilege because once inside, they have access to the rest of the puzzles. In the briefcase, the players find a locked wallet, phone, and computer. They also find a watch and several pieces of paper and a couple of pens (Fig. 2).

At the start of the game, the items in the briefcase appear to be separate, unrelated items. The players must use trial and error to piece together and find the other six

Fig. 1 Birthday card

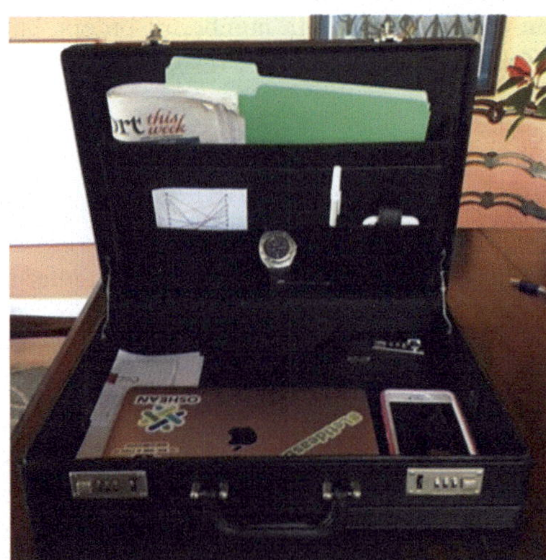

Fig. 2 Open briefcase

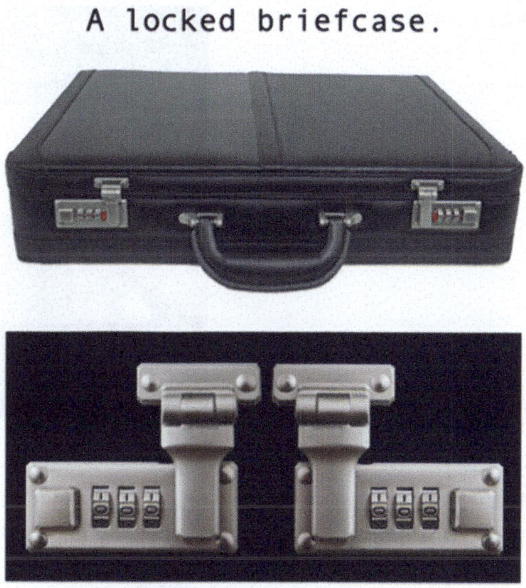

Fig. 3 Locked briefcase

Fig. 4 Wristwatch

puzzles before successfully solving the mystery. The puzzles in the briefcase must first be discovered by the players.

Puzzle 2: A Locked Wallet

For this puzzle, players must unite the wristwatch (Fig. 4) and the locked wallet (Fig. 5) found in the briefcase. The puzzle is solved by trial and error and realizing that there is a three-digit number written on the back of the watch. People often write

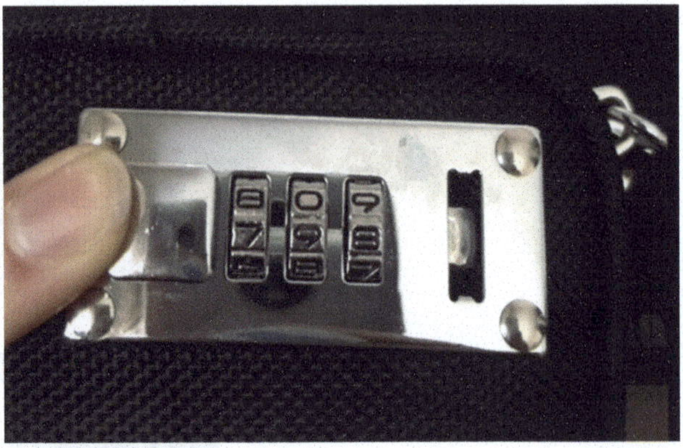

Fig. 5 Locked wallet

their passwords down, so they don't forget them. This is an opportunity to discuss password safety. Is it safe to write your passwords down? Does it matter if you are at home, school, or work?

The wallet holds three clues that are needed in later puzzles. A flash card that explains the Caesar's cipher (Fig. 7), a list of codes on yellow legal pad paper (Fig. 8), and a scrap of paper torn from a spiral notebook with an alphabet substitution puzzle (Fig. 14) written on it.

Puzzle 3: Laptop Break-in

A locked laptop is found in the briefcase. For this puzzle, no clues are explicitly given. To unlock the laptop, players must try the most commonly used passwords, and "password" is the fourth most common password of 2019 according to CNN [18].

On the home screen of the laptop, there is a letter (Fig. 9) and an e-card (Fig. 11) that are needed for later puzzles. To keep the players from going off track, it is important to limit the items that can be explored on the laptop.

Puzzle 4: Phone Phreak

To break into the phone, players need to solve the password hint that appears on the phone's screen saver window (Fig. 6). The screensaver shows a Caesar cipher decal with encrypted text, hinting that a Caesar cipher will unlock the phone. If the players do not know how the Caesar cipher works, a flashcard is found when the locked wallet puzzle is solved (Fig. 7). Once the phone is hacked, the phone's background graphic gives an explanation of a rail fence cipher (Fig. 10) that is needed in Puzzle 6.

Fig. 6 Screensaver

Fig. 7 Caesar cipher

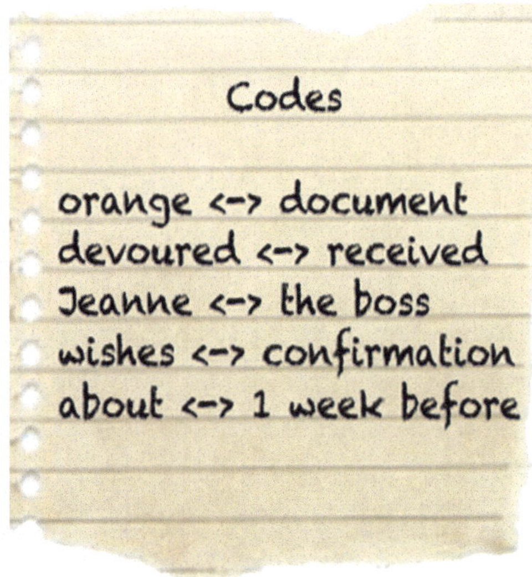
Codes

orange <-> document
devoured <-> received
Jeanne <-> the boss
wishes <-> confirmation
about <-> 1 week before

Fig. 8 Coded letter key

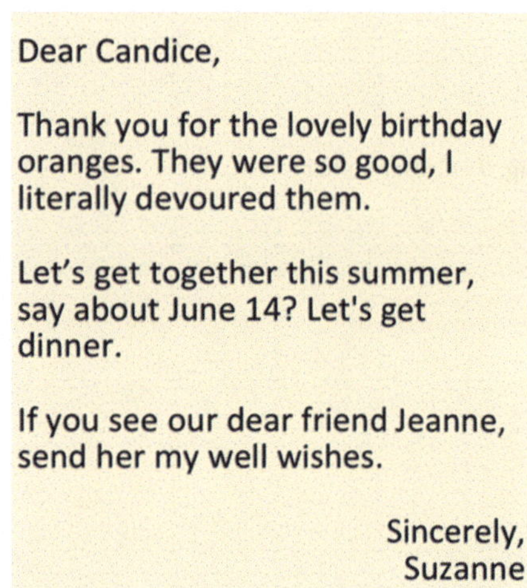
Dear Candice,

Thank you for the lovely birthday oranges. They were so good, I literally devoured them.

Let's get together this summer, say about June 14? Let's get dinner.

If you see our dear friend Jeanne, send her my well wishes.

Sincerely,
Suzanne

Fig. 9 e-Letter

The phone phreak puzzle serves as a lead in for a postgame discussion on the history of classical ciphers and symmetric versus asymmetric cryptography.

Puzzle 5: Orange We Having Fun?

For this puzzle, players need to notice that the codes found in the wallet (Fig. 8) are also words used in the e-letter found on the desktop (Fig. 9). Players substitute the code words in the letter to piece together when the crime is committed. This

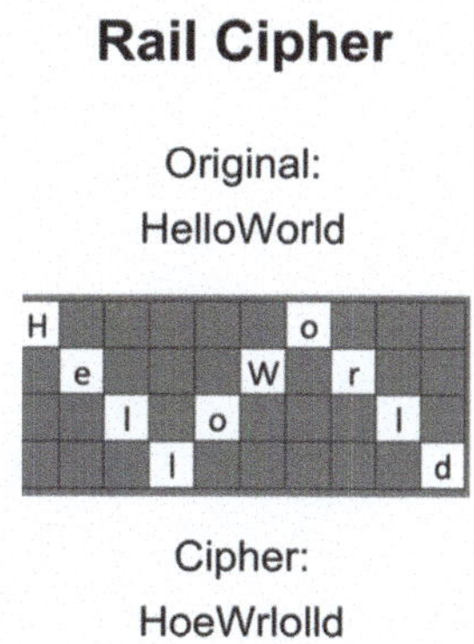

Fig. 10 Rail fence cipher

Fig. 11 Birthday e-card

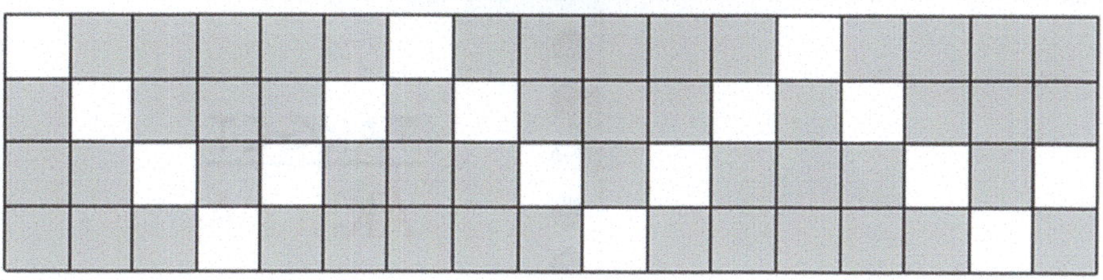

Fig. 12 Blank rail fence cipher

puzzle is an example of steganography. The letter alone reads like an innocent letter between friends. But given the clues, a more sinister hidden message is revealed.

Puzzle 6: Ride the Rails

To solve this puzzle, players must pair the rail fence cipher clue (Fig. 10) with the e-birthday card that talks about riding the rails (Fig. 11). Piecing these two clues together shows where the crime is committed. A blank rail fence cipher is provided without explanation in the briefcase (Fig. 12).

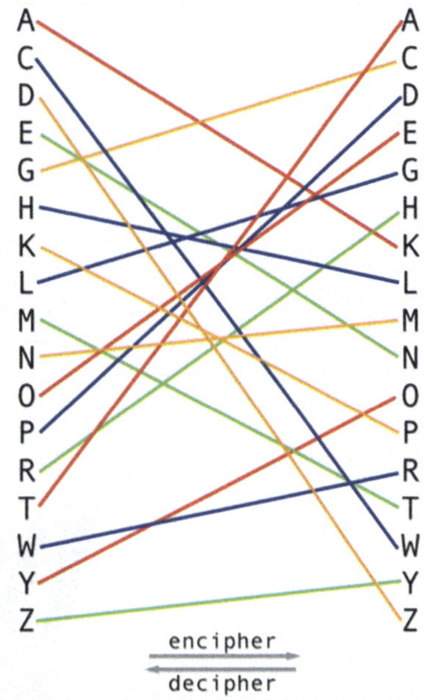

Fig. 13 Alphabet cipher

Fig. 14 Alphabet puzzle

Puzzle 7: Alphabet Soup

The final puzzle is solved by using a specific alphabet substitution cipher that is given in the briefcase (Fig. 13) to unscramble letters on a scrap of paper that is found in the wallet (Fig. 14). This is an example of a monoalphabetic substitution cipher. Together, this puzzle reveals who committed the crime.

Fig. 15 Who–When–Where card

Who: ______________________________

When: ____________________________

Where: ___________________________

7 Game Setup and Hiding Clues

In this section we explain how to set up the game and where to hide each clue. There are four basic steps to setting up the game:

1. Wallet and Watch Setup: Place a version of the alphabet puzzle (Fig. 14), the Caesar cipher key (Fig. 7) and the coded letter key (Fig. 8) in the wallet. Set the wallet combination to "798." Write "798" on the back of the wristwatch (Fig. 4).
2. Phone Setup: The phone needs to have a lock-screen image similar to Fig. 6. The password on the phone needs to be "enigmamachine" (without quotes). When the phone is unlocked, the players need to see an image similar to Fig. 10 to get the rail fence cipher clue.
3. Laptop Setup: Any type of laptop, tablet, or hand-held device that can store files can be used. The device needs to have the password, "password." On the device's desktop, a version of the e-birthday card (Fig. 11) and the e-letter (Fig. 9) should be stored in plain sight.
4. Brief Case Setup: Once you have all the components ready, put the watch in a briefcase pocket and the locked wallet and locked laptop on the floor of the briefcase. Include a version of the alphabet cipher (Fig. 13), the blank rail fence cipher (Fig. 12), and a blank Who–When–What card (Fig. 15) in the pockets or in folders. Add a few pens and blank notebooks for players to use, candy, and any other items to add fun to the game. The briefcase then is ready to close and be locked with the combination "013088."

8 Playing the Game

To start the game, give the players (three players work best) the locked briefcase (Fig. 3) and the birthday card from grandma (Fig. 1). When they are ready to start, read the mission similar to the script below (Fig. 16). Set a 30-min timer (if you wish) and watch the fun.

After the game, a postgame discussion concerning what was learned from each puzzle will further the comprehension of the material.

Suzanne is a foreign sleeper agent. She has been plotting to attack a VIP. We have been following her actions, but we are not yet sure how her target is, when she plans to attack and where it will all go down. Your mission is to uncover these three pieces of intel: who, when and where. We have created a diversion that should keep Suzanne from the room for approximately 30 minutes. Use this time to go through her things and gather this much needed information. We are counting on you!

Fig. 16 The mission

9 Conclusions

There are several future goals of this project. The most immediate goal is to measure whether this game (a noncompetitive game) increases self-efficacy and/or interest in studying cybersecurity. Another short-term goal is to discover ways to extend the game pedagogy to add more security concepts. A longer-term ambition is to create a semester-long course that consists of a comprehensive set of games, puzzles, and activities that could substitute for a customary Introduction to Cyber Security first-year undergraduate course.

Acknowledgments Work described in this chapter was partially supported by NSA/NSF grant H98230-19-1-0166 and Rhode Island College Committee on Faculty Scholarship (CFS) Reassigned Time (RT) grant Spring 2020. The authors would also like to thank Dr. Aberdeen Siraj of Tennessee Tech University, Professor Emeritus Shelly Heller from The George Washington University, and Dr. Ashley Podhradsky from North Dakota State University for all their support on the project.

References

1. GenCyber. www.gen-cyber.com/. Accessed Feb 2020
2. Teach, Learn, and Make with Raspberry Pi—Raspberry Pi. raspberrypi.org/. Accessed Feb 2020
3. Welcome to VirtualBox.org! Oracle VM VirtualBox. www.virtualbox.org/. Accessed Feb 2020
4. Our Most Advanced Penetration Testing Distribution, Ever. Kali Linux, 1 Apr 2020. www.kali.org/. Accessed Feb 2020
5. Workshop 4.4. 2018 WiCyS Workshop—WiCyS—Women in Cybersecurity. WiCyS. www.wicys.net/2018-workshop. Accessed Feb 2020
6. ASE '18. *USENIX*, 10 Nov 2018. www.usenix.org/conference/ase18. Accessed Feb 2020
7. Escape the Room. https://girlscybercamp.seas.gwu.edu/home/resources/. Accessed Feb 2020
8. S. Mello-Stark, *Thinking Outside the Box—Using Escape Room Games to Interest Teachers and Students in Cybersecurity*. 2019 Innovations in Cybersecurity Education, National CyberWatch Center. Innovations in Cybersecurity Education Award-Winning Submission in Instruction for 2019, 2019. Accessed Feb 2020
9. S. Singh, *The Code Book: The Science of Secrecy from Ancient Egypt to Quantum Cryptography* (Anchor Books, New York, 1999). ISBN: 0-385-49532-3. Accessed Feb 2020
10. T. Barr, *Invitation to Cryptology* (Prentice Hall, Upper Saddle River, NJ). ISBN: 0-13-088976-8. Accessed Feb 2020

11. M. Bishop, *Computer Security: Art and Science* (Addison-Wesley, Boston, MA, 2003), pp. 343–344. Accessed Feb 2020
12. T.M. Jones-Wilson, Teaching problem-solving skills without sacrificing course content: marrying traditional lecture and active learning in an organic chemistry course. J. Coll. Sci. Teach. **35**(1), 42–46 (2005). Accessed Feb 2020
13. L. B. Nilson, (2010) Teaching as Its Best, A Researched-Based Resource for College Instructors, 3rd edn. Jossey-Bass, A Wiley Imprint, San Francisco, CA. Accessed Feb 2020
14. G. Kress, C. Jewett, J. Ogborn, T. Charalampos, *Multimodal Teaching and Learning: The Rhetoric of the Science Classroom* (Continuum, London, 2006). Accessed Feb 2020
15. M. Gondree, Z.N.J. Peterson, T. Denning, Security through play. IEEE Secur. Priv. **11**(3), 64–67 (2013). https://www.computer.org/csdl/mags/sp/2013/03/msp2013030064-abs.html. Accessed Feb 2020
16. G. Jin, M. Tu, T. Kim, J. Heffron, J. White, Game-based cybersecurity training for high school students, in *SIGCSE '18: 49th ACM Technical Symposium on Computer Science Education*, 21–24 Feb 2018, Baltimore, MD, USA. ACM, New York, NY, USA, 6 pages. https://doi.org/10.1145/3159450.3159591, http://www.nsfmaneuver.org/document/ACM_SIGCSE2018_Ge.pdf. Accessed Feb 2020
17. M. Bashir A. Lambert, J.M.C. Wee, B. Guo, N. Memon, An examination of the vocational and psychological characteristics of cybersecurity competition participants. *2015 USENIX Summit on Gaming, Games, and Gamification in Security Education (3GSE 15).* https://www.usenix.org/conference/soups2016/workshop-program/wsiw16/presentation/wee. Accessed Feb 2020
18. R. Picheta, The most commonly hacked passwords, revealed. *CNN*, Cable News Network, 23 Apr 2019. www.cnn.com/2019/04/22/uk/most-common-passwords-scli-gbr-intl/index.html. Accessed Feb 2020

Information Visualization as a Method for Cybersecurity Education

Antonio González-Torres, Mónica Hernández-Campos, Jeferson González-Gómez, Vetria L. Byrd, and Paul Parsons

1 Introduction

Cybersecurity is employed to protect against unauthorized access to computational resources and data of individuals and organizations. Learning about cybersecurity is challenging and students can benefit from nontraditional approaches to conveying the complexity of cybersecurity information. One such approach is using Information Visualization (InfoVis) to visually represent content and make learning more accessible and engaging. Using visualizations in teaching can have benefits, including reducing cognitive load, supporting reasoning through graphical constraining [1], enabling the interactive exploration of concepts and phenomena [2], and improving group decision-making [3].

InfoVis offers mechanisms for the visual representation of abstract information, including dynamic and complex phenomena (e.g., cybersecurity data), and facilitates an iterative process of learning through interaction with data [4]. Through iterative and interactive process of working with data, students can analyze both historical and real-time data, explore different perspectives of cybersecurity phenomena, identify patterns and relationships, and draw conclusions [5]. The application

A. González-Torres (✉)
Costa Rica Institute of Technology, Cartago, Costa Rica

ULACIT, San José, Costa Rica
e-mail: antonio.gonzalez@tec.ac.cr; agonzalez@ulacit.ac.cr

M. Hernández-Campos · J. González-Gómez
Costa Rica Institute of Technology, Cartago, Costa Rica
e-mail: mohernandez@tec.ac.cr; jgonzalez@tec.ac.cr

V. L. Byrd · P. Parsons
Purdue University, West Lafayette, IN, USA
e-mail: vlbyrd@purdue.edu; parsonsp@purdue.edu

K. Daimi, G. Francia III (eds.), *Innovations in Cybersecurity Education*,
https://doi.org/10.1007/978-3-030-50244-7_4

of InfoVis to several fields [6], such as knowledge visualization, cybersecurity [5], and software evolution [7] provides strong evidence of its value. For that reason, organizations are considering using it in domains like business intelligence and marketing.

Research has shown that visualizations are effective educational resources for improving computer science education and learning [8–10], partly because students perceive visualizations as fun and engaging, which can enhance the learning process. InfoVis combines two factors that are associated with learning improvement in science; on one hand, it can show both context and detail, showing components of systems within the entire structure. On the other, it displays the behavioral changes and processes within systems [11].

The combination of textual and visual–spatial representations can help students understand complex processes and systems, avoiding misinterpretations that result from textual information only [9, 12]. Furthermore, InfoVis allows learners and teachers to interact with learning content and concept representations in the classroom [13]. Therefore, it acts as a medium to facilitate active learning, which is associated with a better education in higher-order thinking tasks as analysis, synthesis, and evaluation.

The dynamic nature of InfoVis in education promotes the engagement of students in learning activities by themselves and can provide them with feedback about their actions and with knowledge on the process, content, and abilities. The above can have positive effects on the attention and comprehension of participants [9] and encourage and motivate them to keep focused [14]. Hence learners who are actively engaged with visualization technology consistently outperform other students that use visualization passively [15].

Overall, the use of InfoVis can respond to the need for innovative pedagogical environments and techniques to teach cybersecurity content and, at the same time, to engage students in the learning process [13]. In this chapter, we address some cognitive considerations in how visualizations are processed by the human mind according to recent findings and discuss how InfoVis can increase the effectiveness of instructional design.

The rest of this chapter explains a methodological framework based in cognitive and instructional factors (see Sect. 2), discusses a process for the analysis of cybersecurity using InfoVis (see Sect. 3), provides some examples of InfoVis applied to cybersecurity and how it can be applied to education (see Sect. 4), and discusses the main conclusions (see Sect. 5).

2 Cognitive Principles

Teachers and students develop more competent and efficient behavior toward information [16] when they are exposed to the resolution of complex problems that require reflexive answers. In this context, visualizations are a useful tool because they can engage learners in cybersecurity learning activities [8] and challenge them to discover patterns and knowledge not visible at first glance.

Research in the cognitive and learning sciences has demonstrated that learning using visualizations can be more effective due to less working memory processing. Visualizations permit learners to get insight from the relationships between elements easier and faster than verbal and written information, and may also promote better memorability [9, 12].

Cognitive load theory builds upon the widely accepted model of information processing by Atkinson and Shiffrin [17] to support instructional methods for teaching. This instructional approach considers human cognitive architecture for facilitating learning in educational settings. It integrates knowledge on the limited capacity of working memory, the organization of long-term memory, and the interaction between both systems [18].

Instructional procedures based on cognitive load theory are more effective than methods based on traditional procedures [18]. The primary assumption of Sweller is since the amount of information that working memory can hold at one time is limited, instructional methods should avoid overloading learning strategies [9]. According to the findings of the cited authors, the design of visualizations for cybersecurity learning should consider the following factors:

The split attention effect and spatial contiguity principle: A goal of a visualization is to reduce working memory load. So, the placement of visual elements and their text labels should follow the spatial contiguity to avoid split student attention and two separate stimuli sources. The drawback of split attention is that it will demand more working memory resources and will reduce the learning of the content.

The modality effect: The design of visualizations should consider auditory descriptions, rather than a lot of written information. The use of narrations motivates the auditory processing, and so, the visuospatial processing can manage the key visual elements. Narrations should be short to avoid an adverse transient information effect, and its processing occurs separately. Thus, it does not overload the visuospatial processing, and split attention does not happen. Participants who study with supplementary auditory explanations outperformed those presented with supplementary on-screen texts, according to previous studies.

The redundancy effect and coherence principle: The visualization should not include unnecessary redundant information, since the excess of information will require more working memory resources and will affect the learning process. Redundant visual elements hinder learning because they require unnecessary processing in understanding the main concepts.

The signaling principle: Visualizations are more effective when elements are included to cue their essential parts. These signals should not add unnecessary extra embellishments such as color, zooming, or transparencies because they are assets to reduce cognitive load and produce better memorization of the content.

The transient information effect: It predicts that highly ephemeral information will be less effective for learning than less transient information. It is advisable to segment the large-sized visualizations or provide the possibility for students to use the paste facility. The time of display of the information should be large enough for people to process it and not miss it.

Although the previous recommendations are for visualization design, it is necessary to take into consideration not only the quality of the educational resource but the complete planning of the instructional design. Instructional theories recommend not to present direct, explicit information to students, but rather guide them to find the information by themselves. However, there is little confirmation of the effectiveness of minimal guidance for students, but there exists considerable evidence of the importance of explicit instruction [19, 20]. The use of InfoVis for cybersecurity teaching and learning requires a well-organized guide that considers the following:

1. The main concepts behind the visualization
2. Details of the visualization functionalities and its use
3. The interaction that occurs between elements
4. The usefulness of the visualization to solve cybersecurity challenges and the steps to follow

Other recommendations about the design of visualizations [15, 18] that are coherent with what has been mentioned above are the following:

1. Design visualizations according to the topic to teach and the study level in which it will be taught.
2. Provide flexible controls to the visualization for the students to have the ability to execute the tool both forward and backward according to their needs.
3. Add explanations to the visualizations for giving a better understanding of its details and the knowledge that looks to transmit.
4. Implement an appropriate degree of interactivity in the visualizations so that students can browse and discover knowledge more effectively.
5. Permit learners to use worked examples to study because they will perform better on subsequent problems than learners who must solve the same issues, due to a reduction in extraneous cognitive load. The assumption is that worked examples reduce working memory load.
6. Promote collaborative learning for difficult problems where knowledge is spread among two or more people. Collaboratively work facilitates learning and performs better than studying individually.

Teachers must analyze how to implement cybersecurity visualizations when considering the above recommendations and that the purpose of these is to use them in security matters and decision-making but not in learning environments. So, the use of visualizations as an educational resource that allows an efficient and effective learning experience requires following a well-defined instructional strategy [4].

3 Instructional Design Considerations

The first step of instructional design is diagnosis and context analysis. Hence, it is essential to gather information about the learning situation, the environment, the people involved, the abilities of learners, their mastery of concepts, and the level of the course [21]. Then, the second step is the definition of the learning objective. Once the diagnosis of the educational context is complete, it is necessary to define the abilities of the students when the learning experience finishes.

Instructors can decide the scope of the learning strategy using the taxonomy of Anderson and Krathwohl [22], which is a revision of the taxonomy proposed by Bloom [23]. For example, students at the elemental level can remember and explain cybersecurity concepts. In contrast, intermediate-level pupils can apply them to solve problems or connect new ideas in different situations. However, students in upper levels should evaluate options, justify decisions, or create new proposals. So, the use of visualization for cybersecurity as an educational resource could have a broad range of actions depending on the learning goals.

The third step is the definition of the contents. At this stage, the teacher must locate the main contents and the subtopics that emerge from each one. This stage defines the relationships between concepts to properly guide the process.

The fourth step consists of the definition of the learning strategy, which occurs after defining the learning objectives, the contents, and the relation between them. The teacher should consider that the strategy promotes the achievement of the learning objectives. Therefore, the use of visualization can be for classroom demonstrations, laboratory exercises, online learning, homework, pre-class exercises, and reference material [15].

At this stage, it is necessary to remember that students must engage in active learning activities that are easy to understand and which do not overload them of contents. Students should answer strategic questions and even constructing their own visualizations [24] in collaborative teams or individually.

The fifth and final step is the evaluation outcome. In this context, an important recommendation is to create instruments to evaluate the results of the learning process. The assessment provides the instructor with a tool to measure the performance of the strategy and students. Learners should have access to these tools since the beginning of the learning strategy [3].

4 Visualization and Cybersecurity

The function of InfoVis in the analysis of cybersecurity-derived data is to move analysts toward the construction of useful knowledge through an interactive discourse with the data. It should motivate students to engage in a productive and iterative exploration process that allows them to request additional data to the system as they identify interesting patterns. This exploration involves browsing, filtering, and exploring different perspectives of data in one or more visualizations until they obtain the necessary knowledge or consider that is impossible to reach a determinate conclusion, using the data and representations available.

InfoVis facilitates the analysis of historical and real-time cybersecurity data, for example, related to logs, IDS and IPS signatures, email data, and malicious code.

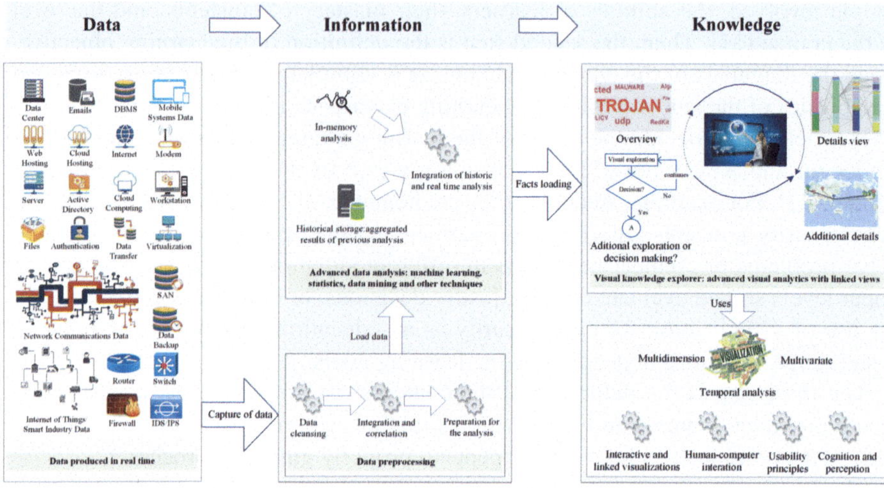

Fig. 1 Big Data visual analytics applied to cybersecurity data

The tools for these types of data include the use of representations as diverse as heatmaps, dispersion graphs, radial layouts, treemaps, parallel coordinates [25], maps, bar charts, and grids [26].

The application of InfoVis to cybersecurity data is a transformation process that could be thought of as a funnel, where raw data is analyzed and filtered in several steps until this is converted into knowledge. The output of the process is a reduction, in terms of volume, of the original input, which contains all the required elements to carry out informed decision-making. This transformation is described in Fig. 1 and is based on the visual analytics process proposed by Gonzalez-Torres [7, 27], which took as a basis the previous definitions formulated by other authors [28–31].

The process mentioned above involves reading data from heterogeneous data sources, then cleaning and correlating data, carrying out the automatic analysis of historical and real-time data, creating visualizations, and studying the relationships between the elements involved in the results. The phases that constitute this process are Extract, Transform, and Load (ETL); Advanced Data Analysis; and Visual Knowledge Explorer (see Fig. 1).

ETL has the function of performing the connection and retrieval of logs from data centers, file servers, DBMS servers, web hosting servers, cloud hosting infrastructure, workstations, routers, firewalls, IPS devices, and switches, and also collects data from network traffic and IoT networks operated by organizations. This phase can take advantage of software tools such as Apache Kafka, Apache Storm, Apache Pig, and Apache Tez. Apache Kafka works as a subscriber system composed of producers and consumers. Producers publish data to "topics" (i.e., a "topic" is a category name to which consumers subscribe), which could be read, cleaned, integrated, correlated, transformed, and filtered by consumers, using Apache Storm, Apache Tez, and Apache Pig [32, 33].

The responsibility of the Advanced Data Analysis phase is to carry out in-memory analysis for real-time data and execute batch processing tasks for historical data. Real-time data might be processed by Apache Spark, which performs analysis tasks without having to write the information directly to the hard disk, decreasing processing times [34], whereas historical data can be handled by the latest version available of Apache Hadoop [32]. The integration of in-memory and historical analysis results is a complex task that has not yet been solved successfully, despite some efforts made in this direction [35]. In this context, the most suitable options nowadays include the use of Spark and MemSQL, which can facilitate the comparison, integration, and query of results of real-time and historical data analysis.

Once data has been processed, the results could be refined in the Visual Knowledge Explorer stage. This phase is concerned with the definition of the visualization designs and the human–computer interaction techniques needed to support students' interactions, and thus, it is responsible for creating the data models, data structures, and visual mappings required to coordinate the data to be displayed.

The function of the Visual Knowledge Explorer is to move students toward the construction of useful knowledge. This stage makes possible a feedback loop between the student and the system: The student explores the visualizations, requests additional data to the system using the available interaction possibilities, and the system provides the required data according to the availability of the proper data models, structures, and visual mappings. The student continues to interact with the system, browsing, filtering, and exploring different perspectives of data in one or more visualizations until she obtains the necessary knowledge or considers that is impossible to reach a determinate conclusion, using the data and representations available.

There are many database options for Big Data, including SQL, Non-SQL, NewSQL, and SQL-On-Hadoop [36]. So, it is essential to study in detail the particular problem at hand, according to the type of data (structured, unstructured, real-time, or historical data), and the features of the available databases for storing and processing large volumes of data, to decide which option to choose.

5 Cybersecurity Learning Through InfoVis

Some factors cybersecurity deals with are security events, malware analysis, network behavior, security metrics, security issues found in the source code of programs, attack patterns, penetration testing, traffic flows, and the relationship between any of these elements. Learning about all these factors requires time and well-prepared instructors. So, InfoVis tools such as the Visual Knowledge Explorer (VKE) [5] and BubbleNet [37] can be of great value for explaining some basic concepts and their connections.

Visual Knowledge Explorer (VKE)

A common task in security analysis is to characterize the composition of network traffic when unusually high traffic rates are coming into the IT infrastructure. The high traffic may be a result of an active attack or maybe a peak in transactions from business branches or partners. The first step in determining the origin and destination of network traffic is to use IP addresses and a map to plot their geographic locations and obtain this information intuitively and quickly, at first glance. If the traffic does not appear to come from a reputable origin, the protocols used (i.e., TCP, UDP, RTP, ICMP, or other) and target ports should be determined, as well as its low-level composition and how traffic has been classified. Furthermore, if the traffic is due to email, the content of attachments should be analyzed to detect malicious code, study execution traces, and identify potential system damages using specialized tools (e.g., Cuckoo).

Inspired by the aforementioned needs of security analysis, and the benefits of InfoVis, VKE makes use of linked views, various interaction techniques, a color code, a parallel node-link (PNL) visualization for characterizing the flows of traffic details (i.e., protocols, IP addresses, ports, and traffic classification), a map to depict the location of the sources and destinations of traffic, and a tag cloud to show statistics on the frequency of the classification of traffic types (see video in link https://youtu.be/ezKU1nWqoVk).

Inspired by the aforementioned needs of security analysis, and the benefits of InfoVis, VKE makes use of linked views (see Fig. 2), various interaction techniques, a color code, a parallel node-link (PNL) visualization (see Fig. 2a) for characterizing the flows of traffic details (i.e., protocols, IP addresses, ports, and traffic classification), a map to depict the location of the sources and destinations of traffic (see Fig. 2b), and a tag cloud to show statistics on the frequency of the classification of traffic types (see Fig. 2c).[1]

The list Fields Available (Fig. 3a) shows the most common fields stored by a Snort IPS and give the possibility to instructors of explaining each one at a time that demonstrate a convenient combination for analyzing traffic. The data displayed by the visualizations is configurable, so students can select the variables to be used by dragging them from a list of available fields to a list of chosen fields, as it is shown in Fig. 3a, b. Figure 3a illustrates a student dragging a variable from the Fields Available list to the Fields Selection list, whereas Fig. 3b display both lists after more variables have been moved from one list to the other.

The PNL visualization characterizes traffic flows using all the variables in the list of selected fields and permits analysts to get an insight of the correlation of variables and to have a clearer perspective on how traffic and data are flowing in the network. PNL depicts the network traffic flows with lines connecting colored squares, which are placed in columns. Columns represent data categories, and colored squares depict variables within a category. The number of columns depends

[1] A demo video is available at https://youtu.be/ezKU1nWqoVk.

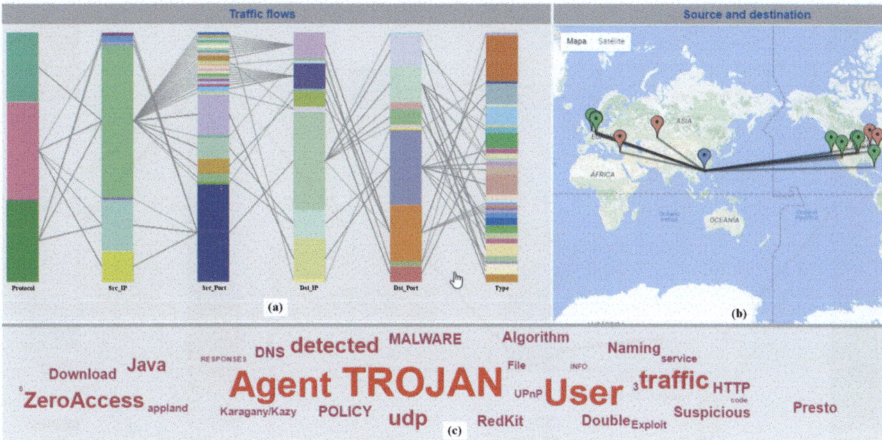

Fig. 2 InfoVis applied to cybersecurity data. (**a**) shows a view of the traffic flow at the protocol level, whereas (**b**) displays how traffic flows between locations, and (**c**) illustrates the type of traffic coursed

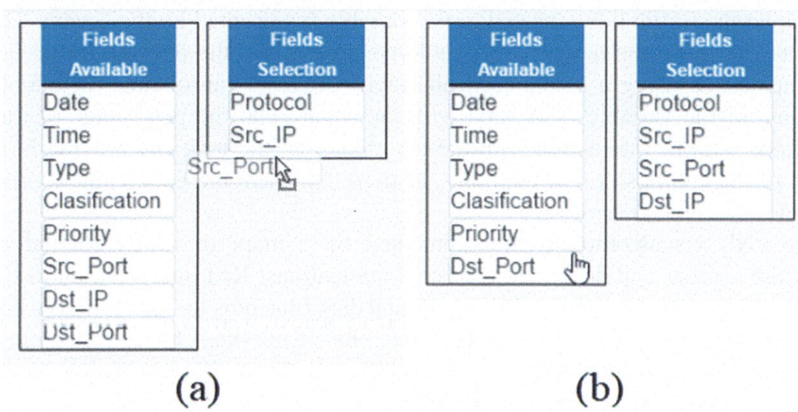

Fig. 3 Selection of fields depicted by the PNL visualization (**a**) shows the traffic fields that users can select to depict, and (**b**) displays the list of data fields to represent for the type of traffic

upon the number of variables selected by students, as described before, whereas the height of colored squares encodes the aggregated value of the corresponding variable. The color of the squares is associated uniquely to a variable. The order in which variables are added to the list of selected fields determines the order of columns in the graphic.

The configuration of PNL in Fig. 4 shows six category columns: Protocol, Src IP, Src Port, Dst IP, Dst Port, and Type. This representation permits to explain a basic concept: traffic flows between a source and destination IP use ports is associated to a layer 4 protocol. The interaction of the visualization permits to highlight specific traffic flows and analyze its composition. The interaction in this visualization starts when students click on a colored square and the tool highlights in orange the lines

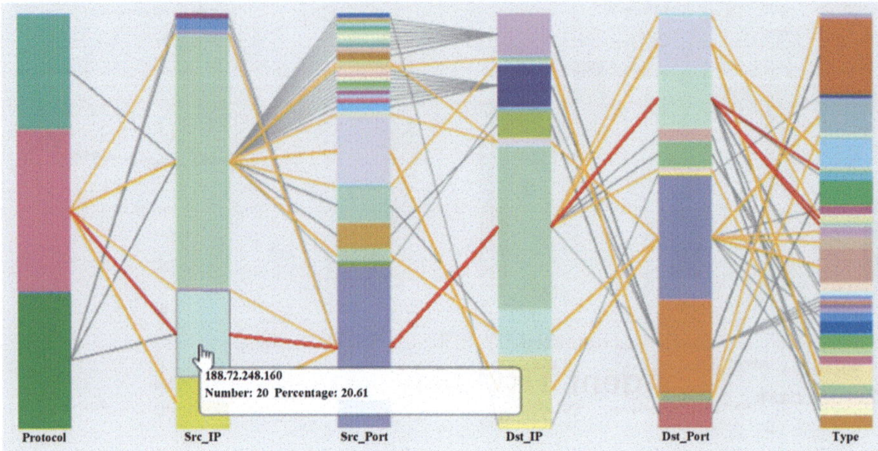

Fig. 4 Parallel node-link (PNL) visualization

that depict the traffic flows related to the variable represented by the square. Then, students can move the mouse over a colored square, and the relevant traffic flows are emphasized using a red line. Furthermore, students can obtain, via a tooltip, the name of the variable associated with the square and the percentage of traffic associated with it. Interaction with the tag cloud and the map can provide further details of the composition of the flows, as these represent the geographic locations and types of traffic cursed.

The PNL visualization is complemented by a map, in which colored pins depict the sources and destinations of communications: Red pins are used to show suspected attackers, green pins for targets, and dark blue pins for locations that could be either a source or target of attacks. The color of the pins changes according to the interactions performed with the PNL and the tag cloud. When selections in these visualizations are made, the colored point in the center of pins is also changed: A red point is used to depict suspected attackers, a green point for targets, and a gray star for those locations that could be either a source or target of attacks.

Furthermore, the tag cloud complements the PNL visualization providing statistical information on the type of suspected threats found in communications (see Fig. 5). The main visual elements of this visualization are the size and color of words (the default color is red), which encode the frequency of threats classified in the Type category (or the category that was chosen to be displayed during the configuration phase). Thus, the higher the frequency of a word in the log data, the greater is its size and the darker is its hue in the tag cloud.

The interaction path students can follow during analysis could start in any of the three visualizations. Thus, several scenarios are possible, and the instructor and students can carry out the study of different factors taking place in cybersecurity attacks. The following scenarios are some practical examples in which an instructor can demonstrate the functionality of VKE.

Fig. 5 Frequency of type of security events detected

Decode source and destinations: A student wants to explore TCP traffic flows and decode source IP addresses and target ports involved in communications. The interaction between the student and VKE could proceed as follows:

1. The student clicks in the magenta square (TCP) in the Protocol category in the PNL visualization.
2. Orange lines encode traffic flows in which the associated variable to that square is participating.
3. The map highlights the involved IP addresses, coloring the pins in orange and the red points in the center of this according to the code codification for source and destination locations, as well as for locations which are both source and destination.
4. PNL shows that 5 IP addresses (Src IP category) are the source of the traffic, which is sent through 7 ports (Src Port category) to 5 destination IP addresses (Dst IP category) and 6 ports (Dst Port category), under multiple traffic types.
5. The student is interested in the light green square in the Src IP category because of its large size and moves the mouse over it, which results in the following actions:

    (a) A tooltip is displayed: the square is associated with the IP 202.45.65.18 and the percentage of packets sent over the network with that IP as the traffic source was 60.82% of the total traffic.
    (b) Red lines are drawn to highlight the traffic flows in which the selected IP address is involved.

An important detail to note is that both categories, Src IP and Dst IP, include a light green-colored square—this signifies that the same IP address has been used as both source and destination. Hence, most of the traffic originated in the communications corresponds to the IP 202.45.65.18 as source (highlighted in red) and destination (emphasized by an incoming orange line). After examining the map, it can be concluded that the traffic destination of this IP address is in the US and that traffic sources are in both Asia and the US.

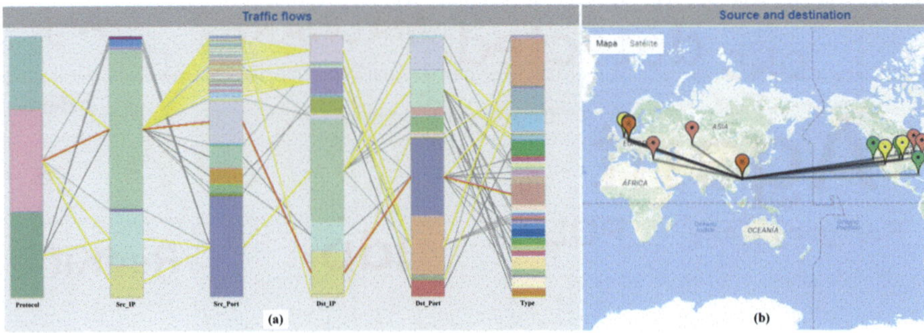

Fig. 6 Traffic flows according to protocols and locations (**a**) shows the view of the traffic flow at the protocol level, whereas (**b**) displays the traffic flows between locations

The above information could be used by an analyst to discern if the incoming traffic is being originated by legitimate branches of the organization or partners or is coming from attackers.

Analysis of suspicious traffic: A student examines the recent traffic activity with VKE, and it comes to her attention that the tag cloud has registered many Trojan and malware events; thus, she decides to investigate the traffic flows and the location of sources and destinations of traffic. Accordingly, the steps involved in this investigation are as follows:

1. The student clicks over the word TROJAN, which is featured in yellow (see Fig. 5).
2. The color of traffic flow lines in PNL and the pins in the map are changed to yellow, as is illustrated in Fig. 6a, b.
3. The student examines PNL and determines that most traffic has been originated by 202.45.65.18 which is her own IP address (represented by the light green square) and sent over to four destinations. Furthermore, the IP address has also received many incoming communications, according to the size of the light green square in the Dst IP category.
4. According to the map, traffic coming into 202.45.65.18 originated in Europe (see the yellow pin in the left side of Fig. 6b, beside the dark orange pin), whereas the outgoing traffic was sent to two different locations in the US. Sources and destinations are identified by a small colored point in the center of the pins, as explained above.
5. Subsequently, the student clicks on the word MALWARE, which is painted in dark orange color in the tag cloud, whereas the corresponding traffic flow is rendered using the same color in PNL and the sources and destination of traffic are represented by pins of the same color in the map.
6. The dark orange line was drawn above a yellow line, indicating that the same traffic parties were sending and receiving Trojan attacks. However, this line brings out a comparison of the magnitude of traffic flows involved with Trojan and malware communications.

7. The dark orange pins in the map contain a gray star in the center, indicating that these locations were both sources and destinations of malware traffic.
8. Based on the information provided by VKE, the student can conclude that she is a source and destination of traffic carrying out Trojans and malware. This may indicate that her infrastructure was taken over by attackers to attack third-party networks. However, an outsider analyst could also conclude that because of the geographic location of the IP address 202.45.65.18, it is an attacker.

BubbleNet

The identification of cybersecurity data patterns, through a proper visualization scheme, can help students understand the types of vulnerabilities and attacks in a real-life environment. In line with the above, BubbleNet [37] is a dashboard to discover and present patterns of real-time data, aggregated by country, day, and hour. Through its five views, BubbleNet permits nonexpert users to learn about Denial-of-Service (DoS) attacks and network scams.

Figure 7 shows the main view of BubbleNet and a demo video of the tool in the following link https://bit.ly/38gMAqF. The circles or "bubbles" represent countries; the bigger the circle, the larger the amount of data associated with the country (i.e., number of records). The heatmap displays critical information: the blue colors represent low deviation from normal data and the red color depicts a high

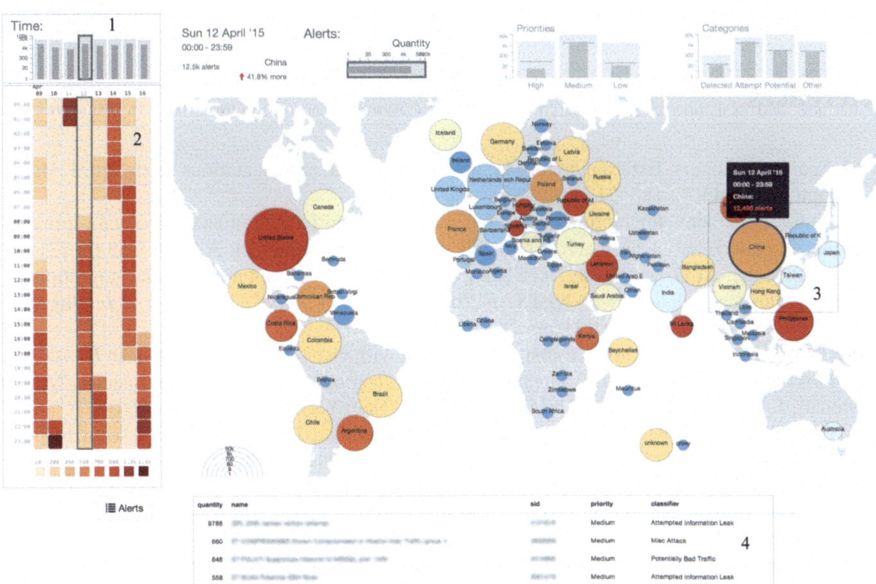

Fig. 7 Interaction with BubbleNet for investigating DoS attacks [37]

deviation. The lateral panel visualizes the patterns for specific days and hours. The visualization permits to aggregate data from the views (time, alerts, priorities, and categories) and study historical data pattern. Therefore, students can easily identify abnormal situations related to possible attacks. The following is a scenario in which students and use the tool to understand DoS attacks.

Investigate traffic latency: A network administrator received complaints about high latency from users and suspects a DoS attack. Figure 7 shows, using numbers, the steps followed for gathering the necessary information in BubbleNet:

1. Check the Time view (upper left corner) to filter dates with higher records. The visualization displays these days using gray-colored bars.
2. The student selects interesting (red) time intervals with high-traffic dates.
3. Then, the map updates and the network administrator select the country or countries with the highest amount of records, by clicking directly on them.
4. The student views the detailed information of the records by checking the Display on-Demand view (bottom). This view contains the number of records, names, ids, priorities, and classifiers.

The examination of each view (Fig. 7) and the careful reviewing of the historical information from the application allow the administrator to determine whether this is a normal situation or in fact a DoS attack.

Acknowledgments The authors wish to thank La Universidad Latinoamericana de Ciencia y Tecnología (ULACIT), Costa Rica, for its support to this research and also to Juan Alvarez Piedra, Osael Josue Jimenez Murillo, and Pablo Andrés Rodríguez Blanco for their contribution with code for VKE.

References

1. M. Scaife, Y. Rogers, External cognition: how do graphical representations work? Int. J. Hum. Comput. Stud. **45**(2), 185–213 (1996)
2. K. Sedig, P. Parsons, Interaction design for complex cognitive activities with visual representations: a pattern-based approach. AIS Trans. Hum. Comput. Interact. **5**(2), 84–133 (2013)
3. N.A. Giacobe, M.D. McNeese, V.F. Mancuso, D. Minotra, Capturing human cognition in cyber-security simulations with NETS, in *2013 IEEE International Conference on Intelligence and Security Informatics*, (2013), pp. 284–288
4. J. Klerkx, K. Verbert, E. Duval, Enhancing learning with visualization techniques, in *Handbook of Research on Educational Communications and Technology*, ed. by J. M. Spector, M. D. Merrill, J. Elen, M. J. Bishop, (Springer New York, New York, NY, 2014), pp. 791–807
5. A. Gonzalez-Torres, V.L. Byrd, P. Parsons, VKE: a visual analytics tool for cybersecurity data, in *2019 International Conference on Security and Management (SAM'19)*, (2019), pp. 56–62
6. A. González-Torres, F.J. García-Peñalvo, R. Therón, A. González-Torres, F.J. García-Peñalvo, R. Therón, Human–computer interaction in evolutionary visual software analytics. Comput. Hum. Behav. **29**(2), 486–495 (2013)
7. A. González-Torres, F.J. García-Peñalvo, R. Therón-Sánchez, R. Colomo-Palacios, Knowledge discovery in software teams by means of evolutionary visual software analytics. Sci. Comput. Program. **121**, 55–74 (2016)

8. G.R. Garay, A. Tchernykh, A.Y. Drozdov, S.N. Garichev, S. Nesmachnow, M. Torres-Martinez, Visualization of VHDL-based simulations as a pedagogical tool for supporting computer science education. J. Comput. Sci. **36**, 100652 (2019)
9. J.C. Castro-Alonso, P. Ayres, J. Sweller, Instructional visualizations, cognitive load theory, and visuospatial processing, in *Visuospatial Processing for Education in Health and Natural Sciences*, (Springer International Publishing, Cham, 2019), pp. 111–143
10. C. Vieira, P. Parsons, V. Byrd, Visual learning analytics of educational data: a systematic literature review and research agenda. Comput. Educ. **122**, 119–135 (2018)
11. R.E. Mayer, J.K. Gallini, When is an illustration worth ten thousand words? J. Educ. Psychol. **82**(4), 715–726 (1990)
12. I. Vekiri, What is the value of graphical displays in learning? Educ. Psychol. Rev. **14**, 261–312 (2002)
13. M.M. North, S.M. North, Dynamic immersive visualisation environments: enhancing pedagogical techniques. Australas. J. Inf. Syst. **23** (2019)
14. D. Schweitzer, W. Brown, Interactive visualization for the active learning classroom, in *Proceedings of 38th SIGCSE Technical Symposium on Computer Science Education*, (2007), pp. 208–212
15. T.L. Naps, et al., Exploring the role of visualization and engagement in computer science education, *Working Group Reports on ITiCSE Innovation and Technology in Computer Science Education*, 2002, pp. 131–152
16. P. Antoniou, L. Kyriakides, A dynamic integrated approach to teacher professional development: Impact and sustainability of the effects on improving teacher behaviour and student outcomes. Teach. Teach. Educ. **29**, 1–12 (2013)
17. R.M. Shiffrin, R.C. Atkinson, Storage and retrieval processes in long-term memory. Psychol. Rev. **76**(2), 179–193 (1969)
18. J. Sweller, *Cognitive Load Theory, Evolutionary Educational Psychology, and Instructional Design* (Springer, Cham, 2016)
19. P.A. Kirschner, J. Sweller, R.E. Clark, Why minimal guidance during instruction does not work: an analysis of the failure of constructivist, discovery, problem-based, experiential, and inquiry-based teaching. Educ. Psychol. **41**(2), 75–86 (2006)
20. R.E. Mayer, Should there be a three-strikes rule against pure discovery learning? The case for guided methods of instruction. Am. Psychol. **59**(1), 14–19 (2004)
21. C.M. Reigeluth, B.J. Beatty, R.D. Myers, *Instructional-Design Theories and Models, Volume IV: The Learner Centered Paradigm of Education* (Routledge, New York, 2016)
22. L.W. Anderson, D.R. Krathwohl, et al., *A Taxonomy for Learning, Teaching, and Assessing. Abridged Edition* (Allyn and Bacon, Boston, MA, 2001)
23. B.S. Bloom et al., *Taxonomy of Educational Objectives. Vol. 1: Cognitive Domain* (McKay, New York, 1956), pp. 20–24
24. T. Mahmood, U. Afzal, Security analytics: big data analytics for cybersecurity: a review of trends, techniques and tools, in *2013 2nd National Conference on Information Assurance (NCIA)*, (2013), pp. 129–134
25. E. Glatz, S. Mavromatidis, B. Ager, X. Dimitropoulos, Visualizing big network traffic data using frequent pattern mining and hypergraphs. Computing **96**(1), 27–38 (2014)
26. H. Shiravi, A. Shiravi, A.A. Ghorbani, A survey of visualization systems for network security. IEEE Trans. Vis. Comput. Graph. **18**(8), 1313–1329 (2012)
27. A. González Torres, F.J. García-Peñalvo, R. Therón-Sánchez, How evolutionary visual software analytics supports knowledge discovery. J. Inf. Sci. Eng. **29**(1), 17–34 (2013)
28. J.J. van Wijk, The value of visualization. Vis. Conf. IEEE **0**, 11 (2005)
29. B. Shneiderman, The eyes have it: a task by data type taxonomy for information visualizations, in *Proceedings 1996 IEEE Symposium on Visual Languages*, (1996), pp. 336–343
30. S.K. Card, J. Mackinlay, B. Shneiderman, *Readings in Information Visualization: Using Vision to Think* (Morgan Kaufman, San Francisco, CA, 1999)
31. E.H. Chi, A taxonomy of visualization techniques using the data state reference model, in *Proceedings of the IEEE Symposium on Information Vizualization 2000*, 2000, p. 69

32. N. Marz, J. Warren, *Big Data: Principles and Best Practices of Scalable Realtime Data Systems*, 1st edn. (Manning Publications Co, Greenwich, CT, 2015)
33. A. Fasale, N. Kumar, *YARN Essentials* (Packt Publishing, Birmingham, 2015)
34. A.G. Psaltis, *Streaming Data: Understanding the Real-Time Pipeline* (Manning Publications Company, Shelter Island, NY, 2017)
35. L. Antova et al., Datometry hyper-Q: bridging the gap between real-time and historical analytics, in *Proceedings of the 2016 International Conference on Management of Data*, (2016), pp. 1405–1416
36. J. Román, The Hadoop Ecosystem Table, 2017
37. S. McKenna, D. Staheli, C. Fulcher, M. Meyer, Bubblenet: a cyber security dashboard for visualizing patterns. Comput. Graph. Forum **35**(3), 281–290 (2016)

Part II
Curriculum Development

How to Prevent Your Smart Home Device from Turning into a Weapon

David Zeichick

1 Introduction to Smart Home Device Security

The term Internet of Things (IoT) first appeared in 1999 and is attributed to the British technologist Kevin Ashton [1]. He described it as physical objects that connect to the Internet via sensors. The term has grown to include the data that is exchanged between devices, stored in the cloud, and then analyzed [2]. Smart home devices are a subset of IoT, referring to IoT devices used in a residence. Here we use the term "smart home devices" instead of IoT since this project is focused on devices found in a home. Examples of smart home devices range from smart light bulbs that can turn on when we enter the room, smart refrigerators that remind us that we are almost out of milk, smart doorbells that call our smart phone and allows us to talk to the person at the front door, to the more bizarre example of soil sensors for house plants that tweets "water me please" when they are too dry [3].

Companies are rushing to meet consumers' growing need for smart home devices. This rush to market by manufacturers has produced serious deficiencies in privacy and security. This same mistake was made 20 years ago when consumers rushed to the Internet to shop and bank online [4]. The Internet was designed to openly share data, which is the complete opposite of what is necessary for secure transactions. Attackers took advantage of the lack of security by creating malware and sniffing unencrypted data with the goal of stealing personal data [4]. The industry reacted by adding security layers and products. Secure Sockets Layer (SSL) was implemented to secure Internet transactions, antivirus programs were developed to rid computers of nasty malware, and passwords were adopted to authenticate users. This solution is not foolproof since it relies on consumers to implement many of the solutions. Unfortunately, most home users are not technically savvy; they do

D. Zeichick (✉)
California State University, Chico, Chico, CA, USA
e-mail: dzeichick@csuchico.edu

K. Daimi, G. Francia III (eds.), *Innovations in Cybersecurity Education*,
https://doi.org/10.1007/978-3-030-50244-7_5

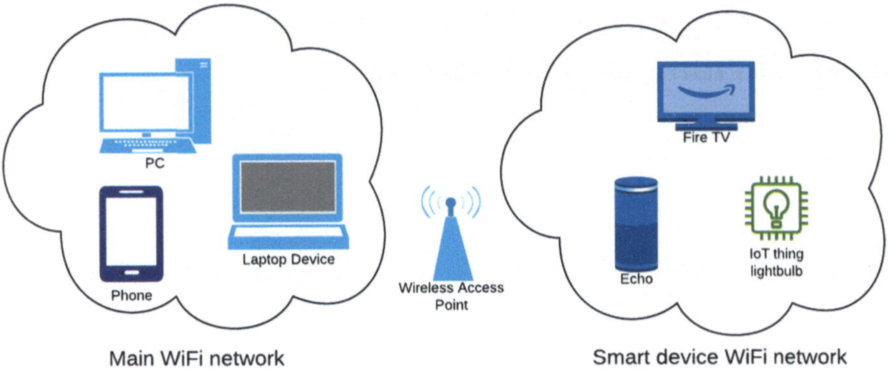

Fig. 1 Separate Wi-Fi networks for smart home devices

not understand how to properly configure their systems or realize the importance of a strong password [5]. The same is true with smart home devices. Most home users can't perform basic security functions which leads to the question: Should they be adding smart devices to their homes [6]?

For those users that have some technical proficiency, making some minor adjustments to their smart home devices will greatly enhance their security. First, changing the default password of their smart home device would prevent their device from being susceptible to the Mirai attack [7]. Second, placing their smart home devices on a dedicated home Wi-Fi network that is separate from their other devices (e.g., laptop and smartphone) prevents attackers that have taken over their smart device from spying on their network traffic (see Fig. 1). Finally, for the student interested in heading off security issues at the network level, this project is for them.

Weaponizing Smart Home Devices

Criminals are turning smart home devices into weapons. In the Fall of 2016, cybercriminals compromised thousands of smart home devices and infected these devices with malware, called Mirai, that hijacked the systems. The cybercriminals then forced these devices to participate in a distributed denial of service (DDOS) attack. A typical DDOS attack involves an attacker commandeering a large number of desktops or laptops. This attack was unique because it was the first large-scale attack leveraging vulnerable networked cameras, digital video recorders, and other smart devices. The Mirai smart home device botnet attack was used against the Domain Name System (DNS) service DynDNS, which made many popular websites, such as Twitter and Reddit, unreachable for several hours [8]. This is accomplished by instructing the thousands of Mirai-infected smart home devices to all send network traffic simultaneously to a victim site, which then renders the victim site unresponsive [9].

Smart home devices are particularly vulnerable to this type of attack since many devices are being produced by manufacturers that contain critical security flaws [10,

11]. Due to the explosive growth in the number of smart home devices purchased for household use, the number of eligible new recruits for the smart home device botnet army is dramatically increasing [12]. Protecting susceptible smart home devices from such an attack is challenging; in comparison to personal computers, the security of smart home devices cannot be solved with traditional solutions, such as antivirus software and automatic patching [13]. This is due to smart home devices' constrained resources including limited power, a slow processor, and minimal storage [14, 15]. Therefore, researchers are now looking at implementing the security of home smart home devices at the network level [16–19].

Smart home devices tend have very predictable network traffic behavior [20]. Unlike laptops, tablets, and smartphones that are used to visit a large variety of websites every day, smart home devices send traffic to a limited number of domains [21]. Smart home devices were found to connect to less than ten servers per day compared to person computers that connected to hundreds of different servers each day (see Fig. 2).

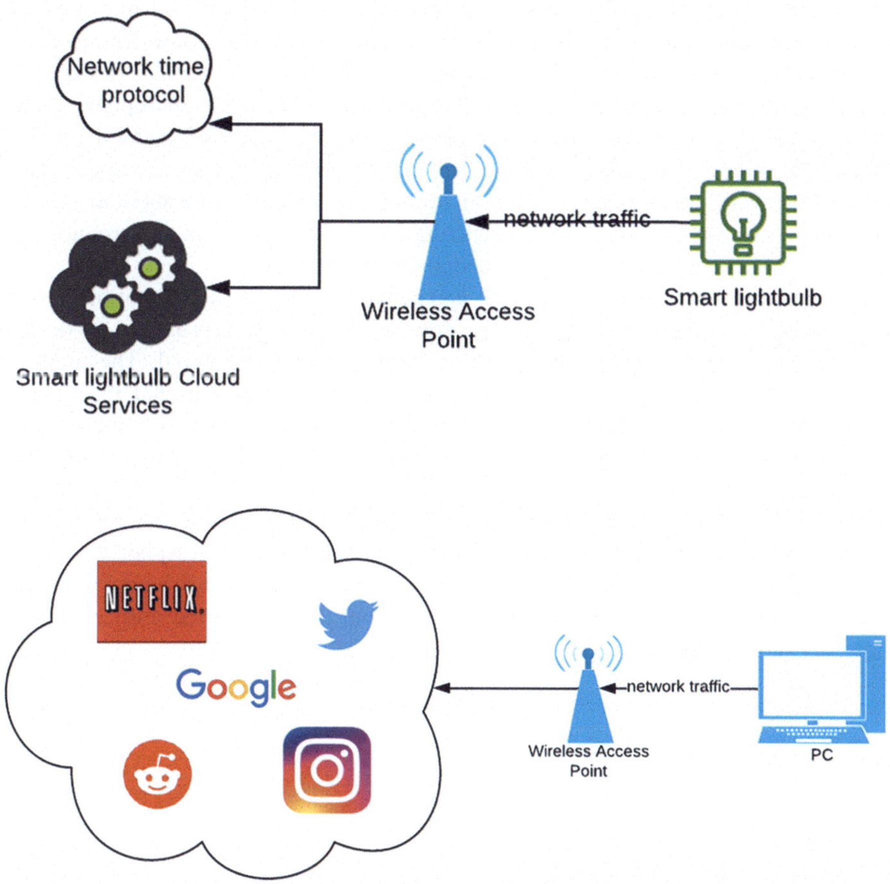

Fig. 2 Typical network traffic of a smart lightbulb versus the typical network traffic of a personal computer

The project proposed in this chapter addresses this issue. Students learn about the configuration and design of smart home devices by creating their own. Next, they analyze the security of their device by analyzing network traffic. Finally, they write a program on their router that will monitor for anomalous traffic, effectively preventing their device from taking place in a DDOS attack. Since smart home devices are appearing in more and more homes, it is very likely that either the student or their parents already have one. The most popular of these devices are smart lightbulbs (Philips Hue), smart thermostats (Google's Nest), home cameras that are either part of a security system or used as a baby monitor, and Amazon's Echo. The student's project mimics the functionality of a motion sensor which is typically part of a home security system.

2 Project Goals

The primary goals of this project are to (1) raise students' interest in computer security, (2) understand the architecture of a smart home device, and (3) introduce the importance of security at the network level. Students are presented with the current real-world issue of smart home devices being hijacked to disrupt access to major websites. This is an issue that most students are aware of, and some were affected by, since some may have noticed that they could not access major sites during the attack. This project provides an opportunity to learn more about the attack and how to prevent the attack from occurring in the future, or at least to reduce the impact of a future attack by securing their smart home device.

Students use a Raspberry Pi, a small inexpensive computer, and a motion sensor to create their smart home device. The concept is to write a program on the Pi that will send a text alert whenever the motion sensor is triggered. The method to implement this functionality includes interfacing with an external third-party website which generates a text. This architecture is typical of smart home device, since smart home devices typically receive input from a sensor and then relay that event to an external website.

Students use a personal mini router to learn about network traffic. First, they utilize a tool on their router to monitor traffic generated by their Raspberry Pi. This establishes a normal baseline of traffic. Next, they write their own program to create an alert whenever their Raspberry Pi deviates from the established norm, and this indicates anomalous traffic.

What to Purchase

1. Raspberry Pi 3 or greater [22]
2. Micro SD card with adapter [23]

3. Mini router with OpenWrt installed (if not installed make sure it is compatible with OpenWrt) [24]
4. PIR Sensor Infrared IR Body Motion Module [25]
5. Raspberry Pi 3 B+ Power Supply [26]
6. USB drive [27]

3 Methodology

Students use a Raspberry Pi 3 Model B, a PIR Motion Sensor Module, and a Cloud application called Twilio to create their own smart sensor (see Fig. 3). The project is divided into four separate phases to reduce the overall complexity. The phases include (1) connecting the sensor to the Pi and writing code on the Pi to react to the sensor being tripped, (2) adding functionality on the Pi so that a text is sent when the sensor is tripped, (3) capturing the network traffic on the router to establish a normal baseline, and (4) writing a program on the router to alert the user when the network traffic deviates from the baseline.

Phase 1: Raspberry Pi Configuration and Sensor Integration

1. Install the Raspbian operating system onto the micro SD card [28].
2. Insert the SD card into the bottom of the Raspberry Pi.
3. Sensor integration: Detailed instructions on how to complete this phase of the project can be found on the Raspberry Pi website [29]. The first step involves connecting the motion sensor directly to the Pi. This allows the sensor to send an alert to the Pi each time it is tripped (see Fig. 4). Students write a Python program on the Pi that receives the alert from the motion sensor and prints a message to the screen.

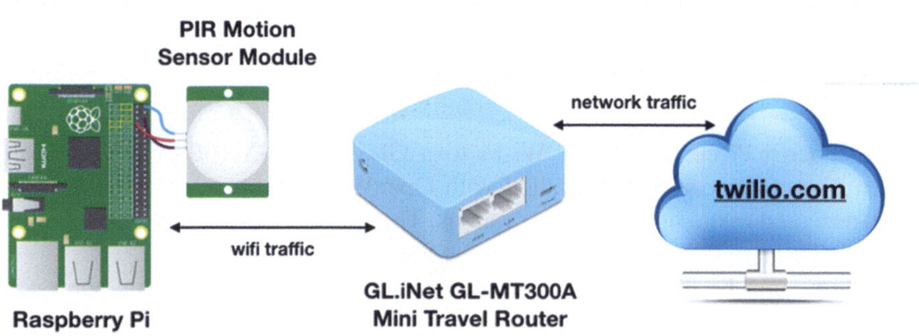

Fig. 3 Project diagram

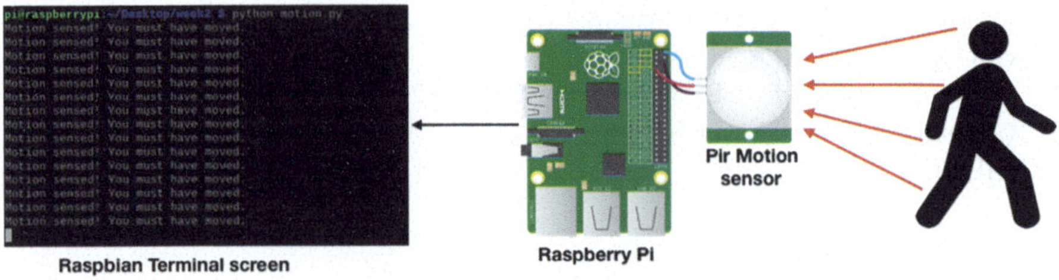

Fig. 4 Python program that prints an alert when the motion sensor is tripped

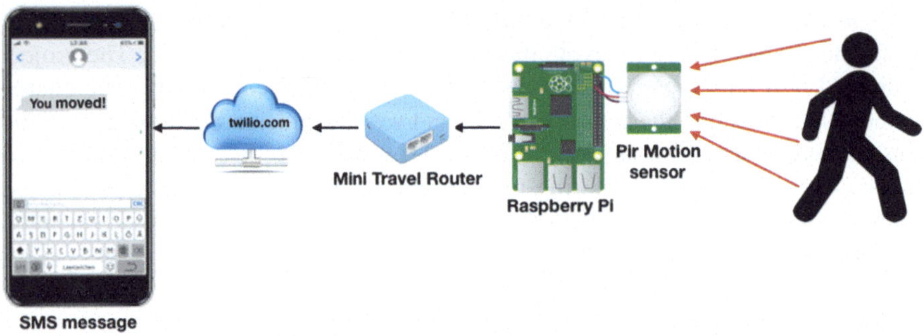

Fig. 5 Send an SMS message when the motion sensor is tripped

Phase 2: Text When Tripped Program

In this phase the students create a Python program to implement functionality to send an SMS message to their cell phone whenever the sensor is tripped (see Fig. 5). This is done with the help of a web application called Twilio. Students set up a free account on Twilio, and Twilio creates a unique code that the students use to authenticate back to the site through the application programming interface (API) [30]. Twilio provides detailed Python examples on how to accomplish this [31].

Phase 3: Capturing Network Traffic

The Raspberry Pi connects over Wi-Fi to a GL.iNet GL-MT300A Mini Travel Router. The router comes preinstalled with OpenWrt, a light-weight Linux distribution designed to run on embedded systems. This configuration allows for students to install applications and run programs that they have created.

Tcpdump can be installed on OpenWrt router itself. Therefore, this approach eliminates the need of having a remote Wireshark or similar listener to analyze the

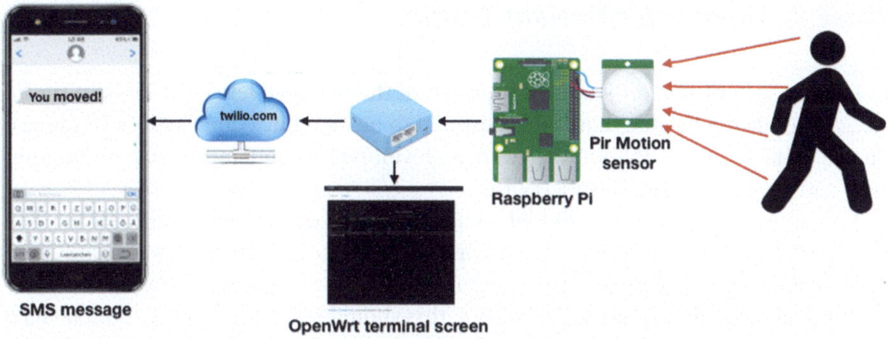

Fig. 6 Capture network traffic using Tcpdump while the Text When Tripped program is running

Table 1 Sample of captured outbound and inbound network traffic

Outbound IP addresses	Inbound IP addresses
192.168.8.1	192.168.8.1
18.212.47.248	239.255.255.250
69.196.229.196	18.212.47.248
52.4.88.97	69.196.229.196
35.167.85.14	35.167.85.14

These IP addresses will be slightly different each time a capture is run since Twilio uses a variety IP addresses for its service

traffic in real time. Tcpdump is a network sniffing tool that runs on OpenWrt and allows for the capture of all network traffic processed by the router. The students launch Tcpdump and then their Python-developed application on the Pi. In order to get a good baseline of traffic, the Python program should run for a couple of hours, with intermittent trips of the motion sensor (see Fig. 6).

1. SSH into OpenWrt installed router and install tcpdump with the commands:
   - opkg update
   - opkg install tcpdump
2. Execute below command to listen on interface (-i) and store captured information to a file (-w) and be verbose while doing so (-v):
   - tcpdump -i any -v -w pcap.cap

The network capture file is then transferred to a student's computer where they import it into Wireshark, a free network analysis tool, to analyze the traffic generated during the capture. Traffic should be seen going to Twilio each time that the sensor was tripped. The destination IPs may vary over time but should remain in Twilio's IP range (Table 1).

Phase 4: Monitor for Deviant Traffic

The next phase of this project is to monitor, at the network level, for deviant traffic (i.e., any outbound traffic that is not destined to an IP address in Twilio's IP range). On the mini router, a Python program is developed to sniff all traffic originating from and destined to the Raspberry Pi.

This concept follows the model of an intrusion detection system (IDS). The idea is that an IDS monitors the network traffic of a host or a network and raise alerts whenever it detects security violations [32]. For example, if you saw a smart lightbulb attempting to connect to Netflix, this should raise a security violation since there is no reason the smart lightbulb should communicate with Netflix (see Fig. 7).

The formal model of an IDS includes the three procedures: (1) feature selection, in which the necessary attributes of the network packet data, P, are identified; (2) profiling, which is the training procedure where P will be run on training data in order to establish the profiling knowledge base, K; and (3) detection which identifies traffic as either anomalous or normal. The feature that we are interested in for this project is the IP address; the training data are the IP addresses found in the first column of Table 2, and the anomalous traffic is any IP address that is not found in the first column of Table 2. Example anomalous IP addresses are in second column of Table 2.

Running Python on the router does take some initial set up. This primarily involves extending the hard drive space of the mini router; the router is inexpensive

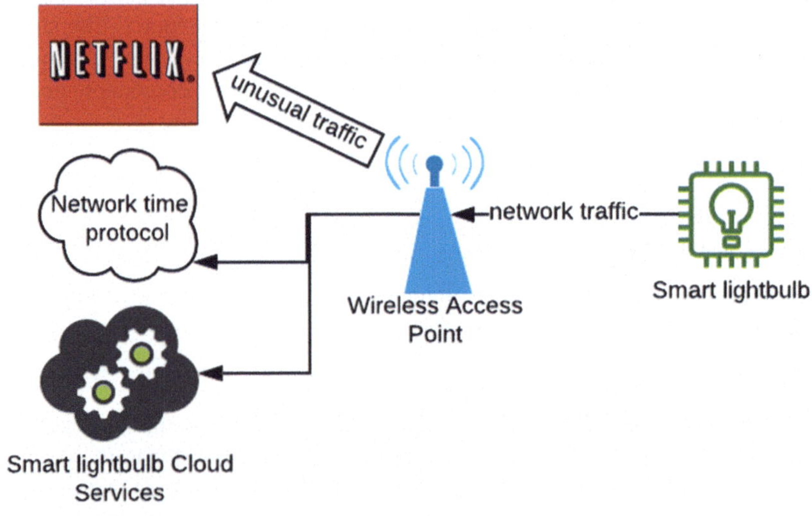

Fig. 7 Example of anomalous traffic from a smart lightbulb

Table 2 Example of expected and anomalous network traffic

Captured outbound IP addresses	Sample anomalous IP addresses
192.168.8.1	34.210.107.195 (Netflix)
18.212.47.248	172.217.6.68 (Google)
69.196.229.196	104.244.42.1 (Twitter)
52.4.88.97	151.101.189.140 (Reddit)
35.167.85.14	Plus any IP not found in column 1

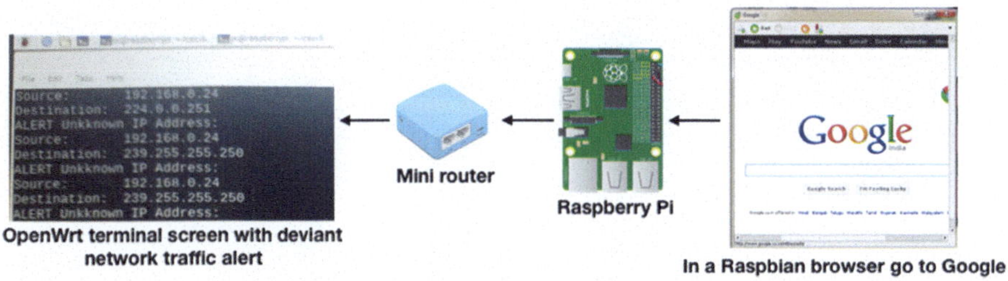

Fig. 8 Display a message when deviant network traffic is detected (in this case the user browsed to Google from a browser on their Raspberry Pi)

and comes with very limited hard drive space. Extending the memory is achieved by inserting a properly formatted USB drive in the router and running the necessary commands to use the USB drive as an overlay filesystem (instructions to do this can be found on OpenWrt's website [33]).

The Python sniffing program should create an alert that will be sent whenever the program detects an IP address that was not seen during the monitoring phase of the project (see Fig. 8). The contents of the alert will indicate that suspicious traffic has been spotted by the router originating from the Raspberry Pi.

It is recommend to use the Scapy Python library which makes capturing network traffic rather painless [34] (see Fig. 9). The first portion of the program creates an array called ipList. ipList contains all of the IP addresses discovered during phase 3, capturing network traffic, in which Tcpdump was used to capture all of the network traffic while the Text When Tripped program was running. The function "deviant" was created to read each packet's IP address, and print if the IP address is deviant. Getting the IP address is accomplished by leveraging the scapy library's getlayer function. In the if statement, the IP address is searched for in the array, ipList. If the IP address is not in the array, then the message "Deviant packet! IP" plus the IP address that was found is printed to the screen. If the IP address is found in the ipList array, then nothing happens, and the program continues to read the next packet.

```
#!/usr/bin/env python3

from scapy.all import *

ipList = [
        '18.208.54.140',
        '18.212.47.248',
        '18.211.224.155',
        '8.8.8.8',
    '8.8.4.4',
    '192.168.8.24',
    '192.168.8.139',
    '192.168.8.1',
    '10.11.55.21',
    '10.11.55.32',
    '10.0.0.1'
        ]

def deviant(pkt):
    pkt = pkt.getlayer(IP)

    if hasattr(pkt, "dst") and pkt.dst not in ipList:
     print("Deviant packet! IP: " + pkt.src + " - " + pkt.dst)


packet = sniff(prn=deviant)
```

Fig. 9 Sample code

4 Summary

Students learn about the typical configuration and design of smart home devices by creating their own. By analyzing the network traffic of their new smart home device, they learn that, compared to laptops and smartphones, their smart home device generates very consistent network traffic patterns; network traffic sent from their smart device is sent over a few protocols to a limited number of destinations. They turn this finding into a solution by creating a program on their router that will monitor for anomalous traffic, securing their device from malicious attacks (i.e., preventing their device from taking place in a DDOS attack).

References

1. K. Ashton, That ‘internet of things’ thing. RFID J. **22**(7), 97–114 (2009)
2. R.H. Weber, Governance of the internet of things—from infancy to first attempts of implementation? Laws **5**(3), 28 (Sep. 2016). https://doi.org/10.3390/laws5030028

3. R. Hammill, M. Hendricks, Gadgets to help tend a garden, *The New York Times*, 24 Apr 2013. https://www.nytimes.com/2013/04/25/technology/personaltech/calling-on-gadgetry-to-keep-the-garden-growing.html. Accessed 30 Jul 2019.
4. S. Shackelford, A. Raymond, R. Balakrishnan, P. Dixit, J. Gjonaj, R. Kavi, When toasters attack: a polycentric approach to enhancing the 'security of things', 2016. https://papers.ssrn.com/sol3/papers.cfm?abstract_id=2715799. Accessed 23 Apr 2017.
5. K. Fu, et al., Safety, security, and privacy threats posed by accelerating trends in the internet of things, Technical Report. Computing Community Consortium, 2017. http://cra.org/ccc/wp-content/uploads/sites/2/2017/02/Safety-Security-and-Privacy-Threats-in-IoT.pdf.
6. K. Walker, The legal considerations of the internet of things, ComputerWeekly.com, 2014. https://www.computerweekly.com/opinion/The-legal-considerations-of-the-internet-of-things. Accessed 30 Jul 2019
7. G. Kambourakis, C. Kolias, A. Stavrou, The Mirai botnet and the IoT Zombie Armies, in *MILCOM 2017—2017 IEEE Military Communications Conference (MILCOM)*, (2017), pp. 267–272. https://doi.org/10.1109/MILCOM.2017.8170867.
8. H. Sinanović, S. Mrdovic, Analysis of Mirai malicious software, in *2017 25th International Conference on Software, Telecommunications and Computer Networks (SoftCOM)*, (2017), pp. 1–5. https://doi.org/10.23919/SOFTCOM.2017.8115504
9. C. Kolias, G. Kambourakis, A. Stavrou, J. Voas, DDoS in the IoT: mirai and other botnets. Computer **50**(7), 80–84 (2017). https://doi.org/10.1109/MC.2017.201
10. I. Andrea, C. Chrysostomou, G. Hadjichristofi, Internet of things: security vulnerabilities and challenges, in *2015 IEEE Symposium on Computers and Communication (ISCC)*, (2015), pp. 180–187. https://doi.org/10.1109/ISCC.2015.7405513
11. G. Hunt, G. Letey, E. Nightingale, The seven properties of highly secure devices, *Microsoft Research*, Mar 2017. Accessed: 10 Feb 2018. https://www.microsoft.com/en-us/research/publication/seven-properties-highly-secure-devices/.
12. V. Sivaraman, H.H. Gharakheili, A. Vishwanath, R. Boreli, O. Mehani, Network-level security and privacy control for smart-home IoT devices, in *Wireless and Mobile Computing, Networking and Communications (WiMob), 2015 IEEE 11th International Conference on*, (2015), pp. 163–167
13. V. Adat, B.B. Gupta, Security in internet of things: Issues, challenges, taxonomy, and architecture. Telecommun. Syst. **67**(3), 423–441 (2018). https://doi.org/10.1007/s11235-017-0345-9
14. J.Y. Kim, W. Hu, D. Sarkar, S. Jha, ESIoT: enabling secure management of the internet of things, in *Proceedings of the 10th ACM Conference on Security and Privacy in Wireless and Mobile Networks*, (2017), pp. 219–229. https://doi.org/10.1145/3098243.3098252
15. E. Vasilomanolakis, J. Daubert, M. Luthra, V. Gazis, A. Wiesmaier, P. Kikiras, On the security and privacy of internet of things architectures and systems, in *Secure Internet of Things (SIoT), 2015 International Workshop on*, (2015), pp. 49–57
16. E. Hodo et al., Threat analysis of IoT networks using artificial neural network intrusion detection system, in *2016 International Symposium on Networks, Computers and Communications (ISNCC)*, (2016), pp. 1–6. https://doi.org/10.1109/ISNCC.2016.7746067
17. Y. Meidan et al., ProfilIoT: a machine learning approach for IoT device identification based on network traffic analysis, in *Proceedings of the Symposium on Applied Computing*, (2017), pp. 506–509. https://doi.org/10.1145/3019612.3019878
18. M. Miettinen, S. Marchal, I. Hafeez, N. Asokan, A.R. Sadeghi, S. Tarkoma, IoT SENTINEL: automated device-type identification for security enforcement in IoT, in *2017 IEEE 37th International Conference on Distributed Computing Systems (ICDCS)*, (2017), pp. 2177–2184. https://doi.org/10.1109/ICDCS.2017.283
19. S. Raza, L. Wallgren, T. Voigt, SVELTE: real-time intrusion detection in the internet of things. Ad Hoc Netw. **11**(8), 2661–2674 (2013). https://doi.org/10.1016/j.adhoc.2013.04.014
20. E. Fernandes, A. Rahmati, K. Eykholt, A. Prakash, Internet of things security research: a rehash of old ideas or new intellectual challenges? IEEE Secur. Priv. **15**(4), 79–84 (2017). https://doi.org/10.1109/MSP.2017.3151346

21. A. Sivanathan et al., Characterizing and classifying IoT traffic in smart cities and campuses. Proceedings of IEEE INFOCOM Workshop SmartCity, Smart Cities Urban Computing, 1–6 (2017)
22. Buy a Raspberry Pi 3 Model B—Raspberry Pi. https://www.raspberrypi.org. Accessed 20 Mar 2020
23. SanDisk Ultra microSD UHS-I Card, *Western Digital Store*. https://shop.westerndigital.com/products/memory-cards/sandisk-ultra-uhs-i-microsd. Accessed 21 Mar 2020
24. GL-AR150 / White—GL.iNet. https://www.gl-inet.com/products/gl-ar150/. Accessed 21 Mar 2020
25. HiLetgo 5pcs HC-SR501 PIR Infrared Sensor Human Body Infrared Motion Module for Arduino Raspberry Pi. http://www.hiletgo.com/ProductDetail/3006354.html. Accessed 21 Mar 2020.
26. Raspberry Pi 3 Power Supply—2.5A (Micro USB). https://www.canakit.com/raspberry-pi-adapter-power-supply-2-5a.html. Accessed 21 Mar 2020
27. SanDisk Ultra Flair USB 3.0 Flash Drive, *Western Digital Store*. https://shop.westerndigital.com/products/usb-flash-drives/sandisk-ultra-flair-usb-3-0. Accessed 21 Mar 2020.
28. Installing operating system images—Raspberry Pi Documentation. https://www.raspberrypi.org/documentation/installation/installing-images/. Accessed 31 Jan 2020
29. Physical Computing with Python—Introduction | Raspberry Pi Projects. https://projects.raspberrypi.org/en/projects/physical-computing. Accessed 31 Jan 2020
30. Twilio | Try Twilio Free. https://www.twilio.com/try-twilio. Accessed 31 Jan 2020
31. SMS API—Twilio Text Messaging for Mobile & Web Apps. *Twilio*. https://www.twilio.com. Accessed 31 Jan 2020)
32. B.B. Zarpelão, R.S. Miani, C.T. Kawakani, S.C. de Alvarenga, A survey of intrusion detection in Internet of Things. J. Netw. Comput. Appl. **84**, 25–37 (2017). https://doi.org/10.1016/j.jnca.2017.02.009
33. Rootfs on External Storage (extroot). https://wiki.openwrt.org/doc/howto/extroot
34. P. Biondi, Scapy, http://www.secdev.org/projects/scapy/

Puzzle-Based Honors Cybersecurity Course for Critical Thinking Development

Mitchell Buchman

1 Introduction

The Advanced Cybersecurity Experience for Students (ACES) was launched by the University of Maryland in 2013. The program includes a living–learning program for freshman and sophomores, which leads to an Honors College Citation in Cybersecurity.

A number of seminar courses were a part of the ACES program, and these seminar courses covered a range of cybersecurity-related subjects such as reverse engineering, the policy implications of cybersecurity, and also included an initial offering of a course titled "Methods for Solving (and Not Solving) Puzzles," which was developed in 2015 when cybersecurity was a fairly new interdisciplinary course of study. This chapter examines how and why the Puzzles course was developed, along with an examination of some lessons learned from teaching this course to the students in the Honors ACES program.

The focus of the course was to get students to learn how to frame and solve unstructured problems, for which they do not have a set of instructions with explicit steps to follow. Although it may not have been apparent at first to the students, the puzzles selected for use in this course all had mathematical bases, which could provide a language for understanding the mechanics of the puzzles and attempting to solve them [1, 2]. The puzzles were often entertaining, as they had nonobvious solutions. As long as the students felt that there was some hope of being able to find a solution for each puzzle and not be derailed by frustration, they maintained interest. The educational background of the students as well as the other courses in their undergraduate studies often gave the students the tools to solve these unstructured

M. Buchman (✉)
ManTech International, Herndon, VA, USA
e-mail: mitchb@ieee.org

K. Daimi, G. Francia III (eds.), *Innovations in Cybersecurity Education*,
https://doi.org/10.1007/978-3-030-50244-7_6

problems. The gap addressed by this course was building the students' skills to go from the statement of the unstructured puzzle to the identification of which of the many tools in their skills tool bag to pull out for a particular puzzle and how to apply those skills.

Within the context of this chapter, tasks which are done for enjoyment where the tasks do not have an obvious answer to someone trained in the relevant field are called puzzles. Puzzles which have a competitive aspect can be called games. These games may be a competition against one or more people or with a solo player measured against an amount of time.

In the process of describing the educational use of particular puzzles in this chapter, it becomes necessary to pull the curtain back, revealing the nonobvious answers to some puzzles, making them trivial to solve. Reading this chapter may of necessity "ruin" those puzzles for the reader.

2 Course Development

The initial offering of the Puzzles course helped to build experience with the course material at the same time that the ACES program was growing out of its infancy. Following the initial offering, the original pair of instructors became unavailable to continue teaching the course. The author of this chapter held meetings with the ACES leadership and agreed to teach the next several offerings of the course, which helped to coalesce ideas for taking the course forward. At that point there was a significant redevelopment of the course material to make it cohesive and to align with the desired program outcomes.

Curriculum Guidelines

This course was being developed just prior to the establishment of the CSEC2017 Joint Task Force on Cybersecurity Education [3]. Using hindsight, this course addresses many of the essential concepts that were identified in the curriculum guidelines, which would be published more than 2 years after the first offering of the puzzle course. What was present at that time was a need to address a perceived deficiency in student preparedness for the ACES program, as well as feedback received from potential employers of graduating students, which raised the same issues. The students in the ACES program excelled at following explicit instructions, but they exhibited significant difficulty when they were asked to engage in activities where those instructions were absent. This appeared to be independent of the students' secondary school system.

In order to address this perceived deficiency, the Puzzles course was developed as a course to foster student critical thinking skills, in the context of cybersecurity. Again, with the value of hindsight, the Puzzles course largely aligned with the knowledge areas which would ultimately be developed in the later cybersecurity curriculum guidelines.

At the time that the course was being developed, there were no courses that could be found which could serve as a model for the educational objectives, to foster student critical thinking skills, in the context of cybersecurity, while using mathematics as the language of understanding solutions. In the absence of prior examples, we set out to develop such a course.

Course Learning Objectives

Given the goals of the course, we developed the following course learning objectives.

Upon completion of this course, students were expected to be able to do the following:

- Interpret unstructured problems that they had not previously seen and identify the relevant issues which are necessary to solve the problem.
- Determine the best type of approach to take when trying to solve a new problem.
- Find the correct solution to a problem in the face of a solution which is counterintuitive.
- Identify strategies for repeated puzzles and games.
- Recognize when external factors may necessitate a shift in strategy.
- Evaluate a potential solution to a problem, in the presence of biases, which they have attempted to solve and determine if their solution is valid.

Course Learning Outcomes

Students would understand how to approach solving unstructured problems without having step-by-step instructions.

Alignment with Program Objectives

The program objective for this course was singularly to enhance students' critical thinking skills, which had been identified as lacking across all incoming students.

Although all of the students had met the prerequisites for the program, their critical thinking skills were falling short of expectations.

Table 1 Course schedule

Week	Topics
1	Course introduction, puzzle survey
2	Combinatorics
3	Combinatorics
4	Number theory, information theory, hamming codes
5	Introduction to cryptography
6	Enigma machine, public key cryptography
7	Unexpected outcomes, mid-term review
8	Mid-term exam, algorithmic puzzles
9	Logic puzzles, algorithmic puzzles, multiple approaches
10	Graph theory, graph theory, Latin squares
11	Latin squares/sudoku, constraint propagation, more squares
12	Critical thinking and lateral thinking, impact of biases
13	Game theory, Rubik's cube
14	Game theory, challenging assumptions, Rubik's cube
15	Real-world problems, Rubik's cube, review for final exam
16	Final exam

Table 2 Course grading

Activity	Portion of total grade (%)
Homework	40
Mid-term exam	20
Final exam	30
Class participation	10

Course Schedule

The course was structured as a 15-week one semester course, which met 3 days a week for sessions that lasted 50 min. Each classroom session focused on a particular topic and began with three to five small puzzles, from various topics, to get the students warmed up to be thinking critically. The schedule is listed in the table above (Table 1):

Course Grading

Students grades in the course were apportioned as shown in Table 2:

Homework problems were designed to ensure that the students had a familiarity with the mechanics of working particular categories of puzzles which had been introduced in the classroom. Students were encouraged to form study groups to work through the homework problems in an effort to get them to discuss these problems and reinforce the learning process.

Exams, on the other hand, were individual in-class efforts, where the focus was on demonstration of the understanding of concepts and to push the students to extend the rote effort of the homework exercises into a more complex demonstration of putting the pieces together. During classroom exam preparation sessions, students were given problems to walk through to ensure that the style of the exam problems would not be a surprise. These exam preparation problems had a similarity to the exam problems, but they were not the exact problems.

Classroom Sessions

Each 50-min classroom sessions began with the presentation of two to four small puzzles, taking no more than 5 min. These served to get the students warmed up and shift their mindset to be more open for the main portion of the classroom session. The remainder of each classroom session was used for lecture and both individual and group interaction of the students, many of which will be described in this chapter.

Puzzle Selection

To the degree practical, puzzles were changed from year to year. However, it can be extremely challenging to develop equivalent puzzles which can achieve an equivalent educational outcome. Some puzzles could be changed by rewriting the narrative but retaining the puzzle structure. Some puzzles could not be easily changed and, therefore, sometimes puzzles were retained across years when it was not practical to develop an equivalent puzzle.

3 Course Topics

This section presents descriptions of some of the course topics, along with a description of some of the puzzles used during those sections. There is also a short discussion of how some of the puzzles were used in the course.

Following instruction on the basic principles for some sets of puzzles, alternate methods were presented. Sometimes the students would be challenged to identify alternate solution methods, as they did in the building height puzzle which is described later in this chapter. Students were then tasked to identify the pros and cons of these alternative approaches.

Students were also challenged to understand the implications of having evolved or alternate constraints added into familiar puzzles. In one case, a simple 3×3 square tic-tac-toe game (also known as naughts and crosses) was changed to a

4×4 square board so the students could assess the implications to the strategies which they had developed in the original 3×3 version. Students were also asked to evaluate the implications of using a two-dimensional tic-tac-toe with a $3 \times 3 \times 3$ playing area. In each of these cases, the students were asked to analyze the game to develop winning strategies, tailored to the constraints.

Another related aspect of student analysis was the determination of solvability. Students were challenged to learn that not every puzzle had a solution and the students learned some approaches which could be used to determine solvability for some puzzles.

Where possible, physically interactive activities were designed into every classroom session. This was done to reinforce the learning experience [4] by increasing the student level of participation in the education process.

The discussion of the course topics, below, contains a sampling of the puzzles used in various parts of the course, as a thorough examination of all of the puzzles used and a description of their role in this course would be of sufficient length to warrant its own book.

Course Introduction

Corner to Corner

The course began with a simple puzzle [5] called Corner to Corner, which was created by Martin Gardner, a prolific author of puzzles. In Corner to Corner, students are asked to determine the length of a diagonal line within a drawing. Students almost always fail to correctly solve this puzzle. Then a small change is made to the puzzle and they almost all instantly solve the puzzle.

This puzzle was used to set the stage for the course and to get the students to see that the way that they perceive a problem can have significant impact on their ability to understand and solve the problem.

Building Height and Barometer

In the second puzzle of this introductory section, students were shown a description of the physics of a mercury barometer and then they were shown a drawing of a multistory building. They are asked to determine the height of the building using a barometer.[1]

The instructor described an approach which involved using the barometer to measure the atmospheric pressure at the top of the building and at the bottom of the

[1] The origin of this puzzle is from Alexander Calandra in an article in "Current Science, Teacher's Edition," 1964. The puzzle has been republished many times by others, often with mistaken later attribution.

building. Although it was not explicitly pointed out at this early point in the course, the demonstration of the physics behind the operation of a mercury barometer was meant to introduce a bias in student thinking. This was pointed out later in the course during the section on biases and this early classroom exercise was pointed out as an example of the impact of the presentation of the puzzle or problem statement and its impact of student understanding of the problem.

The next step was to take the students through the nontrivial math to determine the difference in the building height, based on the atmospheric pressures measured. This was shown to be a valid, but difficult approach to solving the building height problem.

A second approach was presented, which involved taking the barometer to the top of the building and measuring the time that it would take for the barometer to fall and crash on the sidewalk below. The known acceleration due to gravity can be used to determine the height of the building, making certain assumptions. Once again, this was a valid approach and, in this case, it was somewhat easier to solve than the differential atmospheric pressures method.

A third approach was described to the students. This one involved finding the superintendent of the building and offering to present him a gift of the barometer if he can provide an answer to the height of the building. This was an arguably more elegant and still valid solution.

Presenting the problem this way gave the students permission to think about this problem in creative ways. Once the students were in an open-minded frame of mind, they were divided into small groups and were asked to identify as many ways that they could to solve the building height problem. This has always resulted in a high level of energy within these groups as they identified numerous creative solutions. The students then presented their lists to the class to see the range of ideas identified by the other groups.

Nim

The introductory session wrapped up with students playing each other the game of Nim [6]. This is an amazingly simple game with only a small set of rules and a rapid outcome. It also has a rich body of analysis [7–15], which supports revisiting this game later in the course in various contexts, as well as providing an opportunity for students to research this simple game further on their own.

The game begins with three piles of stones as shown in Fig. 1. In a simple variant of Nim, the piles contain 3, 5, and 7 stones. Players take turns picking up stones from a single pile on their turn. They must pick up at least one stone during their turn and may pick up between 1 and the total number or stones remaining in that pile. Play alternates between two players. The player who makes the last possible move to pick up stones wins the game. The game can also be played to have the player who is forced to pick up the last stone as the loser, which is referred to as a misère game.

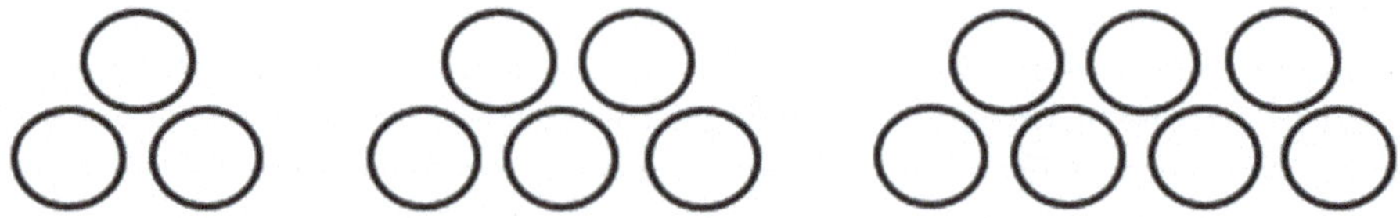

Fig. 1 Example of the setup for a game of Nim

Table 3 Milk jug transfer puzzle state table

	8-L jug	5-L jug	3-L jug
Starting condition	8	0	0
	3	5	0
↓	3	2	3
Desired ending condition ↑	1	4	3
Ending condition	4	4	0

The students were challenged to identify a winning strategy for the game. This game was revisited later in the course when the constraints of the game were further changed and the students again revisited the identification of a winning strategy, if any exist.

Puzzle Survey

Three Milk Jugs

One puzzle examined in the survey of puzzles was the milk jug transfer problem. In this puzzle, there are three milk jugs of capacity 8, 5, and 3 L. The 8-L capacity jug is full and the other two jugs were empty at the start of the puzzle. There is no way to measure except by using the jugs themselves. The goal is to divide the milk so that there are 4 L of milk in one jug and 4 L of milk in another jug.

This puzzle was helpful in demonstrating a structured approach to puzzles. By building a table of the state of each jug, the initial and final conditions of the jugs could be written into the table and students were then able to work from the outside in. They were able to work the problem from the starting and ending conditions, and try to find a way to get those two paths to meet. This is the initial mention of the backtracking approach to solving puzzles. This can be seen in Table 3 where steps are listed moving forward from the initial condition and moving backwards from the final goal condition.

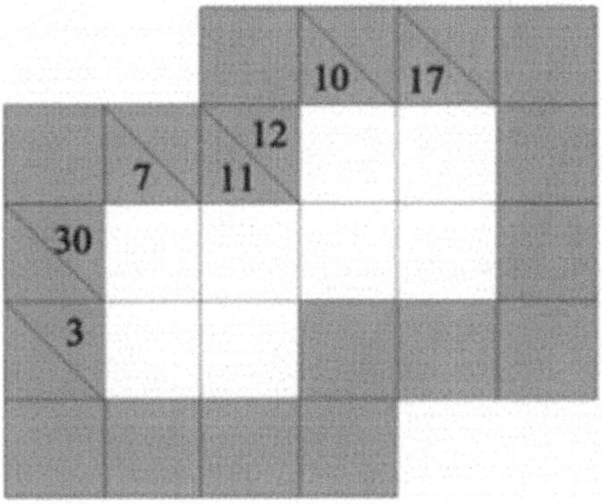

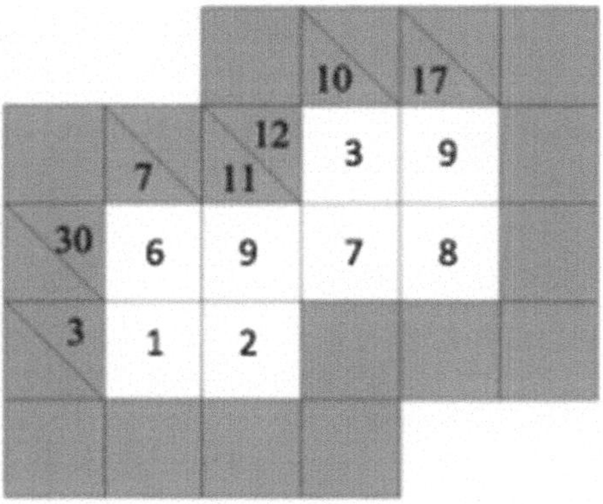

Fig. 2 Example of an unsolved and a solved Kakuro puzzle

Kakuro

Students were then introduced to Kakuro, the first of a set of two-dimensional puzzles, popularized by the Japanese publisher Nikoli. Kakuro is essentially a mathematics addition crossword puzzle, with a grid of squares. Some of those squares contain clues, others contain white space for answers, and the remaining squares are shaded and are not used for that puzzle. This and other two-dimensional puzzles offered opportunities to introduce multiple methods [16] for solving each of these puzzles.

The object of Kakuro is to fill all of the empty squares using the numbers from 1 to 9, so the sum of the horizontal numbers in a row equal the clue in the diagonal box to the left and the sum of the vertical numbers in a column equal the clue in the diagonal block above. Additionally, numbers cannot repeat in a row and they cannot repeat in a column. An example of a small blank Kakuro puzzle and its solution are shown in Fig. 2.

In the process of learning to solve Kakuro puzzles, students were introduced to the concept of enumerating the full range of combinations [17]. The enumeration of two-digit solutions, as seen in Table 4, is small but still demonstrates enough ambiguity to make the puzzle interesting. When the solution extends to six digits, the number of choices grows significantly, further adding to the challenge in solving the Kakuro puzzle.

Shikaku

Another two-dimensional puzzle developed by Nikoli is the Shikaku puzzle, where some of the squares in a grid are numbered and the objective is to divide the large grid into rectangular areas, such that each area only contains one number and that

Table 4 Enumeration of 2-digit Kakuro solutions

Sum	Addends	Addends	Addends	Addends
3	1 + 2			
4	1 + 3			
5	1 + 4	2 + 3		
6	1 + 5	2 + 4		
7	1 + 6	2 + 5	3 + 4	
8	1 + 7	2 + 6	3 + 5	
9	1 + 8	2 + 7	3 + 6	4 + 5
10	1 + 9	2 + 8	3 + 7	4 + 6
11	2 + 9	3 + 8	4 + 7	4 + 7
12	3 + 9	4 + 8	5 + 7	
13	4 + 9	5 + 8	6 + 7	
14	5 + 9	6 + 8		
15	6 + 9	7 + 8		
16	7 + 9			
17	8 + 9			

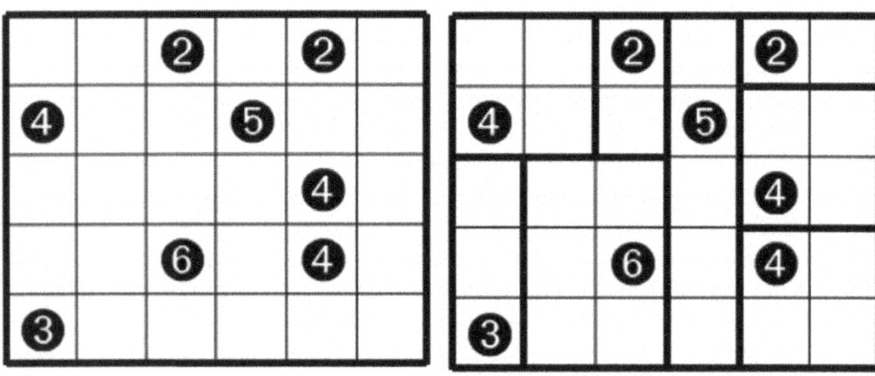

Fig. 3 Example of an unsolved and a solved Shikaku puzzle

number corresponds to the number of blocks in that area. Examples of a blank and a solved Shikaku puzzle are shown in Fig. 3.

Other puzzles from Nikoli were used throughout this course and included Slitherlink, Heyawake, Maysu, Nurikabe, and Sudoku.

Combinatorics

Tile Swap Puzzle

The combinatorics section of this course focused on permutations and drew from course notes from Dr. Jamie Mulholland at Simon Fraser University. These materials have since been collected into a book [18] by Dr. Mulholland.

A good amount of time was spent introducing Dr. Mulholland's Tile Swap puzzle, which provided a good foundational tool that was used at many points throughout this course.

A review of probability, with an emphasis on conditional probability, made certain that the students had the necessary tools to evaluate puzzles with dependent sequential states, where some of the states may have already been traversed, when viewed as a state-transition diagram. Students were also presented with an understanding of permutation cycles.

15 Puzzle

This provided sufficient background to enable the students to venture into the 15 Puzzle [19], which had its origins in the late nineteenth century. This puzzle was a very useful instructional tool. In addition to the 15 Puzzle being a physical tool which students can manipulate, there is a rich body of published analysis to draw from, concerning many aspects of the game. An example of the layout of the 15 Puzzle is shown in Fig. 4.

Students seemed to appreciate the opportunity to simply play with the puzzle for a while, but this was best done in a classroom setting so that they didn't simply mimic steps found on the Internet.

Initial analysis of the movement of the tiles was done using the representation of two-cycle permutations. Initially this was focused on the bottom right corner of a solved puzzle and grew out from there.

The students were then introduced to the concept of solvability, determining if a particular puzzle configuration is solvable. When examining the puzzle for solvability, with the blank in the lowest row, the puzzle is solvable if the number of inversions is even. Even with repeated instruction, the solvability test presented a challenge for some of the students.

Fig. 4 Example of a solved 15 Puzzle

Logic Puzzles

Logic puzzles rely on mathematical deduction to find a solution, where deduction is the process of reasoning based on one or more statements linking premises and conclusions, from specific observations to more generalizations [20, 21].

Students were provided a background on mathematical logic before the introduction of logic puzzles. They were introduced to several techniques for solving logic puzzles which included a chart method, symbolic translation, picture methods, and mathematical methods. Logic puzzles presented to the students started with small-scale easy puzzles and evolved to puzzles which contained more elements and were more challenging to solve.

One example of a challenging logic puzzle used in the Puzzles course is known as the Einstein puzzle. This puzzle contains descriptions of five houses, in five different colors. The owners of the houses drink a certain type of beverage, smoke a certain brand of cigar, and keep a certain pet. Information is described within the puzzle which partially constrains the solution, but there is sufficient information missing that it is challenging, but possible, to find the complete answer. This type of problem calls for a chart method.

Algorithms

Algorithmic puzzles involve clearly defined procedures for solving the problem [22]. They seek to design solutions to puzzles in ways that accept a wide variety of input and scale the utility and applicability of the solution. Algorithms for puzzle solutions are potentially reusable across a number of puzzles. This contrasts with solutions for logic puzzles, where reasoning is designed around the constraints and idiosyncrasies of each puzzle.

Algorithmic design techniques were then presented, including brute force, divide and conquer, decrease and conquer, transform and conquer, greedy approach, dynamic programming, backtracking, and branch and bound. Puzzles which exercised each of these techniques were then presented to help build a foundation.

9-Dot Puzzle

Following the introduction of the types of approaches to algorithmic puzzle solving, exhaustive techniques were compared with non-exhaustive techniques. Students were then provided with a quick counterexample of the utility of brute force approaches with 9-dot puzzle. The puzzle involves taking a square 3×3 grid of dots and attempting to draw four straight lines, without lifting their pencil from the paper, while passing at least once through each dot. An example of the unsolved and the solved 9-dot puzzle are shown in Fig. 5.

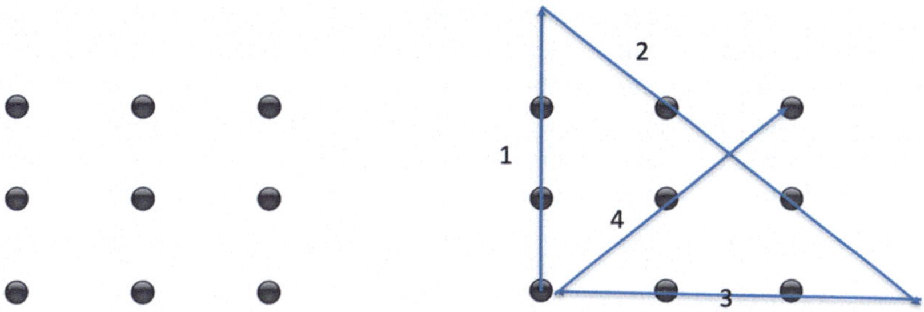

Fig. 5 Example of the 9-dot puzzle before and after being solved

Once the solution was presented to the students, they all quickly saw that they had been enforcing artificial constraints on the puzzle. They also saw that there was no way to apply a brute force approach to this puzzle and achieve a solution.

Information Theory

Card (Hat) Game

With the introduction of basic information theory, students were presented information on data encoding, XOR, parity checks, repetition codes, and Hamming codes. These tools enabled them to engage in the Card (Hat) game. Two students would sit in chairs, facing each other, in the front of the classroom. Each player held a card to their head, showing the face of the card to the other player, but without looking at their own card. Players need to guess the color, red or black, of their own card. Both players would win, if and only if at least one of them guessed correctly. This game can also be played by having players wear colored hats, which is why it is being referred to as the Card (Hat) game.

Students then played a few rounds of the game to get a sense for how it played out. The students then examined the probability of winning.

The game was then modified to introduce a third player and students examined the impact of the additional player on the probability of winning.

Rules were then further changed to permit players to choose red, black, or pass. Under these conditions, the players would all win if at least one player guessed correctly, no player chose incorrectly, and not all players choose to pass.

After playing the game a few rounds, students attempted to determine a winning strategy. At this point, the concept of cooperative strategies was introduced.

For an individual player, the probability of guessing correctly is 50%, where a single correct guess is enough to win, but a single wrong guess is enough to lose. So, an ideal strategy would seem to have all players pass, except for one player. This would result in the 50% probability of a win. Students examined the question if they could do better than that.

What if the players cooperate and the guesses are not uniformly distributed? They were then guided to the cooperative strategy that a player should pass if they see both a red and a black card. If the player sees two cards of the same color, they guess the opposite color. This strategy yields a 75% probability of a win.

With this knowledge, the students then compared the Card (Hat) game with data transmission protection to find commonalities with error coding techniques.

King's Poisoned Wine

Another interesting puzzle involves a king who has 1000 wine bottles and he discovers that one of the bottles has been poisoned. The king has no idea which bottle has been poisoned but needs to make use of the bottles for a celebration tomorrow. Also, the poison doesn't act for 12 h. The king has ten prisoners who can be used as tasters to identify the poisoned wine. How can the king figure out which bottle was poisoned, in time to use the good bottles for the celebration?

The solution of this puzzle involves binary numbers and coding. By assigning binary identifiers to the bottles and to the prisoners, each prisoner can drink a sip from every bottle where the bottle ID bit matches the binary ID of the prisoner. So, the first prisoner will drink from every other bottle, the second drinks from bottles 2, 3, 6, 7, 10, etc. The combinations of the prisoners that die will identify the poisoned bottle. This helped to convey the idea that concepts from one domain can apply to seemingly unrelated problems. In this case Information Theory and data encoding provided a useful technique to solve this puzzle.

Cryptography

The introduction to the cryptography section of the Puzzles course began with the introduction of the Caesar Cipher, a simple substitution cipher, which used a fixed offset between the alphabets for the plain text and the cipher text. This was followed by discussion of the Atbash Cipher, which mapped the plain text alphabet to the cipher text alphabet, by substituting the first letter in the alphabet ("A") with the last letter in the alphabet ("Z"). The letter "B" was substituted with the letter "Y", and the pattern continued.

The students were also introduced to Morse Code and the Freemason's Cipher to broaden their perspective on options for enciphering.

The process of breaking ciphers was introduced, along with the technique of plain text cribbing. Although most of the students had a vague sense of the letters which are used more frequently than others, they saw that mathematical analysis of large bodies of text through letter frequency analysis could produce a probability of the occurrence of each letter in the plain text. Further examination revealed the frequency of occurrence of two-letter sequences.

Zodiac Killer Ciphers

Letter frequencies were then put to use with the real-world example of the Zodiac Killer, from the late 1960s and early 1970s. A person claiming to be the Zodiac Killer, a suspected serial killer, sent four cryptograms (or ciphers) to the press in the San Francisco Bay area. The students in the Puzzles course worked in classroom as a group to decrypt the only one of those enciphered letters to have been decrypted. This was the letter which had been sent to the *Vallejo Times Herald*, which bore a postmark of July 31, 1969.

In the course of breaking this cipher, the students encounter five symbols which did not break out correctly. This was used to point out to the students that not all data that they encounter in their careers will be valid. In this case there could have been errors in the encoding process (human error on the part of the person who wrote the letter) or it could have been a deliberate attempt on the part of the author of those letters to make it more difficult for the police to decrypt the letters.

Students seemed to take to the challenge and enjoyed working on a problem that was not a contrived exercise.

Love in Kleptopia

As an introduction to secret and public key cryptography, students were presented with a puzzle titled "Love in Kleptopia" [23], which examined how to safely send a valuable object through the mail, with the use of padlocks and keys. Working through this problem lead to the introduction of information exchange protocols. In this case it was an easy transition from padlocks and metal keys to encryption and information keys.

The basic concepts of secret key and public key cryptography were presented, since they underlie many of the aspects of computer security. This sequenced well to then introduce Diffie–Hellman, key management, and symmetric key cryptography. Following an examination of the limitations of symmetric key cryptography, students were presented with an introduction to public key cryptography.

Enigma Machine

The instructor was privileged to be able to repeatedly borrow an Enigma machine for use in the classroom. The Enigma machines were used to protect the military and diplomatic communications of the Germany and its allies during World War II. The students were presented with a description of the operation of the device and were walked through the mathematics underlying the design of the evolving family of Enigma machines.

They were enthusiastic about the opportunity to use the device to create and then decrypt messages of their own choosing. The students not only created the messages, but they set the rotors within the machine, keyed in the plain text message, and keyed in the enciphered text. Giving the students a chance to operate a rare piece of cryptographic history helped to make this a memorable experience.

Critical Thinking

Critical thinking skills are seen a foundational competency [24] for cybersecurity practitioners. The complexity of the problems encountered require approaches which are progressive, risk driven, integrated, collaborative, flexible, and professional.

Students were given instruction on Structured Analysis Techniques as a way to deal with incomplete and ambiguous information, as problems will sometimes have missing or concealed information. Structured Analysis Techniques facilitate the identification of relevant and diagnostic data, as it provides a systematic approach to consider a range of alternate explanations and outcomes to avoid eliminating potentially relevant hypotheses.

As we gain more experience, we can counterintuitively become more susceptible to mental models where we overlook, reject, or forget important information which does not align with our assumptions and expectations. The risks of mindsets are that we can perceive what we expect to see and new information in assimilated erroneously into existing mental models.

Students were then introduced to a number of biases which affect the way that we process information. These included perceptual biases, biases in estimating probabilities, biases in evaluating evidence, and biases in perceiving causality.

Students were then introduced to historical strategic assumptions that were not challenged. They were presented with the example from World War II [25], with the assumption that Japan would avoid all-out war because it recognized the US military superiority. The reality was that given that the US superiority would only increase, Japan might view a first strike as the only way to knock America out of the war.

There was also a presentation on the Analysis of Competing Hypotheses as a way to avoid picking the first solution that seems satisfactory instead of going through all of the possibilities to arrive at the best solution.

Along with these structured approaches to critical thinking, students were also presented with puzzles that required them to slow down and reason their way through problems, like the one below where they were asked to determine the degrees of rotation that the upper nickel will have gone through when it returns to its starting position.

Counterintuitive Puzzles

Many puzzles are challenging to solve. But by the time that the solution is obtained, it is clear that the solution is valid. In contrast, sometimes puzzles may have fairly simple processes to arrive at the solution. Sometimes after solving the puzzle, it may seem that the solution is invalid. This class of puzzles is counterintuitive puzzles. It becomes necessary for the problem-solving student to learn to balance intuition and objective assessment.

Rope Length Puzzle

One of the puzzles with the most consistently unexpected answers is found in the Rope Length Puzzle. The puzzle takes various forms but seems to have the most dramatic impact when examined in the extreme.

In the Rope Length Puzzle, a length of rope is laid along the surface of the Earth's equator. Simplifying greatly, we assume that the cross section of the Earth at the equator is a circle. The rope is then lengthened by adding 3 ft. to the rope. The question is how much did the distance from the surface of the Earth to the rope change?

For the interactive portion of this puzzle, students were given a short length of rope to wrap around their waist. They were then told to add 3 ft. to the length of each rope, just like was done virtually for the Earth. The students were asked to decide if the change in the height of the rope above the equator is greater than, less than, or equal to the change in the height of the rope above the students' waist.

The counterintuitive aspect to this puzzle is that the change in height is independent of the radius and the change in height is thus equal for the Earth's equator and the person's waist. The point was made that intuition can be helpful, but doing the math is a good check on intuition.

When the students examined the mathematics for the problem, they found that the term for the original radius dropped out and was no longer relevant to the final solution. They started with the understanding that the circumference of a circle is equal to $2\pi r$, where r is the radius of that circle.

Referring to the diagram of the Rope length puzzle, shown in Fig. 6, with 'x' equal to the added length of rope, we find that

$$c1 + x = 2\pi\ (r1 + h)$$

Substituting for the circumference of the inner circle, we get:

$$2\pi r1 + x = 2\pi\ (r1 + h)$$

and

$$2\pi r1 + x = 2\pi r1 + 2\pi h$$

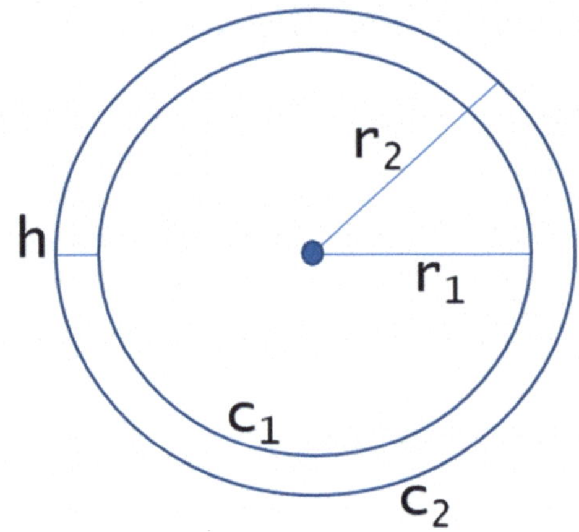

Fig. 6 Rope length puzzle

This simplifies to.

$$x = 2\pi h \text{ or } h = x/2\pi$$

We can see from the final term that the size of the original circle is no longer relevant to the outcome of the question of the change in height, which is simply a proportional relationship between the extra amount of rope and the height of the new circle above the old circle.

This problem was used to help balance the student's sense of dependence on intuition and a sense of being able to rush through problems without needing to work through the underlying math.

Biased Coin Toss

Sometimes taking a problem to an extreme set of conditions can help to understand the problem, as can be seen in this puzzle where a student finds himself as the head referee for the Super Bowl football game. The student is sitting in the referee's locker room just before the game and is passing time by flipping the commemorative coin, of which there had only been one manufactured. The student notices that tails have been coming up more times than heads, and he counts 80 tails and 20 heads. The student hears a knock on the door, and he realizes that it is time for them to go out onto the field. He is convinced that the coin is biased, but also fairly certain that he is the only person who knows that the coin is biased. He needs to decide if he can they use this coin for the coin toss and be confident that the outcome of the coin toss will be fair?

Quickly working through the conditional probabilities, he realizes that the probability of one team calling heads is 50%. Taking into account the coin toss outcome probabilities, as seen in Table 5, they see that the outcome is still fair,

Table 5 Biased coin toss outcome probability

	Toss heads (%)	Toss tails (%)
Call heads	10	40
Call tails	10	40

although it still makes them nervous, since this is somewhat counterintuitive as the coin is significantly biased.

This problem was then taken to the extreme by assuming that the outcome of the coin toss was now certain to land on tails every time. Can a coin toss with this extreme condition still be fair? Students tended to puzzle over this for less time than they did for the original biased coin, since they had already gone through that exercise. As long as the referee was the only person who knew about the biased coin, the outcome of the conditional probability of calling heads or tails, combined with the certainty of the outcome of the toss, will yield a fair outcome.

This problem again reinforced the need to strike a balance between trusting their intuitions and to working out the underling math before reaching a conclusion.

Industrial Batch of Pudding

As an industrial chef, you are mixing a batch of pudding, which when initially mixed will be 100 lb, of which 99% of the weight is water. The recipe calls for the pudding to be served when enough water has evaporated to the point that the amount of water reaches 98% of the total weight of the final mixture. The question is, how many pounds of pudding will you have when the pudding is ready to be served?

The mathematics here is straightforward. It is simply that the outcome is unexpected.

Two Children—One a Boy

Another counterintuitive puzzle is a logic puzzle which asks that if I have two children and one of them is a boy, what is the probability that the other child is a boy?

This puzzle involves enumerating the possible combinations, as male & male, male & female, female & male, and female & female. Of those combinations, the female & female combination is invalid, since one child is a boy, while the other combinations are valid. This yields the unexpected probability of one-third.

Coin Rotation Puzzle

Yet another puzzle with a counterintuitive result involves rotating coins. As shown in Fig. 7, two coins are placed on a flat surface. One is kept stationary and the other is rotated along the surface, around the fixed position coin. The question is how

Fig. 7 Coin rotation puzzle

many degrees of rotation does the moving coin experience by the time that it goes around the stationary coin and reaches its starting position.

Real World Problems

Washington DC Sniper

There is often more than one way to solve a problem, and in real-world problems, extraneous and distracting information is present.

The Structured Analysis Technique of Key Assumptions Check was used in the classroom to examine the puzzle of the DC sniper, a real-world event from 2002. The class worked as one group to build a table of Key Assumptions from news stories which had been published over the timeline of this event.

The initial assessment was that the shootings were the work of a single, white male who had military training and was driving a white van. Through the exercise of applying Structured Analysis Techniques to this problem, students were able to identify the need to avoid jumping to conclusions, be receptive to new leads, and to seriously consider later contradictory information.

Students also walked through the example of the *Challenger* space shuttle launch failure, highlighting the need to avoid the normalization of deviance in solving problems where historical data and existing behaviors are present. This was another case of pointing out how biases can impact the outcome of an analysis.

Additional Topics

The Puzzles course also had sessions which included topics such as graph theory, game theory [26], and the impact of biases on problem-solving.

A partial list of some additional puzzles included in the Puzzles course contains several cryptarithmetic puzzles, the Towers of Hanoi, river crossings, Kant's Clock, fake coin detection puzzles, traveling salesman, 4-knights, chessboard coverage, Rubik's Cube, match-stick puzzles, Konigsberg Bridges, and lateral thinking puzzles.

4 Lessons Learned

Some of the students were already familiar with some of the puzzles presented in the course. Experience in this course indicated that while some students were familiar with more of the puzzles than others, none of them were familiar with more than two or three of them and therefore did not detract from the course.

Students were provided information in class, warning them that various online sources of information on the 15-Puzzle use differing notations to describe the same problem. This still provided some confusion to some of the students who either failed or were unable to heed the warning.

Despite the similarity of exam preparation problems to actual exam problems, some students complained after the exams that they were not taught how to solve the exam problems. This is explicitly the goal that the course was set up to address. These complaints were an expression of the reluctance of some students to accept that their academic success would be dependent on learning to grow beyond following a list of steps. The number of students expressing this concern was small, but it did give insight into what others were experiencing. It was also offset by the larger number of students who expressed enjoyment in the exercise of taking the exams.

One of the challenges in teaching lessons on the Rubik's Cube is that in order to allow students to practice and learn they need to be able to reset the cube to a known condition. That requires them to be able to get incrementally proficient at a fairly consistent pace. Introduction of the Rubik's Cube in the first few weeks of a 15-week semester course would give better time for students to progress to the point where they are able to independently solve the cube by the end of the course. This would also permit instruction of algorithmic concepts, tied to the physical manipulation of the cube.

With students' ready access to Internet-connected resources, they should be expected to be able to find many classic puzzles. This makes it challenging to keep exams fresh, since it can take a significant amount of effort to create new puzzles which cause the student to exercise a particular approach, while keeping the puzzle solvable in the time constraints of an in-classroom exam. An example of this is the

Fig. 8 SEND + MORE = MONEY Puzzle

```
   S E N D
+  M O R E
 ---------
 M O N E Y
```

cryptarithmetic problem “SEND + MORE = MONEY” [27], as shown in Fig. 8. While there are numerous cryptarithmetic problems, this particular puzzle has a good balance of solvability while not being trivial.

Students will research homework problems on the Internet, and they needed to learn that not everything that they find on the Internet is correct. In this case, the highest-ranking solutions were taken out of context, tripping up approximately one-third of the students.

5 Conclusion

This course addressed an observed deficiency of the students in the ACES program. They needed enhanced critical thinking skills to meet the expectations of the other courses in the program. Pre-dating comprehensive cybersecurity curriculum guidelines, this course used a careful selection of puzzles to align with the learning objectives, enhancing the students’ ability to work through problems, in the absence of explicit step-by-step instructions.

By changing the rules and constraints of familiar puzzles or games, students were pushed to learn how to examine if new problems could be interpreted as extensions of what they already learned to understand.

Although quantitative post-course surveillance was not conducted, similar to related efforts [28], students expressed a sense that the course was a positive step toward helping them in approaching problems. The students were also able to successfully demonstrate their knowledge through mid-term and final examinations.

While this course was overtly a puzzle-oriented class, it used the puzzles as the gateway to an undergraduate math survey, while teaching the students to build their set of critical thinking skills, in order to prepare them in their cybersecurity education and careers.

References

1. G. Polya, *How to Solve It* (Princeton University Press, Princeton, NJ, 1945)
2. A. Schoenfeld, Learning to think mathematically: problem solving, metacognition, and sense making in mathematics, in *Handbook of Research on Mathematics Teaching and Learning: A Project of the National Council of Teachers of Mathematics*, (Macmillan, New York, NY, 1992)

3. Cybersecurity Curricula 2017: Curriculum Guidelines for Post-Secondary Degree Programs in Cybersecurity, ACM, IEEE, AIS, IFIP, December 31, 2017. https://dl.acm.org/citation.cfm?id=3184594. Accessed Mar 2020
4. L. Black, Interactive whole class teaching and pupil learning: theoretical and practical implications. Lang. Educ. **21**(4), 271–283 (2007). https://doi.org/10.2167/le679.0
5. M. Gardner, *Entertaining Mathematical Puzzles* (Dover, New York, 1986)
6. B. Charles, Nim, a game with a complete mathematical theory. Ann. Mathematics, 2nd series **3**(1–4), 35–39 (1901–1902)
7. D.J. Davis, The Game of NIM MAT Exam Expository Papers 9, 2006
8. U. Larson, in *Wythoff NIM Extensions and Certain Beatty Sequences*, ed. by Chalmers University of Technology and University of Gothenburg, (December 2008)
9. B.S. Verhovsky, Winning strategies and complexity of Nim-type computer game on plane. Int. J. Commun. Netw. Syst. Sci. **3**, 793–800 (2010)
10. L. Rougetet, Machines designed to play Nim games. Teaching supports for mathematics, algorithmics and computer science (1940–1970). History and Pedagogy of Mathematics, Montpellier, France, July 2016. https://hal.archives-ouvertes.fr/hal-01349260. Accessed Mar 2020
11. P. Frankl, N. Tokushige, The game of n-times Nim. Discret. Math. **260**, 205–209 (2003)
12. C. Allen, V. Ponomarenko, *Fibonacci Nim and a Full Characterization of Winning Moves*, January 26, 2014. https://pdfs.semanticscholar.org/b91d/c51f3a23976c44914a376fcdfa7db2aecc48.pdf. Accessed Mar 2020
13. A.S. Fraenkel, M. Lorberbom, Nimhoff games. J. Combin. Theory Ser. A **58**, 1–25 (1991)
14. E. Duchene, A.S. Fraenkel, V. Gurvich, N. Bao Ho, C. Kimberling, U. Larsson, W. Wisdom. http://www.wisdom.weizmann.ac.il/~fraenkel/Papers/WythoffWisdomJune62016.pdf. Accessed Mar 2020
15. U. Larsson, A generalised diagonal Wythoff Nim. Integers **12**, 1003–1027 (2012)
16. S. Salcedo-Sanz, E.G. O.-G. Sancho, A.M. Perez-Bellido, A. Portilla-Figueras, X. Yao, Solving Japanese puzzles with heuristics. IEE Symposium on Computational Intelligence and Games, 2007
17. S. Panov, S. Koceski, Deterministic and metaheuristic approaches to solving Kakuro puzzles, in *2nd Mediterranean Conference on Embedded Computing*, (2013)
18. J. Mulholland, Permutation puzzles: a mathematical perspective. Self-Published, 2019
19. J. Slocum, D. Sonneveld, The 15 puzzle book: how it drive the world crazy. Slocum Puzzle Foundation, 2006
20. L. Hufkens, C. Browne, A functional taxonomy of logic puzzles. IEEE Conference on Games, 2019
21. B. Parhami, Use of logic puzzles to promote techeracy for non-science/engineering students. 2018 IEEE 9th Annual Information Technology, Electronics and Mobile Communication Conference, 2018
22. A. Levitin, M. Levitin, *Algorithmic Puzzles* (Oxford University Press, Oxford, 2011)
23. P. Winkler, *Mathematical Mind Benders* (CRC Press, Boca Raton, FL 2007).
24. S. Nowduri, Critical thinking skills and best practices for cyber security. Int. J. Cyber-Secur. Digit. Forens. **7**(4), 391–409 (2018)
25. Center for the Study of Intelligence (U.S.), *A Tradecraft Primer: Structured Analytic Techniques for Improving Intelligence Analysis* (U.S. Central Intelligence Agency, Center for the Study of Intelligence, Washington, DC, 2009)
26. S.T. Hamman, K.M. Hopkinson, R.L. Markham, A.M. Chaplik, G.E. Metzler, Teaching game theory to improve adversarial thinking in cybersecurity students. IEEE Trans. Educ. **60**(3), 205–211 (2017)
27. H. Dudeney, *Strand Magazine*, vol. 68, July 1924
28. S.C. Karacal, J.A. Barker, J. Van Roekel, Experiences with a freshman engineering problem solving and reasoning course. American Society for Engineering Education PEER Annual Conference, Seattle, Washington, 1998

Ideologies and Issues for Teaching Blockchain Cybersecurity in Management and Computer Science

Kenneth David Strang, Ferdinand Che, and Narasimha Rao Vajjhala

1 Introduction

Cybersecurity has become a high-demand topic in management science and computer science higher education at universities around the world. The demand stems from the workforce need to educate and train managers and cybersecurity professionals about risk management and decision-making, especially in the cybersecurity domain. Additionally, the demand also emanates from technology companies and government departments, who need specialists trained to design, program, and debug contemporary cybersecurity systems. There is an increasing demand for cybersecurity professionals with a steady yearly growth rate of more than 6% expected until 2020 [1]. The demand for cybersecurity professionals across various domains, including accounting, finance, and logistics, is also in the rise apart from the core computing and information systems disciplines. Close to 70,000 cybersecurity incidents were reported in 2014, and a consistent increase of 10% was observed over the next fiscal year as well [2]. A shortage of two million cybersecurity professionals was forecasted for the 2019–2020 fiscal year [2]. The demand for cybersecurity education is increasing in developed as well as emerging nations. The reasons for the growing demand in cybersecurity education is connected to the fact that many countries are engaged with global trade and with the advances in telecommunication a significant proportion of global trade is increasingly conducted via the Internet [3]. Furthermore, most of these countries have a growing young-aged middle class with money to spend on computer science

K. D. Strang
APCC Research, Long Island, NY, USA

F. Che · N. R. Vajjhala (✉)
American University of Nigeria, Yola, Nigeria
e-mail: ferdinand.che@aun.edu.ng; narasimha.vajjhala@aun.edu.ng

K. Daimi, G. Francia III (eds.), *Innovations in Cybersecurity Education*,
https://doi.org/10.1007/978-3-030-50244-7_7

and management science education [3]. Igonor et al. [4] state that higher education institutions are not meeting the industry needs in the context of cybersecurity education and training. Igonor et al. [4] point out there is a lack of emphasis on the psychological aspects in the current cybersecurity curricula and also the students are not adequately trained with the required technical and soft skills.

A primary reason that the demand for cybersecurity education has increased is due to global cybersecurity breaches. Cybersecurity breaches have made it imperative to update and enhance the protection methods to address the new developments in technology [5]. In March 2019, a database containing over 2.4 million identity records on government officials and politicians from every country was leaked online [6]. In the same month, Facebook admitted that it had not properly secured over 600 million users' passwords since 2012 and that the sensitive data had been accessed by its employees [6]. In the very next month, April 2019, it was discovered that two third-party applications that hold sensitive Facebook datasets, over 540 million records, were exposed online [6]. Then in May 2019, WhatsApp, owned by Facebook, disclosed that its over 1.5 billion users had been left vulnerable to spyware designed by the NSO Group, an Israeli government surveillance agency [6]. These days many businesses generate a vast amount of sensitive data about their customers, data which is extremely attractive to malicious hackers, particularly because much of the data is stored in centralized repositories.

The decentralized nature of blockchain and its immutable ledger technology present great opportunities to secure and protect sensitive customers' data in a way that is also very unattractive for hackers to attack. But companies face major challenges in acquiring the workforce capabilities necessary to adopt and deploy blockchain technology, which promises better data security and privacy. At the same time, companies are increasingly dependent on the Internet to compete and as such must add adequate security to protect their critical infrastructure and sensitive data [7]. Decentralization alone is not enough as it is known that hackers often gain access to companies' networks and systems by attacking weaknesses in their networks or edge devices such as routers and switches [8]. The Internet of Things (IoT) is a paradigm encompassing all the devices and techniques under which every vital object in our daily life, including wallets, watches, refrigerators, cars, and others will be connected to each other and the Internet [9]. IoT systems are particularly vulnerable to cybersecurity attacks because the technology is pervasive and open. A large number of users, especially those with lack of technical knowledge and expertise, are unaware of the security risks posed by the various IoT devices that are extensively used in homes, for instance. Most of these devices are not configured to meet the minimum deterrents for cybersecurity threats.

Although IoT started way back in 1998, there has been significant progress only in the last decade with the proliferation of mobile computing, ubiquitous computing, and wireless sensor networks [9]. The number of IoT devices is expected to reach 7 trillion wireless devices, serving over 7 billion people [9]. Cybersecurity is a significant challenge for IoT devices [10]. IoT devices and components have low power and computing capabilities. Hence, the traditional complex security mechanisms and schemes would not be suitable for IoT devices [9]. Rahim et al.

[5] define cybersecurity awareness as the degree of the users' understanding of the importance of information security to protect the organizational data and network. According to Shammar and Zahary [9], hackers are likely to take advantage of the weaknesses within the IoT objects if cybersecurity in IoT is not managed properly. This is likely to result in disrupting the global IoT network.

The main security challenges in IoT include data confidentiality, privacy, and trust [9]. Data confidentiality is essential to ensure that the data is accessed and modified by authorized users and objects. Proper access control mechanisms need to be created and maintained to ensure that only authorized objects and users have access to IoT devices [9, 10]. IoT devices, especially those used in critical applications, including healthcare, raise privacy issues related to personal and sensitive information. Also, as wireless channels are used for communication between the devices, there are many security vulnerabilities involved in the transmission of the information between the devices [9]. Trust is also an important security challenge for IoT devices as most of the smart objects make decisions themselves, presenting trust challenges. Hence, risk management is needed to ensure cybersecurity attacks are minimized.

Cybersecurity attacks could be minimized if proper trust relationships are set up between the objects as well as the people using and accessing the devices. However, decision-makers in management science as well as programmers in computer science must be educated to perform risk management and to design IoT systems to be resilient against cybersecurity attacks [10]. Blockchain technology is the more important new development in IoT and cybersecurity systems. Therefore, blockchain must be incorporated into management science and computer science higher education.

In the next two sections, we discuss what is covered in blockchain technology in higher education, and then we explore what strategies universities are using to deliver this education.

2 Blockchain Technology in Higher Education

Rapid changes in technology affect not only what we teach but also how we teach [11]. Higher education universities teach blockchain by covering the theory and application as summarized below. Teaching approaches tend to be generally consistent in explaining what blockchain is, exploring its theoretical and mathematical foundations, the building blocks, as well as the technological potential of blockchain [12, 13]. Lending et al. [11] recommend the inclusion of technical teaching cases and tips, involving the use of blockchain technologies as well as other available technologies that are relevant to the information systems practice. On the other hand, even though the application of blockchain in finance, particularly cryptocurrencies, continues to be dominant in the blockchain applications landscape, teaching approaches are more fragmented in the exploration of the impact of blockchain and its other potential applications in areas such as supply chain and logistics, government, healthcare, energy trading, IoT, law, education, and of course cybersecurity [12–15]. Cybersecurity curricula should focus on both technical aspects as well as the business dimensions [2]. The focus on cybersecurity is usually more on the technical dimension, while the business and leadership aspects are often sidelined.

Blockchain, considered as one of the most disruptive technologies, is the core technology behind Bitcoin and the nearly 2000 other cryptocurrencies that are currently available over the Internet [16, 17]. Blockchain technology was developed by Satoshi Nakamoto in 2008 as the underlying technology for the Bitcoin cryptocurrency [18]. Every transaction in a blockchain is fully auditable [17]. Hence, the blockchain technology is increasingly being used across various business sectors, including logistics, banking, pharmaceuticals, and information technology [19]. Several accountancy firms use blockchain technology, and Big Four accounting firms are implementing several blockchain projects and have shown interest in adopting this technology [18]. Zalan [20] states that blockchain may account for 10% of the World's GDP by 2025. The market value of the cryptocurrencies is around $100 billion, and increasing significantly [16]. Some of the applications of blockchain technology include cryptocurrencies, such as Bitcoin. Blockchain technology is now used in a wide range of fields, including digital identify, medical records, financial institutions, and education [21]. Banks and investment funds have also invested more than $1 billion in blockchain technologies in 2016 [18]. Zalan [20] terms blockchain as a transformative technology rather than a disruptive technology that can transform several areas, including financial services, cybersecurity, logistics, and healthcare.

Blockchain addresses two key aspects of business on the Internet, namely, trust and transactions [22]. Data breaches have caused numerous problems for e-businesses and e-commerce transactions over the last two decades. Blockchain can address the privacy and breach of trust issues that have negatively impacted businesses dealing with online transactions. The basis for blockchain technology involves three concepts, namely, the transaction, the block, and the chain [23]. The transaction is an operation in the ledger, the block is responsible for recording all the transaction data over a span of time, and the chain is responsible for managing the set of blocks in a chronological manner reflecting the changes of state in the ledger. The block is cryptographically identified by a Merkle root hash, and every block is chained to its immediate predecessor block [24]. Trust is maintained in blockchain through a free market of independent nodes using a consensus of majority technique, and public visibility of the complete transaction is also ensured through the distributed nodes [24]. Shrestha et al. [25] define blockchain as a distributed and decentralized database of all transactions executed among participating nodes. Blockchain technology uses a distributed ledger to share and record information through a peer-to-peer network [18, 26]. The maintenance and validation of the ledger transactions take place in a decentralized and collective manner by the members of the department [26]. Data verification and content management are easier with blockchain technology because of the use of distributed public ledgers [16]. The blockchain technology uses a cryptographic-based distributed ledger for enabling trusted transactions [19]. The ledger entries are made in a chronological manner and in blocks using a cryptographic hash [18]. Trust is ensured in transactions through a tamper-proof ledger. The immutability of the transactions is ensured through a combination of sequential hashing as well as cryptography [27]. Data is stored in the form of multiple blocks that are connected with each other through a network, and any newly generated block is attached to the previous chain of blocks [26]. The most important feature of this chain of blocks is the permanency of the data in the blocks, which means that any changes can be accounted for in the chain [17, 26]. Blockchain technology coupled with other technologies, such as IoT, can create a permanent and shareable record creating efficiencies for the global economy [17].

One of the key features of blockchain technology is that it runs on top of the current stack of Internet protocols [22]. Blockchain uses cryptographic techniques to ensure that data of a transaction cannot be tampered and also ensures that transactions can be backtracked and verified [23]. The blockchain technology relies on a decentralized and distributed peer-to-peer network. This allows the nodes in the network to exchange data on a trust system improving the efficiency of the data exchange [23]. The security of the data in a blockchain is ensured through the adoption of an asymmetric encryption algorithm [23]. Blockchain technology allows completely secure, transparent, and immutable financial transactions. Another important aspect of blockchain technology is that is built on a distributed network, which means that there is no single point of failure risk, as even if one of nodes is affected, the other nodes in the distributed network maintain the integrity of the ledger [22]. Blockchains also eliminate the need for third parties as the value flows directly from the sender to the receiving party.

Blockchain technology provides key functionalities, including transactional validity, immutability, privacy, and immediacy [28]. The validity of a transaction is done through a timestamp which ensures that there is no duplication of transaction. The timestamp-based approach is implemented by two algorithms using a proofing mechanism to prevent any duplication or fraud in the transactions [28]. Blockchain technologies use mechanisms, including proof of service, consensus, and proof of stake to achieve transactional immediacy [28]. The blockchain infrastructure has five main layers, namely, the data layer, network layer, core layer, contract layer, and application layer [29]. The applications are hosted on the top layer, that is, the application layer. The security issues related to the remaining four layers need to be examined.

A block is a basic unit of a blockchain network used for recording information about a transaction. A block in a blockchain network has two components, namely, the header and the body [29]. The header includes information, including the size of the block, the version of the bloc, the hash information for the previous block, target block, and the node hash value as well as the timestamp [29]. The body of a block includes the number of the transactions as well as the record of the transactions. The first block in a blockchain network is known as the genesis block. This block has all the information needed for all the nodes in the network, including the cryptographic hashes of the records [25]. Transactions are validated based on a set of consensus rules and placed in a block after validation. The blocks of transactions are connected into a chain of blocks cryptographically to ensure high levels of data security [18]. There are four types of nodes in a blockchain network, namely, full nodes, super nodes, light nodes, and mining nodes. The full nodes store the complete history of the transactions on the blockchain network [25]. Merkle Trees are also a core part of blockchain technology [18]. Merkle Trees can take a large number of transaction identification numbers and convert them into a single 64-bit code [18]. This process allows small amount of data to process and verify the transactions and also resolve the memory space problem [18]. Every block in the blockchain has a header and the body. The header includes information on the hash of the previous block while the body has the list of transactions in the blockchain [25]. Blockchain technology also uses self-executing smart contracts extensively. The basis of smart contracts is the predefined business logic that is agreed on by the contracting parties in the transaction [18].

Blockchain technology includes both public and private blockchains. In public blockchains, every transaction is public and users have the option of remaining anonymous. Hence, the networks using public blockchains usually have an incentivizing mechanism so that participants can be encouraged to participate [17]. There is no need for permission from a central authority for public blockchains, while private blockchains involve a certain degree of centralization [15, 25]. This degree of centralization leaves the private blockchains open to a central point of failure [25]. The example of public blockchain includes all the cryptocurrencies, including Bitcoin and Ethereum, where there are no restrictions on participation in the network [16, 17]. Public chains are decentralized and are not prone to the single point of failure problem. Private Blockchains are operated by specific institutions on an independent basis [15, 16]. In permissioned blockchains, the access is controlled either by a consortium of members in the case of a consortium blockchain or by a single organization in the case of a private blockchains [17]. Blockchain technology offers several advantages, including decentralization, distributed security, transparency, and immutability in trustless environments [25].

Thombs and Tillman [30] emphasize on the importance of updating the curriculum in universities to include blockchain and cryptocurrencies. They suggest several topics that should be included in the curriculum for blockchain and cryptocurrencies courses. Among several issues that were identified in this study, was the discussion on fake-blockchain technology [30]. Some of the system designers are substituting the word "database" with "blockchain" to give a superficial blockchain product without any significant redesign. Hence, students would benefit from understanding the difference between traditional databases and the blockchain technology. There should be broad discussion on terms, including, including peer-to-peer, distributed, consensus-based, and cryptographically secure applications [30].

Smart contracts are also an important part of blockchain technology, and should be part of the blockchain curriculum [30]. Thombs et al. give the example of the scripting technology that is used by Bitcoin which is not Turing complete as compared to other cryptocurrencies, such as Ethereum, that support fully Turing-complete scripting. The difference between the support provided for Turing-complete scripting is that scripting languages that do not support Turing complete do not support some of the features, such as iterative loops, which will protect the applications from some of the hacking attacks [30]. These core concepts must be integrated into the blockchain curriculum.

Blockchain technology has disrupted several businesses with several large companies moving quickly to adopt this technology and replace some of the previously used technologies [31]. Hence, universities must include these important topics to integrate early exposure and understanding of these complex issues. In a study carried out by Ryabova and Henderson [31] on integrating cryptocurrency into intermediate financial accounting curriculum, around 88% of the students expressed interest in learning more about blockchain and cryptocurrencies. A similar percentage of students also indicated that the knowledge of blockchain will give them an advantage in their future workplace.

3 Integration of Blockchain in Cybersecurity Education

Craigen et al. [32] describe cybersecurity as a complex challenge that requires interdisciplinary reasoning. Higher education universities teach cybersecurity risk management and how blockchain can be applied to reduce cybersecurity threats. The nature of cybersecurity education is interdisciplinary, with various disciplines including, finance, management, accounting, logistics, etc. Most of these disciplines require varying degrees of cybersecurity knowledge and expertise. Below we discuss some of the controversies and benefits typically covered in cybersecurity higher education.

Cybersecurity awareness is essential for Internet users, especially the young users to combat silent privacy invasion [5]. Blockchain technology strengthens existing secure networks and communications as encryption and hashing are used to store immutable records [19]. A large number of businesses are moving from centralized storage to the decentralized blockchain technology as data privacy is well handled in the blockchain technology [8]. Blockchain market has grown significantly over the last few years and so have the number of malicious attacks on the blockchain system. The increase in the number of security incidents has resulted in increasing demand for studies on blockchain security [26]. Cai et al. [29] state that while several companies and organizations are developing application systems based on the blockchain technology, they are sidelining some of the security issues, especially in the areas of the privacy and security of the blockchain network.

The National Aeronautics and Space Administration (NASA) is using the distributed ledger blockchain technology to strengthen its cybersecurity by preventing denial of service attacks [8]. The decentralized nature of blockchain ensures that there is no single point of entry for hackers. Single point of vulnerability is one of the key factors because of which the current domain naming system is susceptible to attacks from hackers [8]. This vulnerability can be handled through the decentralized blockchain technology as hackers would no longer be able to exploit single points of vulnerability. The domain information can also be stored on a distributed ledger used in the blockchain technology [8]. There is also limited chance for security breaches as the cryptographic access key can be revoked in the event of any security breach [8].

Raban and Hauptman [33] describe emerging technologies, including artificial intelligence (AI), quantum computing, blockchain, and IoT as both having a positive and negative impact on cybersecurity. For instance, AI can possibly be used both for attacking other systems and also for providing enhanced cyber defense capabilities. As large companies move toward adopting blockchain technology, the cybersecurity events would come under the high-impact, low-frequency events. This is because blockchain technology is based on a strong and secure foundation but there are possible vulnerabilities that have not yet been explored as the applications built on the blockchain technology are largely work in progress.

According to Taylor et al. [19], most of the studies on blockchain and cybersecurity focus on IoT, data storage and sharing, network security, data privacy, and navigation and utility of the World Wide Web. IoT devices have proliferated significantly over the last decade. However, there are security issues related to IoT as hackers could use these edge devices to launch cyberattacks [8]. Blockchain technology can be used to address the security concerns in IoT devices, as the devices can form group consensus on what would constitute a normal occurrence within the network [8]. In the event of any suspicious activity, the nodes involved in the activity can be removed or locked down.

Blockchain technology uses a decentralized distributed ledger system that provides an immutable and irreversible record for every transaction [28, 34]. Two important technologies, namely, the distributed ledger technology and cryptography, are combined. Distributed ledger technology involves distributed storage with information stored in multiple locations. The use of cryptography with distributed computing means that the entries in the distributed ledgers are secure. Blockchain is considered a highly reliable technology as it is very difficult to break the encrypted blocks stored in multiple secure locations, and all the transactions on the system have a traceable history. Blockchain technology has a secure basis but as demonstrated by the number of security incidents that Bitcoins had to face in the last few years, a number of security challenges still need to be addressed. For instance, a Japanese Bitcoin exchange, Mt. Gox, as well as InstaWallet Bitcoin wallet service reported losses of over $12 million US dollars because of hacking [35]. A much larger hacking incident occurred with the Hong Kong-based Bitfinex Bitcoin exchange, which reported losses of over $65 million US dollars because of hacking [35]. These incidents expose some of the cybersecurity challenges that blockchain-based technologies and applications face.

Blockchain technology also provides higher security as compared to storing data in a centralized database [35]. The key advantage of blockchain is the unique mix of security and transparency. Blockchain technology ensures security through full encryption of the data blocks [27]. In this way, even if the hacker gains access to a blockchain network and the data blocks, the attacker cannot read the data blocks. Blockchain technology uses encryption keys along with the public key infrastructure

(PKI) technology to provide a higher level of security [27]. There is a risk of compromising transparency in the context of security, but in the case of blockchain technology transparency is assured through the openness attribute [35].

Conte de Leon et al. [36] point out some of the challenges in blockchains in the context of cybersecurity, including the selection of time-resilient strong cryptographic functions. The key issue is that one-way hash algorithms that are cryptographically secure today may not remain secure in the future. For instance, finding the inverse of one-way hash algorithms using the current technology is not computationally feasible, but this could soon be possible. There is a possibility that in the next 20 or 30 years with advances in technology, these algorithms may not remain computationally infeasible [36]. While the underlying distributed ledger system is secure in blockchain technology, there are cybersecurity issues with the correct and secure implementation of these by organizations seeking to implement blockchain technology [36]. This requirement would involve the use of adequate software engineering design and development methods, the use of formal methods for protocol specification, as well as exhaustive testing [36].

However, organizations need to reconsider and reframe their risk management strategies with the implementation of new technology. There is limited literature in the context of blockchain and cybersecurity [19]. Taylor et al. [19] emphasize the need for closer collaboration between the academic researchers and the industry as there is a gap in the context of blockchain cybersecurity research. Furthermore, there may be opportunities to leverage blockchain to improve the credibility and reliability of any data processing that relies on artificial intelligence (AI) in the era of big data [19]. This chapter will add value to the existing literature on these topics and examine why blockchain technology is particularly attractive for blockchain adoption in emerging markets, and how universities can upgrade their course offerings to include blockchain and related technologies [37]. Blockchain technology is one of the key technologies used by Fintech startups, which are increasingly assuming significance in the emerging markets. In the emerging markets, blockchain technology also promises a lot more benefits beyond a sound foundation for Fintech, especially if the data quality tension between trust and truth inherent in blockchain can be bridged [27, 38].

4 Teaching Ideologies for Blockchain and Cybersecurity

We took an empirical approach to investigate what ideologies higher education universities were applying to teach blockchain and cybersecurity. We reviewed the scholarly ideology or rationale for teaching blockchain at universities in emerging economies rather than examine pedagogy underlying the methodology [39]. A search of the peer-reviewed full-text literature using Proquest Central and EBSCO indexes with keywords "blockchain" and "teach," but no other constraints, returned only five relevant peer-reviewed papers [see 12, 22, 40–42]. One paper was from a journal, another was a book section, and the rest were published from peer-reviewed conferences. We assert this limited finding was due to the limited academic interest of the topic within the information technology education domain since more results were returned when we switched from "blockchain" to "object-oriented programming language" in the search keywords. Nevertheless, it is worth examining these contemporary studies about teaching blockchain, which we summarize below, and then we discuss our experiences followed by presenting our proposed conceptual model.

Kursh and Gold [22] published a single case study revealing why and how blockchain was being taught at a USA-based university. Blockchain is relatively new, and this was the first instance of a teaching study emerging in 2016 from the peer-reviewed literature (the other papers were in 2019). In their case, the topic was positioned as an electronic commerce technology innovation within the information management technology curriculum. Their view of blockchain was that it is part of a financial technology enterprise solution—which we assert is a managerial concern in contrast to application design or programming view. They targeted students from business (this was where their degree was granted)—and blockchain was taught in their finance, strategy, entrepreneurship, and information technology courses. They rationalized that teaching blockchain would be in demand for business, technology, and finance college majors because it was developed in 2014 by an anonymous group (Satoshi Nakamoto—a pseudonym for an unknown group) after the 2008 global financial crisis as a secure peer-to-peer digital currency electronic cash system now known as Bitcoin. They noted many banks, insurance companies, and businesses were interested in hiring people with at least a theoretical understanding of blockchain and cryptocurrencies (e.g., Bitcoin). Their emphasis was on entrepreneurship, finance, and cryptocurrency stakeholders. There was little explanation of the pedagogy except that guest lectures were used, a symposium was held, and a student club was formed. It was clear to us that they were covering the managerial aspects of blockchain and not programming.

In his single case study, Liu [41] took the opposite viewpoint by teaching blockchain from an application programming standpoint. Liu [41] developed a tutorial in Java called ChainTutor for helping students learn basic blockchain concepts in computer science and information technology discipline courses at the university level. Liu [41] claimed the targeted audience included stakeholders included decision-makers as well as information technology staff in banking and healthcare companies. Liu [41] classified the topic under the Internet of Things (IoT), which we feel is management information technology rather than pure computer science. His pedagogy was to offer lectures with screenshots and then have students work through examples in the controlled tutor graphical user interface program to appreciate the concepts, but not conduct actual programming, as some concepts of blockchain are not easy for beginners to understand. Liu mentioned that the visual and interactive kinetic delivery tutorial supplemented text-based materials as an effective teaching procedure. In comparison to Kursh and Gold [22], we categorize Liu's [41] approach as less managerial but leaning toward technical project team leaders or application designers but not application programmers.

Rao and Dave [42] published a single case study of teaching blockchain in the computer science and engineering fields as part of their BS/MS programs at Princeton University, Princeton, New Jersey, in the USA. They considered blockchain to be most related to the IoT and cloud computing subjects. Their ideology was clearly to teach basic blockchain within the programming context using a hands-on laboratory-based approach. They provided Raspberry Pi for the teaching delivery platform, which is a small, inexpensive object-oriented programming language that allows students to build IoT applications. They covered concepts such as creating immutable records using cryptographic hash functions, along with transmitting and storing data on the cloud. They provided lectures, tutorials, the sequencing mandatory reading materials, and they specified the graded outcomes but allowed the students to proceed using their choice of methods. As pedagogy, they provided a short theoretical lecture followed immediately by a hands-on lab tutorial with an exercise to complete. We concluded that Rao and Dave [42] positioned blockchain teaching as application development.

Negash and Thomas [40] also published a single case study about teaching blockchain to university students as a nontechnical subject in the business discipline. They described blockchain as a new disruptive technological opportunity for businesses that allows companies to unlock value through trusted and smart peer-to-peer financial transactions where verification and transfer of assets are protected by encryption. They did not provide details for the pedagogy, but they stated their key learning objectives were to educate administrative and infrastructure decision-makers about blockchain enterprise-level or component solutions. As they confirmed, their ideology was to teach blockchain as a managerial decision-making issue rather than a programming or design activity.

IoT devices are gradually expanding to a wide range of applications and appliances, including medical devices and operation of critical infrastructure. Several security problems arise with the use of IoT devices, including insecure web interfaces, lack of adequate authentication, and transport encryption, apart from weak physical security [33]. In the context of cybersecurity, Raban and Hauptman [33] identified that IoT as an emerging technology has an overall net negative impact on cybersecurity while blockchain technology has a net positive impact. The security issues with IoT devices need to be handled before these devices can have a positive impact but this process is likely to be long considering the numerous cybersecurity issues involved.

Dettling [12] published a book chapter in which he explored the challenge that educators face in developing appropriate syllabi and curriculum for teaching blockchain to an audience of business management students, given that teaching approaches are currently fragmented. He recognized that any such blockchain curriculum would have to evolve with the developments in blockchain. Dettling's ideology was to teach blockchain as a managerial decision-making issue, but he took a nuanced position that without an understanding of the basic principles of blockchain it is not possible to effectively assess fundamental managerial decisions about the impact and implications of deploying blockchain [12]. According to Dettling [12], it may be necessary to rethink the syllabi of foundational business management courses such as business and financial mathematics to ensure that business management students acquire the relevant preparation to better assimilate blockchain technology in the managerial context. Dettling put forward a suggested approach to developing blockchain topics for teaching business management students, which focused on business impact, the blockchain building blocks, and the applications of blockchain [12]. We concluded that even though Dettling [12] took a nuanced position he nevertheless positioned blockchain teaching as a managerial decision-making issue.

We also reviewed many non-peer-reviewed sources, and one stood out as interesting. Zeltsinger [43] explained how he designed and taught blockchain at the college level to introductory application design and programming students. He admitted the topic was too broad to include in a single course, so he elected to cover blockchain architecture, specifically the Ethereum Virtual Machine (EVM) environment and protocol (which is a new paradigm emerging from the Bitcoin system). He stated he taught the topic using an application programming standpoint.

There were other related but relatively older sources, including Bicak et al. [44] whose conference paper focused on the development of cybersecurity curriculum for graduate students of universities. Bicak et al. [44] recognized that cybersecurity is a broad field which makes it extremely difficult to cover and draft a curriculum that strikes the right balance between achieving comprehensive coverage and meeting the demands of employers who demand that cybersecurity graduates hit the ground running in relevant but specialized areas. Nevertheless, Bicak et al. [44] adopted the view that to meet the workforce needs cybersecurity education programs should take incorporate both formal education as well as practical training, leading to certification. Therefore, higher education institutions would do well to enrich cybersecurity education offerings by embracing a multidisciplinary approach and nontechnical ideologies.

Mouheb et al. [45] conducted a review and analysis of curriculum design approaches for cybersecurity education. They found that the approaches to teaching cybersecurity were fragmented due to the multidisciplinary nature of cybersecurity, even though the approaches share a common objective to impact the requisite knowledge and skills needed by the workforce demanded by the market. Mouheb et al. [45] found that some approaches designed programs with differing mixes of technical knowledge and practical skills to suit students' career objectives in academia, industry, or government. We concluded that the different approaches to cybersecurity program design were mostly done from a technical standpoint. However, the approaches had a common objective to bridge an apparent disconnect between academia and industry, where universities are perceived as educating meanwhile industry demands trained graduates to meet acute workforce needs. In developing cybersecurity programs it is important to strike a pragmatic balance between the core knowledge necessary and the workforce skills needed in the various disciplines. The process of developing the core knowledge involves curriculum revision and updating the curricula to meet the requirements of the industry taking into consideration the rapid technological advancements.

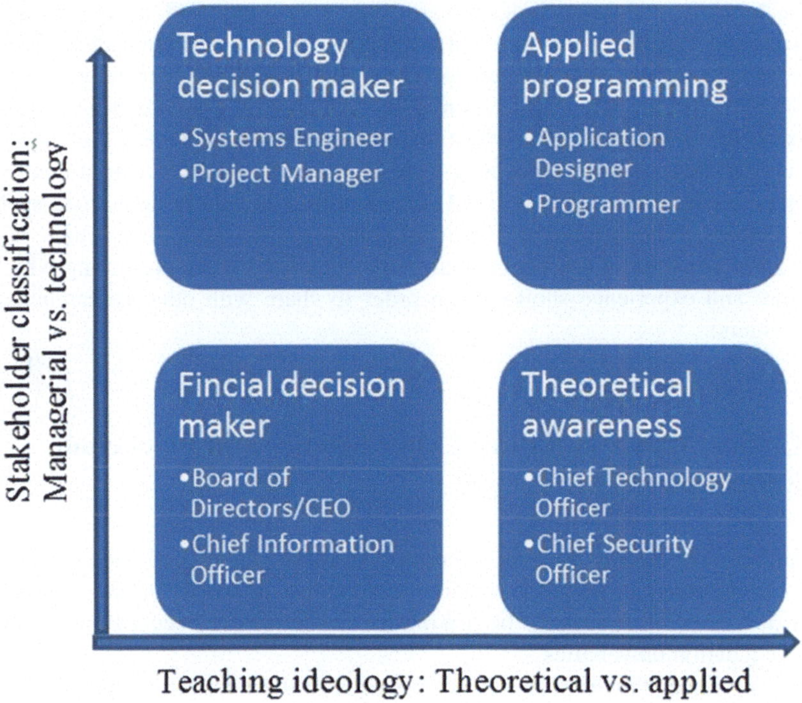

Fig. 1 Conceptual typology

We developed the conceptual typology in Fig. 1 to synthesize the scholarly ideology or rationale for teaching blockchain at universities based on the literature and from our experiences. This model can be used by decision makers at any level to design programs at universities or by consumers to analyze what to study when pursuing a degree. This model will have tremendous utility for academic and entrepreneurial stakeholders. The conceptual typology in Fig. 1 has two dimensions: stakeholder classification and teaching ideology. All universities and educational institutions provide products to consumers, for example, students who become the key stakeholders making decisions about which degrees or courses to purchase and or expend effort toward.

The stakeholder dimension is comprised of the industry position and career goal for the student. Stakeholders include all the parties that have an influence over the project, and can affect or be affected by the project. For example, a stakeholder in the scope of this model relevant to learning blockchain would likely range from application programmers (e.g., maybe using EVM, Python, or other object-oriented programming language), through application designers who model but do not write programs. The other end of the spectrum would include managerial decision-makers, including bankers, insurance investors, CEOs, CIOs, who determine the need then the criteria to purchase or implement blockchain solutions.

The teaching ideology dimension of Fig. 1 is comprised of the pedagogy, lecturing and delivering materials, or andragogy, meaning helping adults learn to learn [46], along with the delivery procedures. The delivery procedures refer to audio/lectures with visual slides, text materials, interactive graphical object-oriented lab exercises, or innovative combinations of these. Another way of looking at the teaching ideology is whether it is applied in a kinetic way (even in a controlled prototype environment such as EVM) or whether it is wholly theoretical such as guest lectures, symposiums, or traditional pedagogy. We do not assert there is one best magic quadrant in our conceptual model—rather we are presenting what the literature and experience shows us, in order to share with other researchers and stakeholders.

5 Critical Issues in Future Cyber Science and Blockchain Education

We further investigated the risk management controversies and issues emerging from cybersecurity at higher education universities. In particular, we focused on the critical issues that seem to be driving the design of cybersecurity programs in higher education universities.

One key issue is the critical balance between distributed technology ease of use and trust. The distributed ledger nature of blockchain technology makes it potentially very attractive for protecting businesses and other entities from cyberattacks in the emerging economies where there is a tendency to distrust centralized authority and control. The distributed nature of the ledger blocks also ensures that there is no single point of failure.

Blockchain offers decentralized technologies through a trustless distributed ledger technology that enables immutable peer-to-peer transactions with minimized central authority control by "trusted" third-party institutions [37]. With most transactions in emerging markets conducted through local agents, businesses are generating a huge amount of customer data that is very complicated to harvest and to store in centralized repositories.

However, the decentralized nature of the customer data in emerging economies makes it unattractive to hackers but extremely attractive to deploy advanced blockchain technologies, such as the next-generation blockchain technologies used for the Apollo currency [8, 47], to efficiently archive and protect customer data and cryptographically grant or revoke access to third-parties as needed. Businesses in the emerging markets are also more likely to recognize that it may be more cost-effective to reach larger customer segments by deploying small IoT devices that are better aligned with existing social and cultural characteristics of the customers.

To protect against cyberattacks, it is necessary to deploy smart IoT devices that do not rely on management via any central authority but can form group consensus about what is normal activity or data within their IoT network, so that each device is accountable for its actions, messaging is secure, and the groups are able to quarantine any devices nodes that behave unusually [47]. The nature of the cybersecurity challenge in emerging economies is somewhat different from the challenges faced by the developed economies, which are very invested and dependent on huge and complex networks and infrastructure and deeply burdened by a great amount of regulation by various private and public institutions.

Cybersecurity education should reflect the fundamental differences explored above. In developing cybersecurity programs and courses those in charge of the cybersecurity curriculum should recognize the relationship between the curriculum required and the workforce skills needed in the various disciplines [48]. Cybersecurity education curriculum for the emerging markets should emphasize not only on the key security knowledge areas [48] that are essential but also on the relevant skills required, taking into account the nature of the opportunities to develop those practical skills that support the application of critical knowledge.

Interestingly, the Association of Computing Machinery Joint Task Force on Cybersecurity Education (CSEC) thought model for cybersecurity education specifies three dimensions: knowledge areas, crosscutting concepts, and disciplinary lenses. The cybersecurity knowledge areas include data, software, component, connection, system, human, organizational, and societal security [48]. The crosscutting concepts include confidentiality, integrity, availability, risk, adversarial thinking, and systems thinking [48]. The knowledge areas indicate critical knowledge with broad relevance within and across multiple computing-based disciplines such as Information Systems, Computer Engineering, Software Engineering, and others. Similarly, the Accreditation Board for Engineering and Technology (ABET)-accredited cybersecurity program focuses on the problem-solving, designing, and implementation skills of the students [4]. There is ample focus in the cybersecurity curricula on ensuring that students have the required skills, including soft skills and decision-making capabilities [4]. The National Centers of Academic Excellence (CAE) cybersecurity curricula framework has three knowledge units, including foundational, core technical, and core nontechnical units [4]. The foundational unit deals with the basics of cybersecurity principles and the related information technology components. The core technical unit includes basic programming knowledge as well as the foundations of cryptography and operating systems [4]. The core nontechnical unit includes the ethical, legal, and planning aspects of cybersecurity and management.

6 Conclusion

In this chapter, we investigated the scholarly ideology or rationale for teaching blockchain at universities. We (authors) discussed our approaches used for designing and teaching information technology-related courses, including topics such as blockchain and IoT to make sense of the literature review. We elaborated on how our colleagues were designing and teaching blockchain at the university level in the business as well as computer science disciplines. We have explored the need for cybersecurity in higher education in this chapter. We dealt with the key questions of why cybersecurity is required in higher education, and discussed the ideologies that university are applying in the realm of management science as well as computer science education. Our findings indicated that blockchain technology was one of the essential components of modern cybersecurity higher education. We also proposed a conceptual typology that will assist in synthesizing the scholarly ideologies for teaching blockchain at universities based on the reviewed literature as well as our experiences. Our conceptual model can be used by decision-makers and academic administrators at different levels to design and deliver cybersecurity and risk management degrees.

We developed a conceptual typology to synthesize risk management at universities in emerging economies. In this chapter, we investigated the scholarly ideology or rationale for teaching blockchain at universities. Ever since the inception of Bitcoin in 2008, the underlying blockchain technology has come into prominence. We developed a conceptual typology to synthesize the scholarly ideology or rationale for teaching blockchain at universities based on the literature and from our experiences. This model can be used by decision-makers at any level to design programs at universities or by consumers to analyze what to study when pursuing a degree. This chapter will contribute to understanding the risk management strategies that are currently being employed by organizations in the context of blockchain technology. This chapter will also benefit educators and students as it would offer critical insights into understanding the cybersecurity risk management strategies in organizations using the blockchain technology.

References

1. S.J. Roohani, X. Zheng, Using ten teaching modules and recently publicized data-breach cases to integrate cybersecurity into upper-level accounting courses, in *Advances in Accounting Education: Teaching and Curriculum Innovations*, vol. 23, ed. by T. Calderon. Advances in Accounting Education (Emerald Publishing Limited, Bingley, United Kingdom, 2019), pp. 113–125
2. S.C. Yang, B. Wen, Toward a cybersecurity curriculum model for undergraduate business schools: a survey of AACSB-accredited institutions in the United States. J. Educ. Bus. **92**(1), 1–8 (2017)
3. A. Fleury, M. Houssay-Holzschuch, For a social geography of emerging countries: introduction to the themed issue. EchoGéo **21**(5), 1–15 (2012)
4. A. Igonor, R. Forbes, J. McCombs, Cybersecurity education: the quest to building "bridge" skills. ISSA J. **8**(1), 18–26 (2019)
5. N.H.A. Rahim, S. Hamid, L.M. Kiah, S. Shamshirband, S. Furnell, A systematic review of approaches to assessing cybersecurity awareness. Kybernetes **44**(4), 606–622 (2015)
6. S. Turner, 2019 data breaches—the worst breaches, so far. *We Aren't Just Protecting You from Identity Theft. We Protect Who You Are*, February 25, 2019. https://www.identityforce.com/blog/2019-data-breaches
7. R. Gupta, *Hands-On Cybersecurity with Blockchain: Implement DDoS Protection, PKI-Based Identity, 2FA, and DNS Security Using Blockchain* (Packt Publishing, Birmingham, 2018)

8. A. Arnold, 4 Promising use cases of blockchain in cybersecurity, in *Forbes*, ed. (Forbes, Jersey City, 2019)
9. E.A. Shammar, A.T. Zahary, The internet of things (IoT): a survey of techniques, operating systems, and trends. Library Hi Tech **38**(1), 1–62 (2019)
10. A.R. Mathew, Cyber security through blockchain technology. Int. J. Eng. Adv. Technol. **9**(1), 3821–3824 (2019)
11. D. Lending, M. Mitri, T.W. Dillon, Ingredients of a high-quality information systems program in a changing IS landscape. J. Inf. Syst. Educ. **30**(4), 266–279 (2019)
12. W. Dettling, How to teach blockchain in a business school? in *Business Information Systems and Technology 4.0: New Trends in the Age of Digital Change* (Springer International Publishing, New York, 2018)
13. L. Li, X. Wu, Research on school teaching platform based on blockchain technology, in *Presented at the 14th International Conference on Computer Science Education (ICCSE)*, Toronto, Canada, 19–21 August 2019, 2019
14. A.F. Camilleri, A. Inamorato dos Santos, A. Grech, Blockchain in education (EUR 28778). Joint Research Centre (European Commission) 2017. doi: https://doi.org/10.2760/60649
15. J.R. Butcher, C.M. Blakey, *Cybersecurity Tech Basics: Blockchain Technology Cyber Risks and Issues*, pp. 1–5. https://www.steptoe.com/images/content/1/8/v2/189187/Cybersecurity-Tech-Basics-Blockchain-Technology-Cyber-Risks-and.pdf
16. J.H. Jo, S. Rathore, V. Loia, J.H. Park, A blockchain-based trusted security zone architecture. Electron. Libr. **37**(5), 796–810 (2019)
17. Y. Wang, J.H. Han, P. Beynon-Davies, Understanding blockchain technology for future supply chains: a systematic literature review and research agenda. Supply Chain Manag. Int. J. **24**(1), 62–84 (2019)
18. E. Bonson, M. Bednárová, Blockchain and its implications for accounting and auditing. Meditari. Account. Res. **27**(5), 725–740 (2019)
19. P.J. Taylor, T. Dargahi, A. Dehghantanha, R.M. Parizi, K.-K.R. Choo, A systematic literature review of blockchain cyber security. Digit. Commun. Networks **6**(1), 1–10 (2019)
20. T. Zalan, Born global on blockchain. Rev. Int. Business Strat. **28**(1), 19–34 (2018)
21. P.R. Vargas, C.L. Soriano, Blockchain in the university: a digital technology to design, implement and manage global learning itineraries. Digit. Educ. Rev. **35**(1), 130–150 (2019)
22. S.R. Kursh, N.A. Gold, Adding Fintech and blockchain to your curriculum. Business Educ. Innov. J. **8**(2), 6–12 (2016)
23. H. Sun, X. Wang, X. Wang, Application of blockchain technology in online education. Int. J. Emerg. Technol. Learn. **13**(10), 252–259 (2018)
24. I. Purdon, E. Erturk, Perspectives of blockchain technology, its relation to the cloud and its potential role in computer science education. Eng. Technol. Appl. Sci. Res. **7**(6), 2340–2344 (2017)
25. R. Shrestha, R. Bajracharya, A.P. Shrestha, S.Y. Nam, A new type of bockchain for secure message exchange in VANET. Digit. Commun. Networks **6**(1), 1–10 (2019)
26. D. Gountia, Towards scalability trade-off and security issues in state-of-the-art blockchain. EAI Endors. Transact. Secur. Safe. **5**(18), 1–9 (2019)
27. E. Piscini, D. Dalton, L. Kehoe, Blockchain and cybersecurity, December 14, 2017. https://www2.deloitte.com/content/dam/Deloitte/tr/Documents/technology-media-telecommunications/Blockchain-and-Cyber.pdf
28. H. Subramanian, Security tokens: architecture, smart contract applications and illustrations using SAFE. Manag. Financ. **45**(12), 1–14 (2019)
29. Z. Cai, C. Du, Y. Gan, J. Zhang, W. Huang, Research and development of blockchain security. Int. J. Performabil. Eng. **14**(9), 2040–2047 (2018)
30. M. Thombs, A.A. Tillman, Designing 21st century curriculum for Bitcoin and blockchain studies. Int. J. Global Business **11**(1), 67–80 (2018)
31. T.S. Ryabova, S. Henderson, Integrating cryptocurrency into intermediate financial accounting curriculum: a case study. J. of Acco. And Fina. **19**(6), 167–179 (2019)

32. D. Craigen, N. Diakun-Thibault, R. Purse, Defining cybersecurity. Technol. Innov. Manag. Rev. **4**(10), 13–21 (2014)
33. Y. Raban, A. Hauptman, Foresight of cyber security threat drivers and affecting technologies. Foresight **20**(4), 353–363 (2018)
34. R. Rust, Banking on blockchains: a transformative technology reshaping Latin American and Caribbean economies. Univ. Miami Inter-Am. Law Rev. **50**(2), 185–196 (2019)
35. J.H. Park, J.H. Park, Blockchain security in cloud computing: use cases, challenges, and solutions. Symmetry **9**(1), 164–177 (2017)
36. D. Conte de Leon, A. Stalick, A. Jillepalli, M. Haney, F. Sheldon, Blockchain: properties and misconceptions. Asia Pacific J. Innov. Entrepren. **11**(3), 286–300 (2017)
37. G. Guo, Blockchain: an emerging technology perfect for emerging economies, December 14, 2019. https://theactuarymagazine.org/blockchain-an-emerging-technology-perfect-for-emerging-economies/
38. E. Murabito, Trust, Consensus and Truth, December 16, 2019. https://medium.com/swlh/trust-consensus-and-truth-3ba142706432
39. K.D. Strang, Teaching virtual online courses in an era of negative student reviews: mixed methods controlled experiment and feedback, in *Emerging Technologies in Virtual Learning Environments*, ed. by K. Becnel, (Information Resources Management Association (IRMA), Hershey, PA, 2019), pp. 20–37
40. S. Negash, D. Thomas, Teaching blockchain for business, in *Presented at the IEEE Canadian Conference of Electrical and Computer Engineering (CCECE)*, Edmonton, Canada, 2019
41. X. Liu, A small Java application for learning blockchain, in *Presented at the Electronics and Mobile Communication Conference (IEMCON)*, Vancouver, Canada, 2018
42. A.R. Rao, R. Dave, Developing hands-on laboratory exercises for teaching STEM students the internet-of-things, cloud computing and blockchain applications, in *Presented at the Integrated STEM Education Conference (ISEC)*, Princeton, NJ, 2019
43. S. Zeltsinger. *What to Teach when Teaching Blockchain?* 2017. https://zeltsinger.com/2017/04/07/teach-teaching-blockchain/
44. A. Bicak, M. Liu, D. Murphy, Cybersecurity curriculum development: Introducing specialties in a graduate program. Inform. Syst. Educ. J. **13**(3), 99–110 (2015)
45. D. Mouheb, S. Abbas, M. Merabti, Cybersecurity curriculum design: a survey, in *Transactions on Edutainment XV*, vol. 11345, ed. by Zhigeng Pan Adrian David Cheok Wolfgang Müller Mingmin Zhang Abdennour El Rhalibi Kashif Kifayat (Lecture Notes in Computer Science, Springer, Berlin, Heidelberg, 2019), pp. 93–107
46. K.D. Strang, *Effectively Teach Professionals Online: Explaining and Testing Educational Psychology Theories* (VDM Publishing, Saarbruecken, 2010)
47. A. Tan, *How Blockchain can Secure the IoT?* 2018. https://www.computerweekly.com/news/252433944/How-blockchain-can-secure-the-IoT
48. D.L. Burley et al., Cybersecurity curricula 2017: curriculum guidelines for post-secondary degree programs in cybersecurity, in *Joint Task Force on Cybersecurity Education*, IEEE Computer Society, 2017, doi:https://doi.org/10.1145/3184594

Early Work Vis-à-Vis Current Trends in Internet of Things Security

Pabak Indu and Souvik Bhattacharyya

1 Introduction

In the 1960s, the Internet was limited to a few scientists and defence only. Since then, computers faced a physical threat, but as time evolved the physical threat became more of a malfunctioning of the devices. International Monetary Fund says that globally around $100 billion financial loss occurs because of cyber-attacks and in some years this goes up to $250–$300 billion [1]. From a small- to a large-scale organization, private to the public sector, many organizations are increasing their technical assets on a large scale to implement their business process efficiently. Failure of these assets fails their business process. Operational cybersecurity threats are defined as [2] operational threats to the technological assets that have consequences affecting confidentiality, availability, integrity or information system. Now let's find the cybersecurity threats and trends for the upcoming digital world. Phishing, ransomware, crypto-jacking, cyber-physical attack, state-sponsored attacks, IoT attacks and social engineering are the most trending cybersecurity attacks in recent years [3]. Though IoT attacks are one of the most trending attacks, still, IoT has been an integral part of our daily life to maintain our daily routine and activities.

IoT devices play a significant role in improving the quality of life for the elderly and people with disabilities [4]. Some IoT devices are used to help and monitor the different vitals of people, even when they are sleeping [5, 6]. IoT devices are also used to revolutionize physical therapy [7]. The Autism Glass [8] aims to provide support for autistic children, which helps to recognize the emotions of other people

P. Indu
Department of Computer Science and Engineering, Adamas University, Kolkata, West Bengal, India

S. Bhattacharyya (✉)
Department of Computer Science and Engineering, University Institute of Technology, University of Burdwan, Burdwan, West Bengal, India

K. Daimi, G. Francia III (eds.), *Innovations in Cybersecurity Education*,
https://doi.org/10.1007/978-3-030-50244-7_8

in real time [9] and help them react and communicate easily in an appropriate manner. Automatic vehicle endeavours to minimize the hazard scenarios in the shade of safety-centric IoT solution [10], by restricting the driver from deviating from the trajectory path or having a collection with an object or notifying about the nearest fuel station, hospital or a particular station in case of emergency [11].

We also use the IoT devices as a warning system for monitoring different safety parameters in the vein of environmental changes, liquid pipeline pressure assessment, chemical leaks, presence of toxic gas, etc. This warning system might protect the people and their belongings.

The undeniable benefits of IoT devices lead it to a profit-driven business. These result in a scenario where manufacturers looked more into the quantity than the quality, overlooking any kind of security aspects, which attracted plenty of security breaches with a minimum effort. Figure 1 shows us the global IoT attack scenarios .

Let us concentrate on some latest IoT security breaches that happened in recent years.

(a) In January 2019, ZDNet reported [12] about an incident occurring in Arizona; a 14-year-old boy has included a friend in their group chat. The boy could listen to all of his friend's conversations surrounding his mobile device without his friend picking up the call. He exploited the 'Facepalm' bug.
(b) As reported by Larry Cashdollar [12], Silex malware crashed an IoT device's storage, inferred firewall rules and changed the network configuration. The malware was quickly spread among all 1650 connected IoT devices. To recover, the owners manually reinstalled all the devices' firmware. A 14-year-old boy had spread this malware using the pseudonym of Light Leaf on.
(c) In October 2018, ZDNet reported [12] that hackers have again tapped in Alexa and Google Home smart assistants, to sneak on users without users' knowledge. No matter how both Amazon and Google have deployed updates every time,

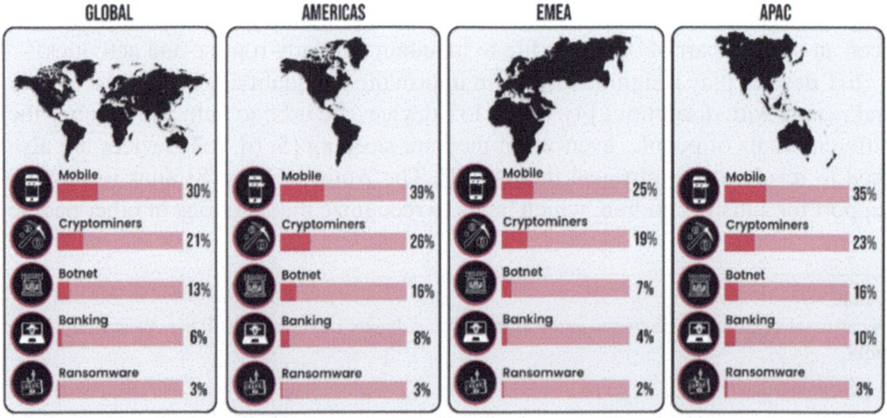

Fig. 1 Global IoT attack scenario (image courtesy [13])

it seems like newer ways to exploit devices have continued to surface by the continuous effort of the advisories.

Such and other security contraventions damage the confidence over IoT paradigm in the consumer market. These various types of attacks not only violate privacy and cause the business disruption but also keep the doorway opened for malicious attacks to cause life-threatening attacks. The US Food and Drug Administration (FDA) has confirmed that security risks prevailed by unauthorized access of IoT medical devices might bring a threat to the patient's life.

All these attacks in the first might look like a failure of the electronic devices, and the problem caused might seem to be just an act of accident or system failure. This can only be revealed if and only if the devices are inspected thoroughly for any kind of suspicious activities.

These risks remain unaddressed by the manufacturers. They only provide some firmware update to the consumers, without changing the design of the product. This updating of the firmware remains the duty of the consumers. Ironically, most of the consumers are not much technically sounded and aware of the fact that they need to update the firmware regularly [15]. Figure 2 describes the increasing rate of the IoT attacks.

As the years pass, the attackers are finding different ways to harm the IoT devices. These extreme situations require highly efficient and well-trained cybersecurity specialists to protect the devices from being misused.

This chapter has been organized as the following: Section 2 discusses the IoT security needs. Section 3 deals with existing IoT security approaches. Section 4 identifies IoT vulnerabilities. Sections 5 and 6 give an insight of innovation in cybersecurity education and future scope of the IoT security in cybersecurity education, respectively. Finally Sect. 7 draws the conclusion.

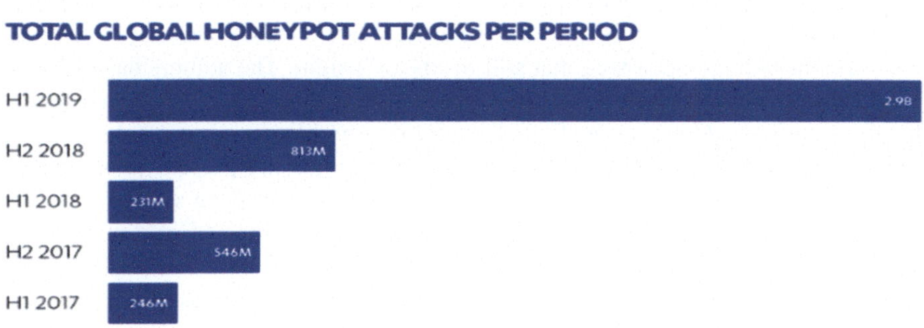

Fig. 2 Year-wise IoT attack statistics (image courtesy [14])

2 IoT Security Needs

To understand the security need of the IoT devices, we need to first understand IoT architecture and the supporting technologies. We then can focus on IoT security as described by different researchers in different ways.

Atzori et al. [16] have subdivided the IoT into two subclasses: (a) Internet-oriented and (b) Things oriented. Gubbi et al. [17] brought light to the IoT-centric application domains and their corresponding challenges. Aneka [18] outlined different security needs for different sensors in IoT applications.

In the analysis conducted by Xu et al. [19] on IoT-enabling technologies and multilayer architectures, shows the different levels of complexity for launching a useful, dynamic IoT device. This launching mechanism takes the IoT security implementation into a different level.

Atzori et al. [20] later have again identified three stages of IoT evaluation: (a) tagged things, (b) web things and (c) social IoT. Further, this chapter will show how different stages have different security needs.

The survey conducted by Perera et al. [21] over 50 different IoT projects shows that privacy and security are taken care in the application level only, but there should be some security protocol in the middle levels for developing a secure IoT device.

3 Existing IoT Security Approaches

Let us dive deep into some research which discussed the security issues of IoT. The IoT uses different approaches and protocols for execution and communication. Therefore, the security solutions adopted by the IoT devices have to differ from conventional IT security solutions. Sicari et al. [22] had surveyed many academic studies and conclude that despite having different and many security solutions, there remains a huge open area that still needs an insight. The authors have further analysed international projects to identify a common solution. But they had found that the solutions address a specific IoT device or a problem.

Mosenia et al. [23] had tried to give a solution that can prevent or minimize the effect of different levels of IoT attacks. They simulated an environment of IoT attacks using the Cisco seven-level reference model [24] and also provided a solution.

There are certain solutions based on IPv6 and wireless personal area network concept [25], in the shade of transportation, routing and application layers as discussed by Granjal et al. [26]. Access control is one of the key features of providing a security solution, especially for distinguishing IoT-tailored. Ouaddah et al. [25] have dived deep into the access control solutions unique from the traditions of access control solutions.

Some security solutions also adopted a distributed architecture for mitigating the attacks on the IoT. Roman et al. [27] in their survey have highlighted the situations where the distributed architecture might also attract various types of attack. Weber et al. [28] also identified the security solutions which cover the legal framework based on IoT devices. Zhang et al. [29] have identified a few areas of IoT using data mining technologies over the Internet where security is still a challenge. They identified areas such as LAN, development flows and applications use many privileges, improper authentication and suspicious environments.

Since the botnet attack has always been vexatious for IoT devices, many security analysts and researchers have performed plenty of investigation on the IoT architecture for the detection of botnet attacks [30–33]. Anagnostopoulos et al. [34] identified the limitations of computations and energy inefficiency after botnet attacks. The authors had also proposed two new commands and central architecture to minimize the cost of the attacks. Burhan et al. [35] proposed a six-layered architecture where each layer handles different attacks.

4 Identifying IoT Vulnerabilities

Going through the existing solutions, we can easily understand that each type of security threat requires different preventions. This takes us to the taxonomy of IoT vulnerabilities as proposed by Nataliia Neshenko et al. [10]. But before jumping into the taxonomy, let us discuss some IoT vulnerabilities so we can understand the categories of the taxonomy.

IoT Vulnerabilities

Before studying details of the IoT vulnerabilities, we must understand the security threats faced by the IoT devices are always multidisciplinary, as shown in Fig. 3.

Deficient Physical Security

Most of the IoT devices are kept in an unattended environment [36] with no or minimum human intervention. As a result, with a very minimal effort, an attacker might access the firmware used by the organization or even physically access the firmware and eventually corrupt it with malicious software or get access to the security protocols or even to the encryption algorithm used to secure the device.

Inadequate Power Storage

Most of the IoT devices rely on limited stored energy available with the device and depend on some manual interventions for restoring the power. The attackers use this

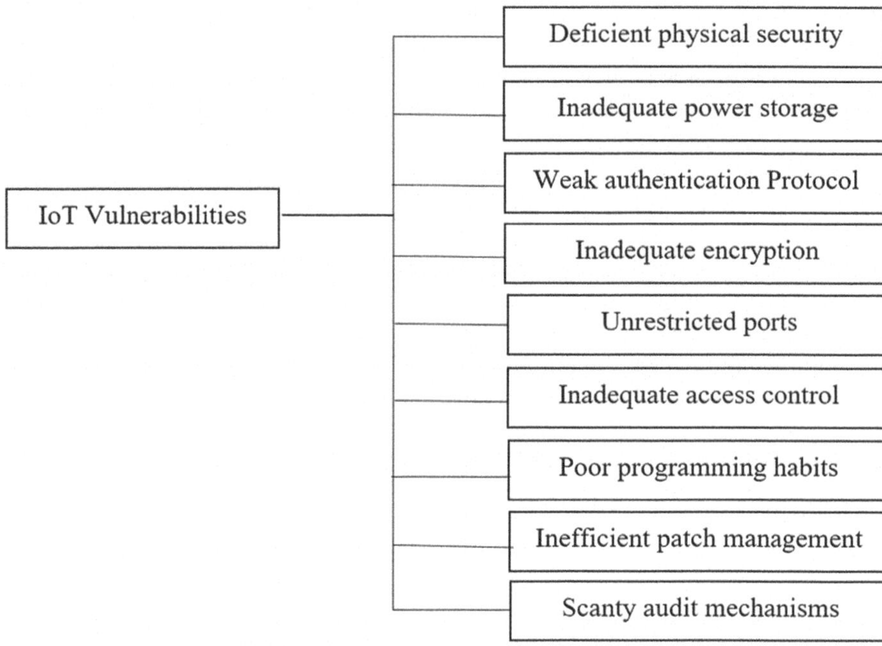

Fig. 3 Various seeds camouflaging IoT vulnerabilities

technological restraint. They use some false instructions or messages to drain up the energy and subsequently make the device unavailable for legitimate users. This type of attack is called DoS attack [37–39].

Weak Authentication Protocol

Efficient authentication protocols come with a complex mechanism and a requirement of higher computational power. It will demand higher power storage for IoT devices. To address the power deficiency, most of the devices sacrifice on the efficiency of the authentication protocol. It results in exposing the authentication keys over the communication medium [40–45].

Inadequate Encryption

Encryption is a prevalent and efficient way of protecting data. The efficiency of the encryption mechanism depends on the algorithm used for the encryption process. But every strong algorithm requires a sufficient amount of resources, which is not always possible because of resource limitations in IoT devices. This results in loss of robustness and efficiency of the algorithm. The attackers exploit this deficiency with a minimum effort [46–51].

Unrestricted Ports

Similar to other computing devices, IoT devices also use different ports to communicate over the network. If some ports become unrestricted, then the communication channel becomes vulnerable [47, 52].

Inadequate Access Control

IoT devices act in collaboration with a cloud platform for various reasons. The complexity of the access control depends on the toughness of the password. But most of the devices do not require any password for connecting with the cloud. The devices also do not request for changing the password in a regular interval [51–57].

Poor Programming Habits

Although most of the IoT devices come up with efficient security protocols, like SSL. Few of the devices do not come up with such efficient security protocol implementation, resulting in unauthorized modification and access of data and even buffer overflow [45, 56, 58–61].

Inefficient Patch Management

As the time progresses, the malicious users find new ways of attacking the devices. So for restricting them, the manufacturers must frequently come up with new security protocols. The end users have reported most of the cases against manufacturers for not coming up with a proper security patch regularly or at all having any mechanism for updating the security protocols for the device. Some security patches are not at all adequate to address the security needs [56, 58, 62–64].

Scanty Audit Mechanisms

To maintain safety and security, IoT devices must face some strong security audit mechanism and help the system to improve the security. But most manufacturers do not equip the devices with a proper security audit mechanism. This results in ignorance to the IoT vulnerabilities before the actual attack [65–67].

Table 1 Different layers with their potential vulnerabilities

Type of layer	Vulnerabilities
Device layer	Hardware elements
Network layer	Weakness in communication
Software layer	Vulnerabilities related to firmware

Taxonomy Overview

The taxonomy of IoT vulnerabilities can be further classified based on (a) layers, (b) security impression, (c) attacks, (d) remediation method and (e) situation awareness capabilities.

Layers

An IoT device has three basic layers, such as device layer, network layer and software layer, which a malicious user can exploit and gain access on the device.

Table 1 shows the vulnerabilities based on some present research works conducted by various researchers. Furthermore, we shall be discussing the vulnerabilities for each layer.

Device-Based Vulnerabilities

We have already seen that non-supervision of the devices is the reason behind some vulnerabilities. It allows easy access to the device by an unauthorized user [68]. Wurm et al. [69] have successfully demonstrated a series of attacks performed on the IoT devices, such as extraction of passwords, learning sensitive information, performing energy theft on a smart meter and retrieval of many update files by taking advantage of a lack of encryption and vigilance at the device level.

Trappe et al. [37] also discussed security threats caused because of the limited availability of energy resources. The researchers also suggested that devices should maintain energy harvest from both artificial resources and natural resources in such a manner that they can adopt efficient security solutions. But this process also encounters some difficulties because of safety regulation and radio propagation.

Network-Based Vulnerabilities

Network-based vulnerabilities have been a prime research area for many years; such researches gave birth to ZigBee protocol [70], which defines network, security and application layer, on top of IEEE 802.15.4, setting up a low power communication network over the sensor network and control network [71]. ZigBee devices use asymmetric key cryptography approach [72], where the key is shared between different nodes. Further to this approach, Vidgren et al. [40] had also illustrated the use of different pre-defined keys in both communicating devices. Sadly, most of the time, the devices transmit these pre-installed keys without proper encryption. This unsecured transmission increases the chance of exposing the keys. Considering the previous area, Morgner et al. [41] examined the ZigBee Light Link (ZLL)-based illuminating systems for different security vulnerabilities. The device is a touch-based connected lighting system. The protocol was used to establish communication between the lighting device and the remote used to control it. The authors performed different attacks and adopted a tailored testing framework [10] to assess the severing of the attack. They also brought the aspect of key management and physical protection of IoT devices to light. The study identified the main drawback was the sharing of pre-defined keys between manufacturers of devices, bringing key management systems (KMS) in the IoT's context.

Roman et al. [73] identified some frameworks based on KMS, such as key pool framework, mathematical framework, negotiable framework and public key framework. Table 2 describes the KMS implementation in brief in contrast to IoT.

As per Petroulakis et al.'s [74], the correlation between energy consumption, security protocols and power control are expressed in Table 3. This shows the problem of adopting a critical security solution in the IoT device. The excessive critical solution requires a higher amount of power consumption leading to a situation where the device itself might be unavailable.

Software-Based Vulnerabilities

Along with the device and network-based vulnerabilities, software-based vulnerabilities can also be a doorway for the attackers to gain unauthorized access to the device. Researchers such as Angrishi [52] highlighted the area of botnet attacks based on IoT devices, often resulting in DDoS attacks. The researchers also identified 90% of [10] the attack happens because the usage of weak credentials and 10% happens by exploiting software weakness.

A study by researchers Patton et al. [75] analysed CPS. The researchers used a search engine, Shodan [76] to identify the devices deployed in the critical service areas. They also tried to identify the number of devices using default or primitive

Table 2 KMS implementation hurdles

Protocol framework	Implementation hurdles
Key pool framework	Weak connectivity
Mathematical framework	Physical placement of client and server nodes
Negotiation framework	Restricted energy of nodes Different network occupancy of client and server nodes
Public key framework	Poor security protocols

Table 3 Security mechanisms vs. energy consumption

Security mechanism	Energy consumption (%)
Encryption	Increased by 15–30
Channel assignment	Increased by 10
Power control	Increased by 4
All three above	Increased by 230

Table 4 IoT vulnerabilities at different architectural layers

Layers	Vulnerabilities
Device-based	Deficient physical security Inadequate power storage
Network-based	Weak authentication protocol Inadequate encryption Unrestricted ports
Software-based	Inadequate access control Inefficient patch management Poor programming habits (e.g., root user, lack of SSL, plain text password, backdoor, etc.) Scanty audit mechanisms

credentials. Ranging from 0.44% to 40% of the IoT devices deployed in different fields suffer from security threats generated by using primitive credentials.

A study by Cui et al. [77] again testified the above problem uncovering almost 540,000 embedded devices, which use default credentials, deployed in many governments and nongovernment sectors.

The Table 4 shows abstraction of different IoT vulnerabilities at different architecture levels.

Security Impression

Whenever the question of securing any IoT devices comes, the main concern lies with their confidentiality, integrity and availability. Confidentiality refers to the unauthorized access to IoT devices and data communicated to or from the device. As discussed in the previous sections, uncontrolled, unauthorized access to IoT devices might lead to the unavailability of devices. To cover the impact of security aspects, we shall consider the classification of security types as shown in Table 5.

Confidentiality

The IoT devices suffer from leakage of confidential data. The main object of this security mechanism is to restrict the leakage of data by implementing rigorous authentication and access control protocols. Let us consider the research work conducted by Copos et al. [78], who examined a home network and got access to some confidential data by doing a network traffic analysis to the smoke detector

Table 5 Security impression and IoT vulnerabilities at different architectural layers ("Y" marked stands for a significant impact on the vulnerability)

		Security impact		
Layers	Vulnerabilities	Confidentiality	Integrity	Availability
Device-based	Deficient physical security		Y	Y
	Inadequate power storage			Y
Network-based	Weak authentication protocol	Y	Y	
	Inadequate encryption	Y	Y	
	Unrestricted ports	Y		Y
Software-based	Inadequate access control	Y	Y	Y
	Inefficient patch management			Y
	Poor programming habits (e.g., root user, lack of SSL, Plain text password, backdoor, etc.)	Y	Y	
	Scanty audit mechanisms		Y	

and thermostat devices installed in the home. The researchers could easily gain access to the smart home completely. They had also identified, though the device encrypts the communication packets, the packets are of different sizes. So if any attacker examines these network packets, he/she can easily identify the destination IP address. The researchers thus proposed a solution, having the same packet size for all communications.

Ronen et al. [46] have identified the leakage of the Wi-Fi password through smart illuminating devices. Installing such illuminating devices can lead to information leakage. These devices transmit the Wi-Fi password without encryption. The authors H. Wang et al. [79] and C. Wang et al. [80] have shown us that the wearable IoT devices can track and study our body movements and hand gestures, read text messages and sometimes even reveal our credentials.

Integrity

Various unique features of IoT devices lead to loss of integrity of data and software, though there are some mechanisms, such as strict inspecting of access control protocols, hashing, encryption, restricting interfaces, input validation and instruction mechanism, used to maintain the integrity of the devices. Further to this, Ho et al. [66] have studied smart lock systems. In their study, they have found network architecture, trust model, reply activity could easily open the lock, and it does not even log the unauthorized access.

Ghena et al. [81] examined a wireless traffic signal and found that because of lack of encryption, anyone could easily gain access to the signalling system resulting in a disruption of traffic movements. To this problem researchers also suggested for installation of malfunction management unit (MMU), which maintains encryption of the network, updates device firmware, blocks network traffic, changes default credentials, etc.

Tekeoglu et al. [82] evaluated the security protocols of an IP camera and revels that user root access to the file system can cause to modify the file, even deleting the file.

Availability

Since all the IoT devices are not having enough chances of implementing proper protocols, they sometimes lead to the unavailability of their services. Most often, protocols used for maintaining the availability of services are monitoring the access of resources, redundancy mechanism and backup systems and providing security updates.

Researchers such as Costa et al. [38] have discussed two layers of service unavailability with wireless visual sensor networks such as hardware and coverage failure. The hardware failure can occur because of the failure of hardware devices associated with the IoT device, or sometimes energy failure. It gives the coverage from failure and inadequate quality of information transmission.

Schuett et al. [83] researched on an IoT device deployed for critical purposes. They replaced the firmware with very minor modifications, resulting in the device's shutdown or even restricting the owner's access.

In the article [84], attackers targeted a small jewellery shop. The website of the shop was unavailable for a significant amount of time. The unavailability was caused, as the website was getting a 35,000 HTTP requests per second from over 25,000 interconnected CCTV cameras. The investigating company later identified that the request sometimes crossed 50,000 requests per second. These requests had been generated from those IP cameras located over 105 countries.

Attacks

The attackers target the confidentiality, authentication, data integrity and availability for performing attacks on the IoT devices. So far, we have discussed the security gaps in IoT devices. Now to understand the ways of attack, we shall focus on the activities conducted only for harming an IoT system.

Attacks Against Confidentiality

Attacks against confidentiality can also lead to an attack against the authentication. This attack impacts the resources available with the IoT devices. These attacks are performed to gain access to the devices for malicious activities, brute force attack, snooping on IoT devices or camouflaging identity of devices. These are the widely used approaches for causing the attack into the systems.

Dictionary attacks are one of the famous attacks, where the attacker tries to generate a high volume of words for making a potential password and, subsequently, gains unauthorized access to the system. Researchers Kolias et al. [85] explained that the 24/7 online IoT devices attract different malicious users to the system. As discussed in Fig. 4, the malicious user can turn the interconnected devices into an army of malicious bots. The attack is executed in various phases, such as rapid scanning [86] for identifying the target.

Antonakakis et al. [87] evaluated over 1000 malware variants, to learn the detection avoidance techniques of Mirai malware. The researchers identified 1.2 million Mirai-infected IP addresses associated with various deployed IoT devices by inspecting some routable but unused IP addresses to filter Mirai malware. Using honeypot and network telescope, Metongnon et al. [88] found that the infection of Mirai is also present in crypto-currency systems.

The attack described in [90] is conducted in two phases, such as information acquisition and correlation analysis. The attack is sometimes called a side-channel attack. The side-channel attack is used to perform power consumption analysis, revealing the encryption key of the devices. The information acquisition phase allows the malicious user to observe the association between several physical components of the IoT devices based on the power consumption, electromagnetic emission, etc. on different parameters. The information acquired through the previous phase is further studied by establishing a correlation between the different input parameters. In Table 6 are some real-world examples indicating the incidents.

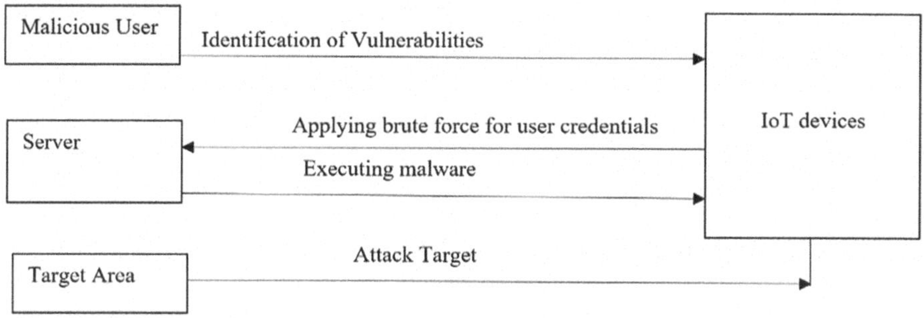

Fig. 4 Attack against confidentiality

Table 6 Some security breaches

Year	References
2016	[87]
	[57]
	[89]

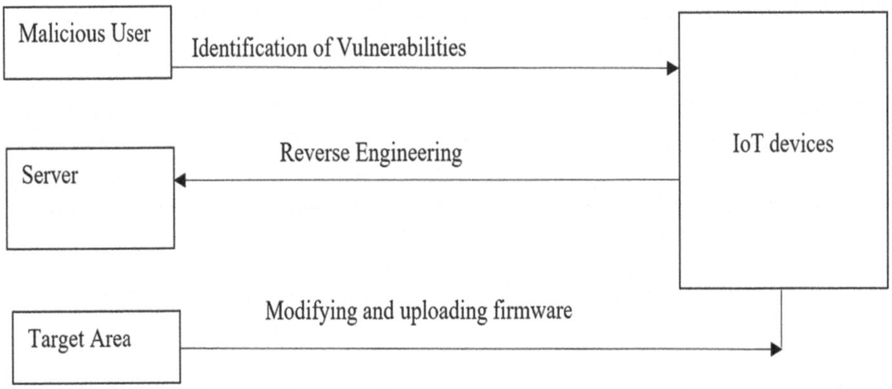

Fig. 5 Attack against data integrity

Table 7 Some security breaches

Year	References
2015	Baby monitor "converses" to children [92]

Attacks Against Data Integrity

Many crucial decisions and applications depend on the data acquired by the IoT devices. Now any attempt of modification might lead the system to an unstable state. Liu et al. [91] tested false data injection attack, where they performed a data injecting attack on power utilities. The research revealed a scenario in which an attacker can inject random measurements to the IoT sensors. In this attack, the malicious users need to tamper with only 1% of the sensors used, which eventually will hamper the performance of an entire power grid.

Another category of attack can be firmware modification, as explained in Fig. 5. In this attack, a firmware is being replaced with malicious firmware, exposing the data to integrity threats.

A research conducted by Basnight et al. [62] put light on data integrity-related attacks. The researchers had chosen on Allen-Bradley Control Logic. The device firmware updates using PLC (Programmable Logic Controller). The unauthorized users can do device alteration at the time when firmware update is taking place. Cui et al. [58] analysed a large number of printers connected with Internet and found that the malicious users can easily exploit over 90,000 unique vulnerabilities using firmware update attacks. This is only possible because update mechanism mostly does not require any authentication protocol. Table 7 shows an example of security breach.

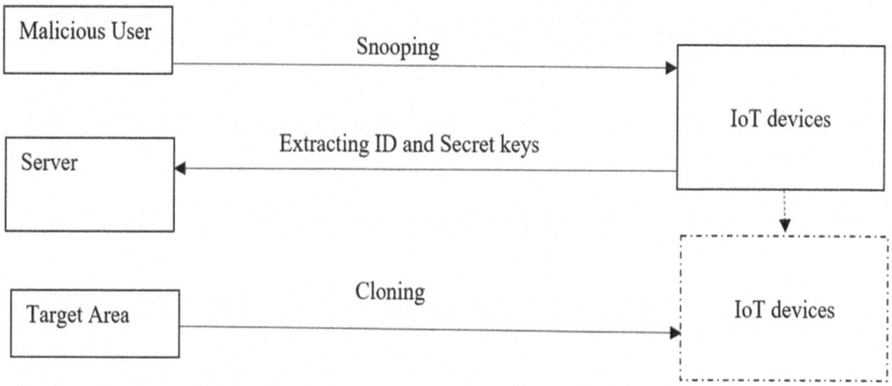

Fig. 6 Attack against data availability

Table 8 Some security breaches

Year	References
2016	Cold Finland [94]

Attacks Against Availability

The primary goal of this attack is DoS (Denial of Services), resulting in restricted access for authorized users to the services provided by IoT devices. Smache et al. [93] formalized this attack as a device captured attack. They implemented a model that will capture a node with a combination of passive, active and physical attacks, as described in Fig. 6.

The attack has three steps: (a) snooping and selecting victim, (b) extracting sensitive information and (c) cloning node.

Zhao [95] has identified that the node capturing can be done by random key generation, as Eschenauer et al. [96] proposed the q-composite scheme. Table 8 shows an example of security breach.

Bonaci et al. [97] applied a framework related to network security issues faced by the IoT devices by analysing network performance and stability of the devices.

Vasserman et al. [39] described the battery draining attack as the vampire attack. The researchers conducted the attack on the energy requirement of message transmission. A message transmission requires significantly more energy from the network. The researchers have found that there are two types of attack: carousel attack and stretch attack. In a carousel attack, the malicious user sends the same message in a loop, such that the node appears in the routing table several times, whereas a stretch attack deals with an unnecessary Long route for message transmission. Both types of attacks result in enormous energy consumption which might lead to drainage of power.

Table 9 shows us the snaps of the vulnerabilities with the corresponding attacks.

Table 9 Snap of the detailed attack survey ("Y" marks the particular vulnerabilities)

Vulnerabilities	Dictionary attack	Side-channel attack	False data injection	Firmware modification attack	Device capture	Battery draining attack
Deficient physical security		Y		Y	Y	
Inadequate power storage						Y
Weak authentication protocol	Y		Y		Y	
Inadequate encryption		Y				
Unrestricted ports	Y					
Inadequate access control	Y					
Inefficient patch management				Y		
Poor programming habits	Y		Y	Y		
Scanty audit mechanisms	Y		Y	Y	Y	

Table 10 The implementation regulation and the areas where it can be implemented

Areas	Implementation regulation
Access control	Firewall
Authentication control	Algorithm authentication
Software assurance	Lightweight
Security protocol	Security schemes

Counter Measures

Based on the remedies towards the attacks of IoT devices, we can categorize the strategies of protection into three categories: (a) access and authentication controls, (b) software assurance and (c) security protocols. Table 10 gives a snapshot of the implementation regulation and the areas where countermeasures can be implemented.

Access and Authentication Control

Studying the previous sections, we can conclude that authentication protocols play a vital role in restricting IoT attacks, but limitation of computational and energy resources makes protocols weak.

Hafeez et al. [43] proposed a secure box approach. The secure box is a platform providing isolation for the devices connected. The platform intercepts any communication from IoT devices to remote locations and verifies via a series of security protocols and raises alarm about any suspicious activity.

Qabulio et al. [98] developed a generic framework for securing mobile wireless IoT devices against physical attacks. In this approach, the authors have deducted the spoofed or cloned nodes by detecting the message transmission direction towards an unwanted destination. The approach depends on the time difference in the inter-equivalent rate to detect spoofed packets. The proposed framework was successfully tested on Contiki OS [99] and COOJA simulation [100].

Software Assurance

Since software associated with IoT devices are also an integral part of an IoT device, in making. Therefore, software security is also important for securing IoT devices. Costin et al. [101] developed an automated framework for dynamic analysis aiming to discover vulnerabilities within embedded IoT farmers. Authors used free penetration tools, Arachni [102], Zed attack proxy (ZAP) [103] and w3af [104], on 2000 firmwear packages, and almost 10% of devices have vulnerabilities, like command injection and cross channel scripting.

Li et al. [105] came up with another concern of traditional code verification, but it lacks domain specificity, which is critical in IoT contexts, especially for embedded medical devices. The authors discussed the problem of code execution delay, which might bring a threat to people's life. So they gave a solution based on a code model checker, such as CBMC [106]. The same technology is used in pacemaker software verification.

Security Protocol

The limited energy resources of IoT devices bring up the challenge of implementing an energy-aware security protocol. In the article published by Balasubramanian et al. [107] designed an Energy-Aware-Edge-Aware (2EA), an architecture, where each node depends on the energy harvesting. In the proposed framework, it creates an energy matrix of every node. If a node lacks energy, it approaches the nearest node with enough source for energy support. This energy resource utilization improved the communication between the parts of IoT, resulting in reduction of packet drop.

Zhang et al. [108] proposed an encloser for each node with a tamper-resistant hardware, but it is a cost-inefficient approach. The researchers proposed a Coverage Interface Protocol (CIP). The authors claimed that the proposed method is tolerant against both physical attacks and attacks originating from a compromised node. The CIP works on two phases, Boundary Node Detection scheme (BOND) and Location-Based symmetric Key management protocol (LBSK). BOND focuses on identifying boundary nodes, whereas LBSK establishes a secure key-based network.

Situational Awareness Capabilities

The availability of countless number of IoT devices opens many kinds of vulnerability issues, under different accessible environments and different parameters. So to provide security solutions is very challenging, and the malicious users will also evolve in their skill set and technologies, making it more challenging. So in providing a framework, we develop four categories, for instance, vulnerability assessment techniques, honeypots, network discovery nodes and intrusion detection mechanisms.

Vulnerability Assessment

The main aim of security is to identify vulnerabilities before they are being exploited and adopting countermeasures. In this context various methods are used, starting from test beds to attack simulation prediction [109] and fuzzy-based assessments. In these aspects, Tekeoglu et al. [56] used different open-source software, such as Kali Linux, OpenVAS, Nessus, Nexpose and bindwalk to produce test beds. In the proposed approach, the network traffic is being captured and analysed. Most of the IoT devices do not lock out the user after the failure of login attempts. Many IoT devices have several unnecessary open ports, and a large number of devices work with outdated software and firmware. In the article designed by Siboni et al. [51]

is described a unique test bed for wearable IoT devices. The architecture of the proposed model contains different functional models, such as a test management module, a security test execution module, a context-aware assessment module and report generation module. They run the test on different devices such as Google Glass and smartwatch.

Honeypot

It is a trap for the malicious user, which does not bring any form to the system. But it reveals the identity of the malicious user and the potential vulnerabilities, exploited to bring harm to the system. Honeypot mimics a specific IoT device, but scalability remains a vital issue.

Pa et al. [110] developed an IoT POT, with the concept of the honeypot, to secure Telnet services. The researchers created an attack simulation, using the existing malware files, and performed a three-step attack, such as intrusion, infection and monetization. The first phase was mostly with the login attempts and found ten commands were mostly used to gain access. In the second stage, the system automatically downloads some malicious software, whereas the third stage was mostly controlled by the attacker conducting DDoS attack and Telnet and TCP open ports scan and spreading malware.

In another study conducted by Guarnizo et al. [111] is designed a scalable high interaction honeypot, a platform for IoT devices. The platform attracts a good amount of malicious traffic by mimicking many IoT devices using worldwide wormholes and few physical devices.

Similarly, Vasilomanolakis et al. [112] proposed a honeypot named HosTaGe used to target malicious activities in the ICS network. It supports the identification of attacks on various protocols, for example, HTTP, SMB, Telnet, FTP, MySql, SIP and SSH, wherein upon their detection, the proposed honeypot generates effective attack signatures and reveals the details of the malicious users.

Network Discovery

The first target of the attackers always has been the network snooping, with IoT devices. Therefore, securing the communication network is an utmost priority. Bou-Harb et al. [113] proposed a solution to make the CPS resistant to the attacks. The CPS attacks are combined from the measurement flow of cyber-physical data to simulate a real attack scenario and discover the flows in the network.

In a study conducted by Fachkha et al. [114], more than 20 heavily used communication protocols were analysed and a pattern of the attackers' intentions to target Internet-facing CPS was drawn. Galluscio et al. [115] have proposed a unique solution to identify unsolicited IoT nodes.

Table 11 Correlation between the vulnerabilities and the situation awareness capabilities ("Y" marks the particular vulnerabilities)

Vulnerabilities	Vulnerability assessment	Honeypots	Network discovery	Intrusion detection
Deficient physical security	Y			Y
Inadequate power storage				Y
Weak authentication protocol	Y		Y	Y
Inadequate encryption	Y			
Unrestricted ports	Y	Y		
Inadequate access control	Y	Y	Y	
Inefficient patch management	Y			
Poor programming habits	Y			Y
Scanty audit mechanisms				Y

Intrusion Detection

An intrusion detection system (IDS) is one of the famous methods to restrict the IoT attack from being executed. Since the IoT devices operate on a very low resource and network environment, the security protocols are not always up to the mark. Under these circumstances, it blocks any compromised node from the entire network [124].

Raza et al. [116] designed an IDS named SVELTE, which maintains the inconsistency of communication between the network and protects against the different scale of attacks. The system is deployed in a 6LoWPAN [124] border router in three phases. The first phase, 6Mapper, recreates a destination-oriented directed acyclic graph, based on gathered information about the network. The second module is responsible for intrusion detection, while the third module acts as a firewall for blocking unwanted traffic.

Yang et al. [117] proposed a scheme that detects FDI attacks in IoT-based devices, mainly surveillance system. The researchers had detected wrong and sequential hypothesis testing to determine malicious nodes. The detection has two-phase frameworks: (a) local false data detection, which uses thresholding to detect false data, and (b) malicious aggregate identifier, which uses the previous phase knowledge to identify malicious nodes.

Table 11 draws a correlation between the vulnerabilities and the situation awareness capabilities.

5 Innovation in Cybersecurity Education

According to an article by the Center for Strategic and International Studies and Intel Security, three out of four security professional survey results indicate that their government has not invested enough in cybersecurity education [118]. This issue of inadequate cybersecurity workforce isn't limited to a few sectors; it goes across from government to education sector as well as industry.

Even though government, industry and education are attempting to address the problem, the entire supply of talent is stressed. Industry is facing a deficit of skilled candidates. Those working as security professionals currently are under steady pressure, as they are required to have continuous training and professional development to keep up with evolving technologies and the threat landscape. Academic institutions want to meet industry needs, but they are struggling to evolve curriculum to keep pace with industry shifts and technological advances.

Realising the Requirement

Cybersecurity-related academic programs have increased significantly around the globe in recent years. The prime reason for this growth can be explained from a study in March 2017 by Frost and Sullivan that said between the years 2013 and 2021, there will be a growth of 350% in the cybersecurity workforce for both the government and private sector units [119]. Currently both sections are facing a significant amount of skill gaps against cyber threats. The National Audit Office (UK) report says that it may take 20 years to fill the current and future cybersecurity skills gaps [120]. To address this epidemic in the digital world, industry, and academia are coming up with many cybersecurity-related programmes. They are also encouraging many professionals to join this field. The National Security Agency of the United States had certified over 160 academic programmes relating to cybersecurity [121]. The University Grants Commission of India had asked the universities to include cybersecurity as a subject in their undergraduate and postgraduate levels [122]. A recent study by IBM shows 56% of students and 44% of educators have rated cybersecurity as a subject of tomorrow, considering the workforce demand will be highest in the coming years [123].

The current trends of IT, such as Cloud Computing, Mobile Computing, and Social Network, observe cybersecurity, as the prime hurdle in their regular operation, as described in Fig. 7.

Apprehending the Trends

The current report of IBM [123] has helped us to identify the following four upcoming trends of cybersecurity.

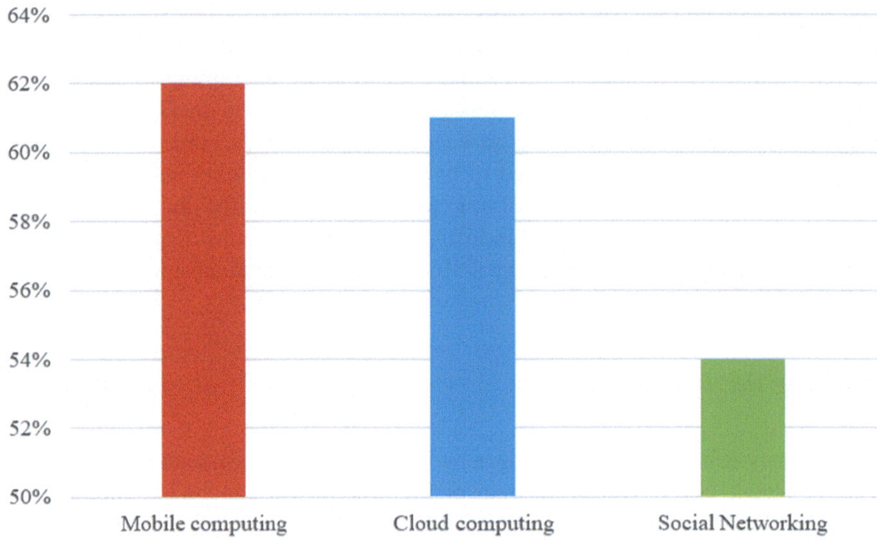

Fig. 7 Cybersecurity as hurdle for current IT trends (image courtesy [123])

Cybersecurity Is Becoming Necessity Day by Day

Cybersecurity is no longer a call of the void. It has become something that impacts our daily life. Constant innovation is a must for success against malicious activities. Any cyber-attack has become more impactful than any terrorist attack, affecting more number of people irrespective of their geographic location.

Increasing Demand for Various Genres of Organizations

Financial sectors, banks, aerospace, defence firms and health care recruit cybersecurity professionals for protecting their data against malicious users. Nowadays, many countries building cybersecurity professionals have become a national priority.

More Information to Secure and More Ways to Attack

In the present days, the attackers have become more silent and efficient and been causing more damage to society. So for countering, the cybersecurity education system has to become more innovative.

More on Practice

The conventional system of academia is focused on the theory aspect of different areas. With cybersecurity, our focus has to differ from the convention and be out of the box. Where the focus is more on the practice, live challenges and learning through experience.

Gap Between Supply and Demand

The demand and supply of the cybersecurity workforce are not converging. Rather the gap is increasing as time passes. The following reasons play a vital role in building the gap.

Cybersecurity Trainers with Inadequate Skills

The organizations require skilled professionals for training of the cybersecurity workforce. But most of the current academia is not having adequately skilled trainers for producing skilled cybersecurity workforce.

Inadequate Resources for the Cybersecurity Area of Study

There are limited numbers of innovative resources available to support the specialized area. As a result, the students aspiring to choose cybersecurity as a profession are looking for proper guidance for flourishing.

Lack of Equipment, Laboratories and Opportunities for the Cybersecurity Aspirants

The students require a specialized laboratory facility with state-of-the-art equipment to become a skilled cybersecurity professional. They also lack enough opportunities for them to get exposed to practical challenges, which can help them to enhance their skills.

6 Future Scope

The current trends and challenges identified through our detailed study have shown us that the cybersecurity programmes are going through a revolutionary phase. To make our future prepared against any cybersecurity attacks, our industry and academia should join hands. The following are the key indicators for developing cybersecurity education [123].

(a) Increase awareness and expertise between people.
(b) Consider cybersecurity education as a global issue.
(c) Develop innovative mind as cybersecurity workforce.

In this development of cybersecurity education, IoT shall play a vital role. As our study shows, IoT has a versatile use from critical task solving devices to making our daily life easy in various forms. Every IoT device is unique and requires separate attention and attracts different levels of security threats. This brings us to a situation where proposing any unique solution is not possible. Previous researchers had identified different levels of vulnerability issues for different devices and proposed different solutions. We conduct a detailed study to understand the different needs of IoT devices in the past and present and to predict the future of the threats faced by IoT devices. We can achieve some future-ready security solutions by combining our past and present experiences or can provide a whole new solution covering multiple vulnerability areas.

7 Conclusion

We summarize the above recommendations and our detailed study of different-scale IoT attacks towards the upliftment of cybersecurity education. We must maintain the intensity of the passion and enthusiasm towards the long-term goal of cybersecurity education. As the attackers are bringing innovative challenges in the field of IoT, the cybersecurity professionals also need to be innovative and enthusiastic towards their skill upgradation and ability to neutralize attacks. Rather, predicting the security gaps and fill them before the attack could take place. If cybersecurity education is not given the required importance, it will directly hamper the trust of technology. As we have already seen from the study, IoT devices intervene and make our life easy in every possible aspect. That is the reason IoT devices require a special place in the cybersecurity education systems. The innovative ideas behind the IoT devices make it more challenging for the cybersecurity workforces. It would take government, academia and industry to play their vital roles towards the upliftment of cybersecurity education for a safer digital world.

References

1. Estimating Cyber Risk for the Financial Sector by Christine Lagarde. https://blogs.imf.org/2018/06/22/estimating-cyber-risk-for-the-financial-sector/access. Accessed 15 Mar 2020
2. J.J. Cebula, M.E. Popeck, L.R. Young, A taxonomy of operational cyber security risks version 2. No. CMU/SEI-2014-TN-006. Carnegie-Mellon University Pittsburgh PA Software Engineering Inst (2014).

3. Top Cybersecurity Threats in 2020 by Michelle Moore. https://onlinedegrees.sandiego.edu/top-cyber-security-threats/. Accessed 15 Mar 2020
4. M.C. Domingo, An overview of the internet of things for people with disabilities. J. Netw. Comput. Appl. **35**(2), 584–596 (2012)
5. M. Chan, D. Est'eve, J.-Y. Fourniols, C. Escriba, E. Campo, Smart wearable systems: current status and future challenges. Artif. Intell. Med. **56**(3), 137–156 (2012)
6. A.G. Ferreira, D. Fernandes, S. Branco, J.L. Monteiro, J. Cabral, A.P. Catarino, A.M. Rocha, A smart wearable system for sudden infant death syndrome monitoring, in *Industrial Technology (ICIT), 2016 IEEE International Conference on*, (IEEE, New York, 2016), pp. 1920–1925
7. I. Bisio, A. Delfino, F. Lavagetto, A. Sciarrone, Enabling iot for in-home rehabilitation: accelerometer signals classification methods for activity and movement recognition. IEEE Internet Things J. **4**(1), 135–146 (2017)
8. Stanford University, The autism glass project at Stanford Medicine (2020). http://autismglass.stanford.edu/. Accessed 29 Jan 2020
9. P. Patel, Autism glass takes top student health tech prize (2020). https://www.openminds.com/market-intelligence/bulletins/autism-glass-takes-top-student-health-tech-prize/. Accessed 29 Jan 2020
10. N. Neshenko, E. Bou-Harb, J. Crichigno, G. Kaddoum, N. Ghani, Demystifying IoT security: an exhaustive survey on IoT vulnerabilities and a first empirical look on internet-scale IoT exploitations. IEEE Commun. Surv. Tutorials **21**(3), 2702–2733 (2019)
11. R. Coppola, M. Morisio, Connected car: technologies, issues, future trends. ACM Comput. Surv. (CSUR) **49**(3), 46 (2016)
12. Top 5 Shocking IoT Security Breaches of 2019. https://www.pentasecurity.com/blog/top-5-shocking-iot-security-breaches-2019/. Accessed 20 Jan 2020
13. Global IoT attack scenario. http://tadviser.com/index.php/Article:Cyber_attacks. Accessed 20 Jan 2020
14. Year wise IoT attack statistics by Forbes. https://www.forbes.com/sites/zakdoffman/2019/09/14/dangerous-cyberattacks-on-iot-devices-up-300-in-2019-now-rampant-report-claims/#51b74e615892. Accessed 20 Jan 2020
15. Canonical Ltd, Who should bear the cost of iot security: consumers or vendors? https://ubuntu.com/blog/who-should-bear-the-cost-of-iot-security-consumers-or-vendors. Accessed 29 Jan 2020
16. L. Atzori, A. Iera, G. Morabito, The internet of things: a survey. Comput. Netw. **54**(15), 2787–2805 (2010)
17. J. Gubbi, R. Buyya, S. Marusic, M. Palaniswami, Internet of things (iot): a vision, architectural elements, and future directions. Futur. Gener. Comput. Syst. **29**(7), 1645–1660 (2013)
18. Y. Wei, K. Sukumar, C. Vecchiola, D. Karunamoorthy, R. Buyya, Aneka cloud application platform and its integration with windows azure, arXiv preprint arXiv:1103.2590 (2011).
19. L. Da Xu, W. He, S. Li, Internet of things in industries: a survey. IEEE Trans. Indust. Informat. **10**(4), 2233–2243 (2014)
20. L. Atzori, A. Iera, G. Morabito, Understanding the internetof things: definition, potentials, and societal role of a fast evolving paradigm, Ad Hoc Netw (2016)
21. C. Perera, A. Zaslavsky, P. Christen, D. Georgakopoulos, Context aware computing for the internet of things: a survey. IEEE Commun. Surv. Tutorials **16**(1), 414–454 (2014)
22. S. Sicari, A. Rizzardi, L.A. Grieco, A. Coen-Porisini, Security, privacy and trust in internet of things: the road ahead. Comput. Netw. **76**, 146–164 (2015)
23. A. Mosenia, N.K. Jha, A comprehensive study of security of internet-of-things. IEEE Trans. Emerg. Top. Comput. **5**(4), 586–602 (2017)
24. CISCO, The internet of things reference model. http://cdn.iotwf.com/resources/71/IoT_Reference_Model_White_Paper_June_4_2014.pdf. Accessed 29 Jan 2020
25. A. Ouaddah, H. Mousannif, A.A. Elkalam, A.A. Ouahman, Access control in the internet of things: big challenges and new opportunities. Comput. Netw. **112**, 237–262 (2017)

26. J. Granjal, E. Monteiro, J.S. Silva, Security for the internet of things: a survey of existing protocols and open research issues. IEEE Commun. Surv. Tutorials **17**(3), 1294–1312 (2015)
27. R. Roman, J. Zhou, J. Lopez, On the features and challenges of security and privacy in distributed internet of things. Comput. Netw. **57**(10), 2266–2279 (2013)
28. R.H. Weber, E. Studer, Cybersecurity in the internet of things: legal aspects. Comput. Law Secur. Rev. **32**(5), 715–728 (2016)
29. N. Zhang, S. Demetriou, X. Mi, W. Diao, K. Yuan, P. Zong, F. Qian, X. Wang, K. Chen, Y. Tian et al., Understanding iot security through the data crystal ball: where we are now and where we are going to be, arXiv preprint arXiv:1703.09809 (2017)
30. E. Bou-Harb, M. Debbabi, C. Assi, A novel cyber security capability: inferring internet-scale infections by correlating malware and probing activities. Comput. Netw. **94**, 327–343 (2016)
31. Behavioral analytics for inferring large-scale orchestrated probing events, in 2014 IEEE Conference on Computer Communications Workshops (INFOCOM WKSHPS) (IEEE, New York, 2014), pp. 506–511
32. Big data behavioral analytics meet graph theory: on effective botnet takedowns, IEEE Network, **31**(1), 18–26 (2017)
33. E. Bou-Harb, C. Fachkha, M. Debbabi, C. Assi, Inferring internet scale infections by correlating malware and probing activities, in *2014 IEEE International Conference on Communications (ICC)*, (IEEE, New York, 2014), pp. 640–646
34. M. Anagnostopoulos, G. Kambourakis, S. Gritzalis, New facets of mobile botnet: architecture and evaluation. Int. J. Inf. Secur. **15**(5), 455–473 (2016)
35. M. Burhan, R. Rehman, B. Khan, B.-S. Kim, Iot elements, layered architectures and security issues: a comprehensive survey. Sensors **18**(9), 2796 (2018)
36. R. Mahmoud, T. Yousuf, F. Aloul, I. Zualkernan, Internet of things (iot) security: current status, challenges and prospective measures, in *2015 10th International Conference for Internet Technology and Secured Transactions (ICITST)*, (IEEE, New York, 2015), pp. 336–341
37. W. Trappe, R. Howard, R.S. Moore, Low-energy security: limits and opportunities in the internet of things. IEEE Secur. Privacy **13**(1), 14–21 (2015)
38. D.G. Costa, I. Silva, L.A. Guedes, F. Vasques, P. Portugal, Availability issues in wireless visual sensor networks. Sensors **14**(2), 2795–2821 (2014)
39. E.Y. Vasserman, N. Hopper, Vampire attacks: draining life from wireless ad hoc sensor networks. IEEE Trans. Mob. Comput. **12**(2), 318–332 (2013)
40. N. Vidgren, K. Haataja, J.L. Patino-Andres, J.J. Ramirez-Sanchis, P. Toivanen, Security threats in zigbee-enabled systems: vulnerability evaluation, practical experiments, countermeasures, and lessons learned, in *System Sciences (HICSS), 2013 46th Hawaii International Conference on*, (IEEE, New York, 2013), pp. 5132–5138
41. P. Morgner, S. Mattejat, Z. Benenson, All your bulbs are belong to us: investigating the current state of security in connected lighting systems, arXiv preprint arXiv:1608.03732 (2016)
42. T. Kothmayr, C. Schmitt, W. Hu, M. Brunig, G. Carle, Dtls based security and two-way authentication for the internet of things. Ad Hoc Netw. **11**(8), 2710–2723 (2013)
43. I. Hafeez, A.Y. Ding, L. Suomalainen, A. Kirichenko, S. Tarkoma, Securebox: toward safer and smarter iot networks, in *Proceedings of the 2016 ACM Workshop on Cloud-Assisted Networking*, (ACM, New York, 2016), pp. 55–60
44. P. Porambage, C. Schmitt, P. Kumar, A. Gurtov, M. Ylianttila, Pauthkey: a pervasive authentication protocol and key establishmentscheme for wireless sensor networks in distributed iot applications. Int. J. Distrib. Sens. Netw. **10**, 7 (2014)
45. A. Furfaro, L. Argento, A. Parise, A. Piccolo, Using virtual environments for the assessment of cybersecurity issues in iot scenarios. Simul. Model. Pract. Theory **73**, 43–54 (2017)
46. E. Ronen, A. Shamir, Extended functionality attacks on iot devices: the case of smart lights, in *Security and Privacy (Euro S&P), 2016 IEEE European Symposium on*, (IEEE, New York, 2016), pp. 3–12
47. V. Sachidananda, S. Siboni, A. Shabtai, J. Toh, S. Bhairav, Y. Elovici, Let the cat out of the bag: a holistic approach towards security analysis of the internet of things, in *Proceedings of the 3rd ACM International Workshop on IoT Privacy, Trust, and Security*, (ACM, New York, 2017), pp. 3–10

48. H. Shafagh, A. Hithnawi, A. Droscher, S. Duquennoy, W. Hu, Talos: encrypted query processing for the internet of things, in *Proceedings of the 13th ACM conference on embedded networked sensor systems*, (ACM, New York, 2015), pp. 197–210
49. B. Wei, G. Liao, W. Li, Z. Gong, A practical one-time file encryption protocol for iot devices, in *Computational Science and Engineering (CSE) and Embedded and Ubiquitous Computing (EUC), 2017 IEEE International Conference on*, vol. 2, (IEEE, New York, 2017), pp. 114–119
50. A. Biryukov, D. Dinu, Y. Le Corre, Side-channel attacks meet secure network protocols, in *International conference on applied cryptography and network security*, (Springer, New York, 2017), pp. 435–454
51. S. Siboni, A. Shabtai, N.O. Tippenhauer, J. Lee, Y. Elovici, Advanced security testbed framework for wearable iot devices. ACM Trans. Inter. Technol. **16**(4), 26 (2016)
52. K. Angrishi, Turning internet of things (iot) into internet of vulnerabilities (iov): iot botnets, arXiv preprint. arXiv:1702.03681 (2017)
53. L. Markowsky, G. Markowsky, Scanning for vulnerable devices in the internet of things, in *Intelligent data acquisition and advanced computing systems: technology and applications (IDAACS), 2015 IEEE 8th International Conference on*, vol. 1, (IEEE, New York, 2015), pp. 463–467
54. P.K. Dhillon, S. Kalra, A lightweight biometrics based remoteuser authentication scheme for iot services. J. Informat. Secur. Appl. **34**, 255–270 (2017)
55. Y.J. Jia, Q.A. Chen, S. Wang, A. Rahmati, E. Fernandes, Z.M. Mao, A. Prakash, S.J. Unviersity, Contexlot: towards providing contextual integrity to appified iot platforms. NDSS (2017)
56. A. Tekeoglu, A.S. Tosun, A testbed for security and privacy analysis of iot devices, in *Mobile Ad Hoc and Sensor Systems (MASS), 2016 IEEE 13th International Conference on*, (IEEE, New York, 2016), pp. 343–348
57. Radware Ltd., "brickerbot" results in pdos attack (2020). https://security.radware.com/ddos-threats-attacks/brickerbot-pdos-permanent-denial-of-service/. Accessed 29 Jan 2020
58. A. Cui, M. Costello, S. J. Stolfo, When firmware modifications attack: a case study of embedded exploitation. NDSS (2013)
59. A. Costin, J. Zaddach, A. Francillon, D. Balzarotti, S. Antipolis, A large-scale analysis of the security of embedded firmwares. *USENIX Security* (2014), pp. 95–110
60. Q. Feng, R. Zhou, C. Xu, Y. Cheng, B. Testa, H. Yin, Scalable graph-based bug search for firmware images, in *Proceedings of the 2016 ACM SIGSAC Conference on Computer and Communications Security*, (ACM, New York, 2016), pp. 480–491
61. H. Elmiligi, F. Gebali, M.W. El-Kharashi, Multi-dimensional analysis of embedded systems security. Microprocess. Microsyst. **41**, 29–36 (2016)
62. Z. Basnight, J. Butts, J. Lopez, T. Dube, Firmware modification attacks on programmable logic controllers. Int. J. Crit. Infrastruct. Prot. **6**(2), 76–84 (2013)
63. C. Konstantinou, M. Maniatakos, Impact of firmware modification attacks on power systems field devices, in 2015 IEEE International Conference on Smart Grid Communications (SmartGridComm) (2015), pp. 283–288
64. B. Bencs'ath, L. Butty'an, T. Paulik, Xcs based hidden firmware modification on embedded devices, in *Software, Telecommunications and Computer Networks (SoftCOM), 2011 19th International Conference on*, (IEEE, New York, 2011), pp. 1–5
65. B. Ur, J. Jung, S. Schechter, The current state of access control for smart devices in homes, in Workshop on Home Usable Privacy and Security (HUPS). HUPS 2014 (2013)
66. G. Ho, D. Leung, P. Mishra, A. Hosseini, D. Song, D. Wagner, Smart locks: lessons for securing commodity internet of things devices, in *Proceedings of the 11th ACM on Asia conference on computer and communications security*, (ACM, New York, 2016), pp. 461–472
67. K. Yang, D. Forte, M.M. Tehranipoor, Protecting endpoint devices in iot supply chain, in *Computer-Aided Design (ICCAD), 2015 IEEE/ACM International Conference on*, (IEEE, New York, 2015), pp. 351–356

68. E. Bou-Harb, C. Fachkha, M. Pourzandi, M. Debbabi, C. Assi, Communication security for smart grid distribution networks. IEEE Commun. Mag. **51**(1), 42–49 (2013)
69. J. Wurm, K. Hoang, O. Arias, A.-R. Sadeghi, Y. Jin, Security analysis on consumer and industrial iot devices, in *Design Automation Conference (ASP-DAC), 2016 21st Asia and South Pacific*, (IEEE, New York, 2016), pp. 519–524
70. S. Farahani, ZigBee wireless networks and transceivers. News (2011)
71. A. Elahi, A. Gschwender, ZigBee wireless sensor and control network (Pearson Education, 2009)
72. P. Radmand, M. Domingo, J. Singh, J. Arnedo, A. Talevski, S. Petersen, S. Carlsen, Zigbee/zigbee pro security assessment based on compromised cryptographic keys, in *P2P, parallel, grid, cloud and internet computing (3PGCIC), 2010 International Conference on*, (IEEE, New York, 2010), pp. 465–470
73. R. Roman, C. Alcaraz, J. Lopez, N. Sklavos, Key management systems for sensor networks in the context of the internet of things. Comput. Electr. Eng. **37**(2), 147–159 (2011)
74. N.E. Petroulakis, E.Z. Tragos, A.G. Fragkiadakis, G. Spanoudakis, A lightweight framework for secure life-logging in smart environments. Inf. Secur. Tech. Rep. **17**(3), 58–70 (2013)
75. M. Patton, E. Gross, R. Chinn, S. Forbis, L. Walker, H. Chen, Uninvited connections: a study of vulnerable devices on the internet of things (iot), in *Intelligence and Security Informatics Conference (JISIC), 2014 IEEE Joint*, (IEEE, New York, 2014), pp. 232–235
76. Shodan R. http://shodan.io. Accessed 29 Jan 2020
77. A. Cui, S.J. Stolfo, A quantitative analysis of the insecurity of embedded network devices: results of a wide-area scan, in *Proceedings of the 26th annual computer security applications conference*, (ACM, New York, 2010), pp. 97–106
78. B. Copos, K. Levitt, M. Bishop, J. Rowe, Is anybody home? Inferring activity from smart home network traffic, in *Security and privacy workshops (SPW), 2016 IEEE*, (IEEE, New York, 2016), pp. 245–251
79. H. Wang, T.T.-T. Lai, R. Roy Choudhury, Mole: motion leaks through smartwatch sensors, in *Proceedings of the 21st annual international conference on mobile computing and networking*, (ACM, New York, 2015), pp. 155–166
80. C. Wang, X. Guo, Y. Wang, Y. Chen, B. Liu, Friend or foe?: Your wearable devices reveal your personal pin, in *Proceedings of the 11th ACM on Asia conference on computer and communications security*, (ACM, New York, 2016), pp. 189–200
81. B. Ghena, W. Beyer, A. Hillaker, J. Pevarnek, J.A. Halderman, Green lights forever: analyzing the security of traffic infrastructure. WOOT **14**, 7–7 (2014)
82. A. Tekeoglu, A.S. Tosun, Investigating security and privacy of a cloud-based wireless ip camera: Netcam, in *Computer Communication and Networks (ICCCN), 2015 24th International Conference on*, (IEEE, New York, 2015), pp. 1–6
83. C. Schuett, J. Butts, S. Dunlap, An evaluation of modification attacks on programmable logic controllers. Int. J. Crit. Infrastruct. Prot. **7**(1), 61–68 (2014)
84. Botnet of 25,000 Cameras Located in 105 Countries Launches Massive DDoS Attacks ByRafia Shaikh. https://wccftech.com/massive-botnet-25000-iot-launch-ddos/. Accessed 22 Jan 2020
85. C. Kolias, G. Kambourakis, A. Stavrou, J. Voas, Ddos in the iot: Mirai and other botnets. Computer **50**(7), 80–84 (2017)
86. E. Bou-Harb, M. Debbabi, C. Assi, Cyber scanning: a comprehensive survey. IEEE Commun. Surv. Tutorials **16**(3), 1496–1519 (2014)
87. M. Antonakakis, T. April, M. Bailey, M. Bernhard, E. Bursztein, J. Cochran, Z. Durumeric, J.A. Halderman, L. Invernizzi, M. Kallitsis et al., Understanding the mirai botnet, in *26th fUSENIXg Security Symposium (fUSENIXg Security 17)* (2017), pp. 1093–1110
88. L. Metongnon, R. Sadre, Beyond telnet: prevalence of iot protocols in telescope and honeypot measurements, in *Proceedings of the 2018 workshop on traffic measurements for cybersecurity*, (ACM, New York, 2018), pp. 21–26
89. Internet of things teddy bear leaked 2 million parent and kids message recordings. https://www.vice.com/en_us/article/pgwean/internet-of-things-teddy-bear-leaked-2-million-parent-and-kids-message-recordings. Accessed 29 Jan 2020

90. Y. Zhou, D. Feng, Side-channel attacks: ten years after its publication and the impacts on cryptographic module security testing. IACR Cryptol. ePrint Arch. **2005**, 388 (2005)
91. Y. Liu, P. Ning, M.K. Reiter, False data injection attacks against state estimation in electric power grids. ACM Trans. Informat. Syst. Secur. **14**(1), 13 (2011)
92. CRIMESIDER STAFF, CBS news, Baby monitor hacker delivers creepy message to child. https://www.cbsnews.com/news/baby-monitor-hacker-delivers-creepy-message-to-child/. Accessed 29 Jan 2020
93. M. Smache, N. El Mrabet, J.-J. Gilquijano, A. Tria, E. Riou, C. Gregory, Modeling a node capture attack in a secure wireless sensor networks, in *Internet of Things (WF-IoT), 2016 IEEE 3rd World Forum on*, (IEEE, New York, 2016), pp. 188–193
94. Metropolitan.fi, Ddos attack halts heating in Finland amidst winter. https://metropolitan.fi/entry/ddos-attack-halts-heating-in-finland-amidst-winter. Accessed 29 Jan 2020
95. J. Zhao, On resilience and connectivity of secure wireless sensor networks under node capture attacks. IEEE Trans. Informat. Forens. Secur. **12**(3), 557–571 (2017)
96. L. Eschenauer, V.D. Gligor, A key-management scheme for distributed sensor networks, in *Proceedings of the 9th ACM conference on computer and communications security*, (ACM, New York, 2002), pp. 41–47
97. T. Bonaci, L. Bushnell, R. Poovendran, Node capture attacks in wireless sensor networks: a system theoretic approach, in *Decision and Control (CDC), 2010 49th IEEE Conference on*, (IEEE, New York, 2010), pp. 6765–6772
98. M. Qabulio, Y.A. Malkani, A. Keerio, A framework for securing mobile wireless sensor networks against physical attacks, in *Emerging Technologies (ICET), 2016 International Conference on*, (IEEE, New York, 2016), pp. 1–6
99. A. Dunkels, O. Schmidt, N. Finne, J. Eriksson, F. Osterlind, N. Tsiftes, M. Durvy, The contikios: the operating system for the internet of things (2011). http://www.contikios.org
100. F. Osterlind, A sensor network simulator for the contikios, *Swedish Institute of Computer Science (SICS), Tech. Rep. T2006-05* (2006)
101. A. Costin, A. Zarras, A. Francillon, Automated dynamic firmware analysis at scale: a case study on embedded web interfaces, in *Proceedings of the 11th ACM on Asia conference on computer and communications security*, (ACM, New York, 2016), pp. 437–448
102. Sarosys LLC, Arachni. Web application security scanner framework. http://www.arachni-scanner.com/. Accessed 29 Jan 2020
103. OWASP, Owasp zed attack proxy project. https://owasp.org/www-project-zap/. Accessed 29 Jan 2020
104. Andres Riancho, w3af—open source web application security scanner. www.w3af.org. Accessed 29 Jan 2020
105. C. Li, A. Raghunathan, N.K. Jha, Improving the trustworthiness of medical device software with formal verification methods. IEEE Embed. Syst. Lett. **5**(3), 50–53 (2013)
106. C. Mellon, Cbmc. bounded model checking for software. http://www.cprover.org/cbmc/. Accessed 29 Jan 2020
107. V. Balasubramanian, N. Kouvelas, K. Chandra, R. Prasad, A.G. Voyiatzis, W. Liu, A unified architecture for integrating energy harvesting iot devices with the mobile edge cloud, in *2018 IEEE 4th World Forum on Internet of Things (WF-IoT)*, (IEEE, New York, 2018), pp. 13–18
108. C. Zhang, Y. Zhang, Y. Fang, Defending against physical destruction attacks on wireless sensor networks, in *Military communications conference, 2006. MILCOM 2006*, (IEEE, New York, 2006), pp. 1–7
109. M. Husa'k, J. Koma'rkova, E. Bou-Harb, P. Celeda, Survey of attack projection, prediction, and forecasting in cyber security. IEEE Commun. Surv. Tutorials **21**(1), 640–660 (2019)
110. Y.M.P. Pa, S. Suzuki, K. Yoshioka, T. Matsumoto, T. Kasama, C. Rossow, Iotpot: a novel honeypot for revealing current iot threats. J. Inform. Process. **24**(3), 522–533 (2016)
111. J. Guarnizo, A. Tambe, S. S. Bunia, M. Ochoa, N. Tippenhauer, A. Shabtai, Y. Elovici, Siphon: towards scalable high-interaction physical honeypots, arXiv preprint. arXiv:1701.02446 (2017)

112. E. Vasilomanolakis, S. Srinivasa, C.G. Cordero, M. Muhlhauser, Multi-stage attack detection and signature generation with ics honeypots, in *IEEE/IFIP Workshop on Security for Emerging Distributed Network Technologies (DISSECT)*, (IEEE, New York, 2016)
113. E. Bou-Harb, W. Lucia, N. Forti, S. Weerakkody, N. Ghani, B. Sinopoli, Cyber meets control: a novel federated approach for resilient cps leveraging real cyber threat intelligence. IEEE Commun. Mag. **55**(5), 198–204 (2017)
114. C. Fachkha, E. Bou-Harb, A. Keliris, N. Memon, M. Ahamad, Internet-scale probing of cps: inference, characterization and orchestration analysis, in *Proceedings of NDSS* (2017), vol. 17
115. M. Galluscio, N. Neshenko, E. Bou-Harb, Y. Huang, N. Ghani, J. Crichigno, G. Kaddoum, A first empirical look on internet-scale exploitations of iot devices, in *2017 IEEE 28th annual international symposium on personal, indoor, and mobile radio communications (PIMRC)*, (IEEE, New York, 2017), pp. 1–7
116. S. Raza, L. Wallgren, T. Voigt, Svelte: real-time intrusion detection in the internet of things. Ad Hoc Netw. **11**(8), 2661–2674 (2013)
117. L. Yang, C. Ding, M. Wu, K. Wang, Robust detection of false data injection attacks for the data aggregation in internet of things based environmental surveillance, Comput. Netw. (2017)
118. Hacking the skills shortage: a study of the international shortage in cybersecurity skills, Center for Strategic and International Studies 2016, McAfee, Part of Intel Security, Santa Clara. https://www.mcafee.com/us/resources/reports/rp-hacking-skills-shortage.pdf. Accessed 17 Mar 2020
119. Cybersecurity Talent Crunch To Create 3.5 Million Unfilled Jobs Globally By 2021 by Frost & Sullivan. https://cybersecurityventures.com/jobs/access. Accessed 15 Mar 2020
120. The UK Cyber Security Strategy: Landscape Review. National Audit Office. February 2013. http://www.nao.org.uk/wp-content/uploads/2013/03/Cyber-security-Full-report.pdf. Accessed 17 Mar 2020
121. National Centers of Academic Excellence. National Security Agency (NSA) Central Security Service (CSS). https://www.nsa.gov/resources/students-educators/centers-academic-excellence/. Accessed 17 Mar 2020
122. Cybersecurity to be Part of India's College, University Curriculum. *The Times of India*. January 17, 2013. http://articles.timesofindia.indiatimes.com/2013-01-17/education/36393726_1_cybersecurity-security-scenario-information-security. Accessed 17 Mar 2020
123. M. Viveros, D. Jarvis. *Cybersecurity Education for the Next Generation: Advancing a Collaborative Approach*. Center for Applied Insights. IBM Corporation (2013)
124. Z. Shelby, C. Bormann, *6LoWPAN: the wireless embedded internet*, vol 43 (Wiley, Hoboken, 2011)

Using a Business Compromise Scenario to Teach Cybersecurity

Andrew D. Wolfe, Jr.

1 Introduction

Computer Science 674, "Database Security," is part of Boston University Metropolitan College graduate curricula for multiple Computer Information Systems and Computer Science degree concentrations. This course has had no prerequisites on information security but covered advanced database techniques. Originally developed in the mid-2000s, by 2017 its obsolete textbook was out of print, and no suitable replacements were available. Assignments and tests had been compromised via online "student coaching" sites. While these factors strongly indicated a need for rework, one central issue had to be wrestled: specifically, how to create effective pedagogy for cybersecurity fundamentals and apply those in an advanced database context. In addition, the course is delivered by both online and in-person modes, posing additional challenges for instructional design.

The overall learning objectives for the course have been that students learn:

- Key concepts of information security
- Primary threats to information security and basic cybersecurity practices
- Concepts of cryptography
- Measures for database hardening
- Administering database users and authentication
- Authorization of database tables and procedures under the "grant" framework
- Row-level authorization of database data using virtual private database
- Mechanics of SQL injection and defenses against it
- Keys to secure database application development and deployment

A. D. Wolfe, Jr. (✉)
Department of Mathematics and Computer Science, Loyola University New Orleans, New Orleans, LA, USA
e-mail: adwolfe@loyno.edu

K. Daimi, G. Francia III (eds.), *Innovations in Cybersecurity Education*,
https://doi.org/10.1007/978-3-030-50244-7_9

The ultimate objective was to restructure the course material and student work in a way that would improve learning outcomes for the on-campus students and online students. In addition to the specific goals regarding database security, I also wanted to provide a way to track students' mastery of the cybersecurity basics in the syllabus. It was important that there be a continuity among the assignments performed by the students, so that these connections would reinforce the concepts and practices from early topics on a recurring basis throughout the course. Finally, I wanted to present the general cybersecurity topics, as well as the specifics of database security, in a fashion that would facilitate the students' ability to recognize these in their professional lives, and also to extrapolate them to help them work with situations they would encounter in the future.

2 Fundamentals in Focus

In this course, we must introduce fundamentals of cybersecurity to the student, who may have little or no prior knowledge of them. As such, the course presents the CIA Triad [1] of *Confidentiality*, *Integrity*, and *Availability*[1] which form the underpinning of information security. Confidentiality refers to whether sensitive information is protected from disclosure to unauthorized people. Integrity refers to the correctness and internal consistency of protected information. Availability refers to the ability of authorized parties to use the protected information in a timely fashion. However, these are often treated as metrics, techniques, or mechanisms in contemporary literature. Such treatments are deficient; in this course, we present these as key security *qualities* which may be realized by many kinds of techniques and mechanisms. While there are excellent theoretical models of security mechanisms, most notably from Bell-La Padula, this course takes a more practical slant. We build from Lampson's "Gold Standard" [2]: Authentication, Authorization, and Audit. Authentication identifies each party seeking to use an information system. Authorization (also called "access control") renders decision as to whether a party can use information as requested. Audit records security-related information. The qualities and mechanisms are the beams and pillars of the Database Security course.

3 Existing Virtualized Laboratory

An important element of the "Database Security" course from its inception was the student laboratory exercises to be performed on an Oracle database. For

[1]The origins of the so-called CIA Triad are very hard to pin down, viz., http://blog.electricfork.com/2010/03/cia-triad.html. While Bell-La Padula engages confidentiality strongly and Clark and Wilson add the element of integrity, the inclusion of availability is obscure, even though it seems to be referenced by Donn Parker in 1988.

the first years, students installed Oracle9i directly on their personal computers. However, in every semester several students would encounter severe installation problems, and we instead decided to use desktop virtualization to provide students with their own database installations. We created a virtual machine "appliance" file that each student would download and import into a virtualization tool such as VirtualBox[2] or VMWare.[3] This approach provided students with a consistent laboratory environment for their assignments and virtually eliminated student problems with installation.

The virtualized environment provided other benefits as well, arising mainly from the ability to add more features to the virtual machine in addition to the database manager. While its Linux operating system was little-known among students, it was possible to cue up the virtualized environment in order to present operating system issues in database security very clearly and accessibly. Also, it was possible to configure a simple web application allowing students to explore the dangers of SQL injection, as with Basit and Chen et al. [3].

4 The Importance of Row-Level Security for Relational Databases

Relational database security mechanisms contain effective means of authentication and audit; vendors typically support much flexibility here. However, authorization—the decision process about whether a user can employ a particular data item—has been a challenge. IBM'S Griffiths and Wade devised the first security approach for SQL relational databases, since followed by all vendors, of granting simple operations on a table to a user. An administrator could GRANT to a user the privilege of inserting rows into the table (INSERT privilege), updating rows (UPDATE privilege), deleting rows (DELETE), and querying rows (SELECT). Such privileges allowed the user operations across all rows in the table. Others later expanded this mechanism to authorize the same operations to "roles," which are groupings of users. However, the fundamentals remained the same:

> *A mechanism which permits the users of a shared database to maintain private data, and which permits them to share a set of privileges on their data with a selected group of other users, or with all users. Subsets of a user's data, derived data, and other transformations of data may be shared by defining a view and sharing that view* [4].

Unfortunately, this approach began to hit limitations. These limitations do not reflect badly on Griffiths and Wade, whose work was published *3 years* before there were any commercial relational databases. As business applications for relational

[2]VirtualBox desktop virtualization can be run under MacOS, Windows, and Intel-based Linux computers. The VirtualBox home page is at https://virtualbox.org

[3]VMWare is a well-regarded commercial virtualization package, available for Windows and MacOS. Its home page is https://vmware.com

databases emerged and expanded, the table became unworkable as the fundamental unit of access control. While the original approach envisioned, perhaps, a limited set of users with very specific business roles, the high usefulness of relational technology led to the extension of database access to more and more end users. This multitude of users would often share the same tables; however, a user's rows in various tables would often require individual privacy. One need only think of a bank account table listing account numbers and account owners to see this issue. Table-level grants could not provide such privacy, insofar as the mechanism would allow access to *all* rows, or *none*, never *some*.

Griffiths and Wade allowed for this need by supporting authorization on database *views* as well as tables. A database view is essentially a stored or predefined database query, against one or more tables, that can filter and manipulate data in those tables. By defining a database view filtering data by the logged-in user, an administrator could individually authorize specific rows of a table to those users who should have access. However, the view-based approach breaks down for database *updates* via views, due to serious implementation challenges. By the late 1990s, it had become clear that "database applications, with large numbers of users, require fine-grained access control mechanisms, at the level of individual tuples [rows in a table], not just entire relations/views, to control which parts of the data can be accessed by each user [5]." This gave rise to implementations of "row-level security" or "virtual private database." Oracle's pioneering virtual private database (VPD) implementation uses transparent query rewriting to limit data access to users on a row-by-row basis [6]. Again, this capability is now provided by many database vendors, both commercial and open-source.

The final circumstance requiring a very strong row-level security component has to do with the ubiquitous implementation practice of *database connection pooling*. A connection is a software construct through which an application authenticates to a database as a particular user and works with the data. All data operations on the connection are restricted to the authorizations of that single authenticated database user. However, web applications serving tens of thousands of users, usually self-registered, could not support individual accounts for each user. Moreover, the overhead of web applications creating and dropping thousands of connections individually proved impossible for a database to support.[4] The technical solution to this problem was "connection *pooling*." In the connection-pooled environment, the application maintains a shared pool of perhaps 250–1000 active database connections. These connections are allocated from the pool to service a user's request and released when the request is completed. Under this approach, the application manages "virtual" end users completely apart from database users; each end user runs under one of a handful of actual database users. This technique also defeats the original Griffith/Wade approach and requires a stronger row-level authorization capability.

[4]Mitre actually classifies unpooled database connections as a security weakness compromising Availability: "CWE-1072: Data Resource Access without Use of Connection Pooling," https://cwe.mitre.org/data/definitions/1072.html

5 Initial Experiment with Business Scenario

Having completed multiple implementations of virtual private database as a practitioner, I was anxious that students master this important technology. However, the "Database Security" course did not emphasize this very strongly and used a relatively simplistic exercise for training students in VPD. Teaching the course on campus in 2013, I replaced this assignment on Oracle's "virtual private database" security feature with a completely new one. I built the new exercise on a business-compromise scenario, the final version of which is described in this paper. While the course previously carried an additional five lab assignments, the new lab replaced one simple assignment with a new assignment in five progressive phases.

While the scenario was successful in the on-campus delivery mode, the additional workload it entailed was burdensome on the students. This also made it difficult for online delivery mode. Moreover, as I made refinements over time, it became clear that there was overlap between the new scenario-based lab and the existing lab assignments covering authentication and authorization. In fact, some new possibilities for addressing useful aspects of database security presented themselves—for example, secure password hashing—as simple additions to the lab.

6 Designing a Business Cybersecurity Scenario

From its first attempt, I was optimistic that a realistic business scenario would effectively provide course continuity and also concretize, for the students, the material we were covering. Students' outcomes with the initial scenario-based lab were very auspicious, as well as their reactions, so I explored ways of extending the scenario in the course. Before long, the aforementioned objectives, things any instructor would likely keep in mind, began materializing against the scenario I had built. But even in the initial formulation, before the scenario-based assignment evolved into the comprehensive course overhaul it became, I applied some key requirements:

- Place the student into a narrative.
- Base the narrative on a real situation.
- Use a small business rather than large.
- Present scenarios with numerous business partners.
- Simplify with only vendor partners or only customer partners.
- Simplify business circumstances.
- Include realistic defects.

A Scenario Places the Student into a Narrative

While obvious, we must define what a scenario is and how it works in the classroom. The scenario tells a story into which we place the student as an actor. The scenario must involve background information and the various steps of events in the story that are relevant to the learning objectives. This allows the student to understand what problems or circumstances must be resolved. In addition, we need to specify to the student the role he is to play, that is, establishing what authority and what resources are available to his "character" to resolve the problem. Note our course learning objectives are central "resources" for the student.

Base It on Real Situations

A key element in using scenarios for educational purposes is *realism*. Knights and fairies are useless, as are interplanetary monsters and lightsabers. Any deviation from reality, or covering gaps with vagueness, compromises the pedagogical effectiveness of a scenario. However, in cybersecurity such deficiencies are particularly risky because of the great attention to detail required of the practitioner. We simply can't tell the puzzled student "oh, it's covered by *XYZ*" and expect the student not to use the same *XYZ* to cover gaps in the coursework he submits.

Make the Business Small

For a 15–18-week course, any sizable business would present a number of complexities high enough to constitute an obstacle to learning, rather than a vehicle. Issues like corporate officers, human resources departments, and organization charts would be a distraction at the least. Certain such scenarios—grocery stores, for example—might turn up some real ratholes as they develop. I began seeking a business that might have only a handful of people, perhaps as few as 5 or as many as 15. In such a business, there would not be a proliferation of roles among employees; we would only need to handle a few.

Design Numerous Outside Partners

A ubiquitous issue I foresaw for my students was dealing with partner organizations in the businesses where they would practice cybersecurity. Even before the Internet age, no business was an island. I therefore wanted to present some sort of "business community" in the scenario I created. The partners must need some legitimate access to the information system of the main business in this scenario. In this way, their interactions provide an important point of interest for students assessing the cybersecurity of the business. Simple vendors and customers buying unitary products or services might not expose these concerns. The partners could be customers in one sense and subcontractors serving customers in another. Alternatively, they might be both vendors and distributors in different circumstances. Differences of business interests between partners, as well as between the partner and the primary scenario business, would also provide depth to the scenario.

Include Vendors or Customers, but Not Both

While all businesses have both vendors and customers, I felt it would be impossible to cover both these "sides" of the scenario business within a semester course. While the cybersecurity issues of both activities mirror each other in some way, the details differentiating them require individualized handling.

Simplify the Surrounding Circumstances

Locating the scenario business in an area of dense population complicates many issues around cybersecurity. For this course I did not want to address issues of physical security such as trespassing, vandalism, and theft; it is sufficient to cover these as part of general cybersecurity practices outside the scenario.

Include Realistic Defects as well as Features

One of the touchiest parts of teaching a scenario is cueing up corrective actions for the students to apply. If the scenario is built to already manifest the best practices taught in the course, there is nothing for them to do. But multiple problems arise if the scenario is too much of a mess. To start with, the students simply will not take it seriously. They may also conclude that the instructor or author of the scenario short-shifted the work. The corrective measures may become too much for a realistic mitigation of problems or may make the scenario look like rebuilding a situation (an application) from scratch—too much work for the student. The system presented in our scenario must work reasonably well for the business depicted, and its problems must look typical, or at least believable, for the scenario. The scenario designer must be able to show how the problems arose realistically in the context of the overall narrative.

Please, No More Schools, Computer Makers or Sellers, or Small-Scale Factories!

Academic computer science is plagued by certain scenarios that are hackneyed to the point of unusability. While many students only find it boring to deal once more with "disk drives, motherboards, and monitors," for others it may become distracting or even annoying. We don't want students to conclude that we spent no time creating the scenario—however authentic the details, they will doubt them and not learn from them.

7 Scenario: Mountain Sports and Game Guides

The scenario I devised for the "Database Security" course was "Mountain Sports and Game Guides." From the course introduction of the scenario:

> *You are working with Mountain Sports and Game Guides, located in a beautiful part of Wyoming that is renowned for its scenery, wildlife, and water sports. This business serves as a supplier for local photographers, outdoorsmen, and boaters and also acts as a broker and scheduler for guides for these activities. These guides are in effect competitors, so it is important not to allow one guide to see another's clients, trips, or even the supplies.*
>
> *The system is used directly by employees but also by the guides and by outdoor sport clients. Each such user has a special set of database scripts he or she runs to check orders, trips,* etc. *These have been found to show each such client the right data, no more, no less.*
>
> *Recently some of the guides realized that, when using the database apart from these scripts, their information was not protected from other guides seeing it. This caused a lot of upset and now we must straighten it out.*
>
> *The original developer of this application left some gaps that we will fill in through the assignments in this course. Some of these are simply bad work, but you may find some bad stuff!*

8 Addressing the Intended Requirements

- **A Scenario Places the Student into a Narrative:** Here we tell the story of a business that, while successful, has a cybersecurity problem that indicates a poor cybersecurity posture for the company's database. The student is placed in the position of an IT staffer or consultant to rectify the problem using learnings from the course.
- **Base It on Real Situations**: Like many other Rocky Mountain states, Wyoming has dozens of licensed hunting and other outdoor guides. While my personal experience using guides is limited, there is ample information available online to provide detail for the scenario.
- **Make the Business Small**: "Mountain Sports and Game Guides" is not clearly described as to its number of employees, but the treatment is intended to indicate

it has about ten employees. The scenario thus excludes things like extensive internal networking, VPNs, and internal servers.

- **Design Numerous Outside Partners**: The scenario uses an outfitter to guides, which gives it a number of potential customers among guides (dozens, as noted above) as well as outdoorsmen who would be those guides' customers as well as Mountain Sports'. While this hypothetical business is very small, the fact that it nonetheless has external partners using its information systems gives it a very realistic "big business" problem.
- **Include Vendors or Customers, but Not Both**: This scenario does not include the supplier side of Mountain Sports and Game Guides. This is not simply an omission of these concerns from the description of the scenario. An important way in which we have excluded that supplier side is by removing inventory and pricing from any assignments. There is very little room for the student to work on this side of the scenario business, and no need, a simplification allowing tighter focus on the assignments as given.
- **Simplify the Surrounding Circumstances**: Setting the scenario in Wyoming reduces a lot of complexity. Even if Mountain Sports is in the largest city, Wyoming's capital of Cheyenne, it's a municipality with only 60,000 people—approximately 1/50 the size of Chicago. This virtually eliminates the concerns about issues like sharing utilities in an office building, gang violence, and traffic.
- **Include Realistic Defects as well as Features:** The student can see and use the application, both through a web interface and through direct database access. The business system operates reasonably well according to its requirements. However, the students can easily look "under the hood" and see the weaknesses in its security implementation.
- **Please, No More Schools...!**

9 Rebalancing Theory and Practice Elements in Course Re-design

Through four or five semesters' teaching with the new scenario-based lab, I began to recognize that this scenario offered me the possibility of reworking the entire course around it. I could reduce the redundant workload of existing labs by moving their learnings into phases of the scenario. In effect, I could have my students go through the assignment as if a database security consultant, progressively applying security practices and principles on behalf of "the client."

To do this, I decided to identify any essential academic or theoretical topics in cybersecurity and present them at the beginning. Learnings that would be applied from the beginning phases of working the scenario can hardly be presented in parallel to those phases. Of these topics, the "CIA Triad" was clearly foremost. In addition, certain aspects of cybersecurity practice, such as defense in depth, could also be handled most effectively before launching the student into the "database security consulting engagement." Afterwards, the mechanisms of authentication, authorization, and auditing provide the progression of hands-on activity for the student.

This rebalancing resulted in a very short academic/theoretical beginning of the course, and thereafter the course takes students the bulk of the learning objectives against the backdrop of practical steps in the business scenario.

10 Parallel Pedagogical Goals in the Sequence of Assignments

After revision, the database security course has progressive assignments presenting the realistic scenario of a small business with an insecure business database and the steps to apply security measures to that database. The sporting guides in the scenario are allowed to use the database, but the database operations have no auditing and almost no authentication or authorization. Each student individually remedies these defects in a stepwise fashion in the virtual machine hosting the scenario. The sequence of assignments takes the students through important information security concepts and mechanisms while correcting the vulnerabilities in the database.

There are three separate parallel threads in the sequence of assignments: the *academic* topics of information security/assurance; the *practical*, that is, the meta-mechanisms employed to achieve information security; and the *specific* application of database security techniques in the scenario business. Once the student has passed beyond the initial academic topics, the practical framework drives the learnings in the course.

Academic: Key Security Qualities

Cybersecurity can be understood, as an academic discipline, through various principles in support of the "key security qualities" of Confidentiality, Integrity, and Availability. These principles include least privilege, defense in depth, and valuation of information assets. As an academic topic, Cybersecurity also includes the theoretical underpinnings of technology, most notably cryptography. In this course, our academic treatment also presents known threats and attacks and how vulnerabilities are discovered, reported, and addressed. Of these, only the "key qualities" are presented to the student in advance of the scenario-based assignments. The other academic topics are raised as the student progresses through the course.

Practical: Key Security Meta-Mechanisms

An important component of formulating the course was what I call the "Security Meta-Mechanisms." This is how I frame Lampson's "Gold Standard of Security": *Authentication*, *Authorization*, and *Audit*. Every implementation of information security includes at least these three security *mechanisms*, or, more precisely, *kinds* of security mechanism. Since these are "kinds" or "categories," rather than specific mechanisms, we term them *meta-mechanisms*. Authentication is how the system positively identifies an end user or agent that is active in that system. Authorization is what most people think of as 'security': the set of rules by which an authenticated user is granted or denied access to an information resource. Audit is the recording of interactions of users with information assets, including those assets supporting Authentication and Authorization. There are many ways these meta-mechanisms may be implemented in a system, but they are always present.

Specific: Database Security Techniques

There are several techniques in database security covered in this course. These include administration of end users and roles, definition of password policies, configuring and hardening a database system,[5] recovery mechanisms, and the definition of grants authorizing users' access to database resources as well as to administrative functions. One of the most important technologies in database security, as noted before, is *virtual private database* ("VPD") or row-level security. While Oracle was an early implementer of VPD, similar capabilities are now available in other commercial DBMSes, including Microsoft SQL Server and IBM DB2. Auditing techniques take on a distinctive flavor, also, in the database environment. Database audits may be enabled by AUDIT commands, by triggers in the database, or (in Oracle and other databases) by declarative audit policies. The key distinction of database audit is the importance of recording SQL commands that mediate data changes and the set-wise, bulk nature of many database operations. Finally, it is important to expose students to secure coding practices[6] as these pertain to database applications, in particular, to avoid SQL injection.[7]

---

[5]See CIS Benchmarks, https://www.cisecurity.org/cis-benchmarks/, specifically "CIS Oracle Database 12c Benchmark v2.1.0—09-18-2018."

[6]OWASP Secure Coding Practices https://owasp.org/www-project-secure-coding-practices-quick-reference-guide/ is our main starting point. However, more comprehensive guides are available from the Carnegie-Mellon CERT site: https://wiki.sei.cmu.edu/confluence/display/seccode/SEI+CERT+Coding+Standards

[7]One of many good descriptions of SQL Injection is from OWASP: https://owasp.org/www-community/attacks/SQL_Injection

11 Quasi-Experiential Learning in Progressively Securing "the Client"

My objective in course design was to take the student through an authentic sequence of activities to secure a relational database. The approach is akin to experiential learning [7]; however, the constraints of accommodating part-time adult learners were intrinsically limiting. BU Met's adult students attend only one evening class weekly, and with online students, the only "common class time" consists of six or seven mandated weekly web conferences. Consequently, it would have been impractical to attempt to derive learning styles (Kolb or Honey-Mumford [8]) from the students. However, the Kolb cycle itself of experience-reflect-conceptualize-test [9] applies as the students take the experience-reflect actions of one assignment into the conceptualization and test of the next. Kolb and Honey-Mumford introduce elements of disorientation and dialectic to learning, which is applied to cybersecurity education by Lowe and Rackley [10]. To some degree, every experiential learning scenario is initially disorienting, of course, but the same limitations precluding learning-style assessment would make intentional repeated disorientation counterproductive. Consequently, student activities were performed against a static test environment with no hidden attackers nor sudden changes.

It was important to engage the question of setting an order for the learning objectives in reworking the assignments. Practical guides and administrative manuals list techniques exhaustively, but did they provide an effective sequence for using them? If so, was that sequence going to be productive in the teaching environment? Clearly, I would need to select one of the parallel tracks as my "master" to drive the assignments and the inclusion of the other tracks. I concluded that neither the academic learnings nor the particular DB security techniques presented a clear progression; both are essentially unordered collections. Now the meta-mechanisms have some sort of ordering, in that Authorization is clearly dependent on Authentication and Auditing "feels" dependent on both. Moreover, there are only three of them. So, I used the meta-mechanisms as the primary organization mechanism, rather than grouping by academic topic or by some sets of database functionality.

One might ask why, instead of starting with Authentication, I began the assignments with Audit. Authentication seems to be required for Audit—to identify who was doing what—as well as for Authorization. However, most DBMSes, including Oracle, have the ability to perform *generic* audits of activity independent of which user performs it. Moreover, one key *technique* I felt must apply early on was database "hardening." What is hardening? Because each DBMS must be adapted to the security needs and environment in which it is deployed, its default installation has intentional security gaps. Therefore, best database security practices include a "hardening" activity: limiting user accounts, adapting authorization, and controlling network access to the database. Hardening also involves *creating audits*. For this reason, I placed Audit first. After that came Authentication and then Authorization.

The following table presents the stepwise progression, in this course, of the topics, meta-mechanisms, and techniques.

Module	Learnings	Meta-mechanisms	Techniques applied to scenario
1	• Principles • Assessment and valuation		• (Installing and) reviewing the "mountain sport server" and its applications and functionality
2	• Purposes of audit • Defense in depth	Audit	• Hardening the server • AUDIT implementation using standard AUDIT commands, by triggers and by policies
3	• Models of users and groups	Authentication	• User administration • Database user provisioning • Implementing virtual users • Implementing database roles
4	• Virtual private database—requirements	Authorization	• Row filtering to support authorization • Implementing virtual private databases using database views
5	• Virtual private database—models	Authorization	• Using declarative virtual private database to filter access by rows • Applying virtual private database to filter sensitive columns
6	• Secure programming • Encryption • Password management		• SQL injection • Encryption lab • Managing passwords for scenario "users" using one-way encryption

At the conclusion of the course, the student has taken a poorly configured Oracle database and:

- Hardened the means of accessing the database
- Designed and activated audits of database operation
- Built a system of authentication using not only built-in database user management but finer-grained management of virtual users
- Set up a model for authorization of access to information assets
- Implemented a virtual private database
- Applied encryption to protecting the passwords of virtual (non-database) users
- Performed penetration exercises against the scenario database using SQL Injection

12 Packaging "Mountain Sports and Game Guides" for Students

"Mountain Sports and Game Guides" has a database tracking both its sportsmen/naturalist customers and independent outdoor guides for whom it serves as a broker. Therefore, for the student to engage this scenario, he has to have direct access to "the database." Moreover, database security requires administrative privileges on the database system at least and usually administrative privileges on the server itself.

To provide the student with the ability to perform these tasks authentically, we packaged the "database server" as a virtual machine. The virtual machine was created using the free/open-source VirtualBox virtualization application. The following software and configuration were included:

- Oracle Linux 6 operating system with Gnome-based end-user graphical interface
- Java Development Kit 8 ("1.8")
- Tomcat web application server
- An Oracle19c database server owned by a user named "oracle"
- A database schema "vpd674" containing the tables for "Mountain Sports and Game Guides," populated with several dozen sample users and approximately 18 months of sample data
- Four simple Java web applications installed into Tomcat
- Oracle SQL Developer database client
- Custom SQL scripts for each "user"
- Custom user configuration for *bash* shell and Gnome Desktop

The virtual machine was exported using Open Virtualization Format (version 1). This format is portable and can be imported by VirtualBox for the student's use and can also be imported using VMware Workstation. While this format is an open standard and thus should be usable by other virtualization tools, the only thorough verification of the export "appliance" file was performed with VirtualBox. Some successful tests were performed using VMware, but we have performed no tests on any other virtualization system.

By using this packaging, we intended to provide an expeditious and straightforward introduction to the scenario and the laboratory exercises. The following steps, with screen shots, illustrate a typical sequence for the student.

The student downloads the appliance file to his computer and imports it into VirtualBox (Fig. 1).

Once the import is completed, the student starts the imported virtual machine and can log in as "oracle" (Fig. 2).

To enhance a sense of realism, the login presents a message "to the Mountain Sports partner" (Fig. 3).

After closing the readme window within the VM, the student has a conventional Linux desktop (Fig. 4). Note there are shortcuts on the desktop for the SQL Developer tool, a terminal window, and the "Mountain Sports web site".

Fig. 1 Import settings for the course virtual machine appliance

The SQL Developer tool is already pre-configured to connect to this database; however, the student will have to configure additional connections for various laboratory exercises (Fig. 6).

The virtual machine is also configured to expose various ports to the student's host computer (Fig. 7).

These ports allow the student to use the tools he has installed, and with which he will presumably be comfortable, to access the scenario virtual machine. Here we see the virtual machine desktop in one window and a browser in my host Macintosh using the "Mountain Sports" site (Fig. 8).

The links provide access to pages that are susceptible to SQL injection, like the following (Fig. 9).

The "search for animals" results show the student the SQL that is executed, allowing fairly easy SQL injection experiments (Fig. 10).

After starting the SQL injection assignment on this page, we present the student with assignments requiring more injections to attempt. In the final steps, the pages to inject require the student to work without being able to see the SQL being run.

When the student opens the "Web Site for Mountain Sports", he sees a simple but realistic home page (Fig. 5), with links to various functions.

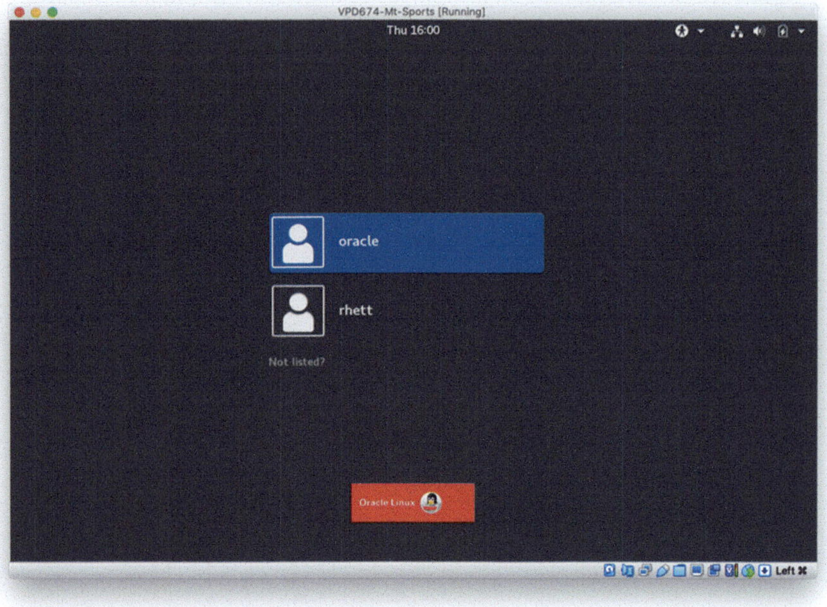

Fig. 2 Virtual machine login window

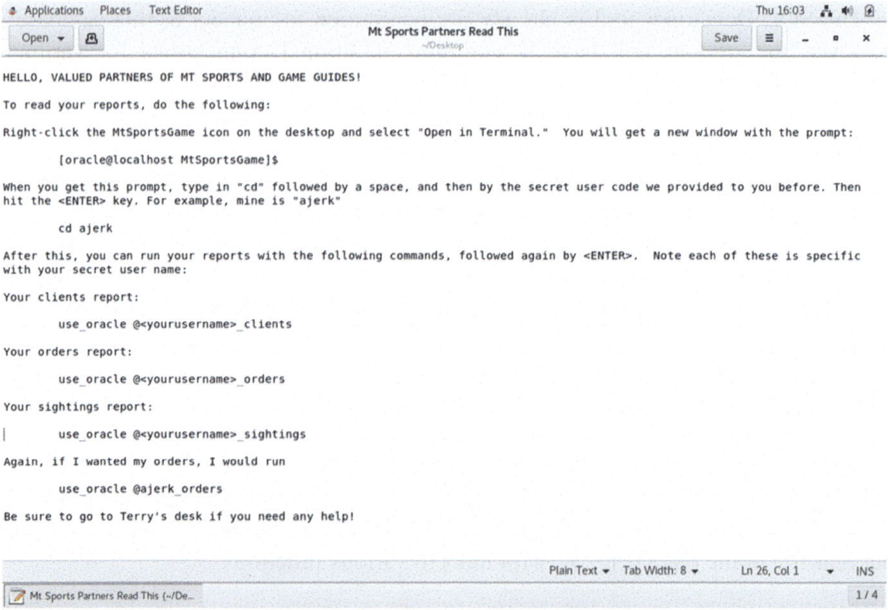

Fig. 3 A plausible "read-me"-type file for the scenario "partner"

Fig. 4 The "oracle" user's desktop in the virtual machine

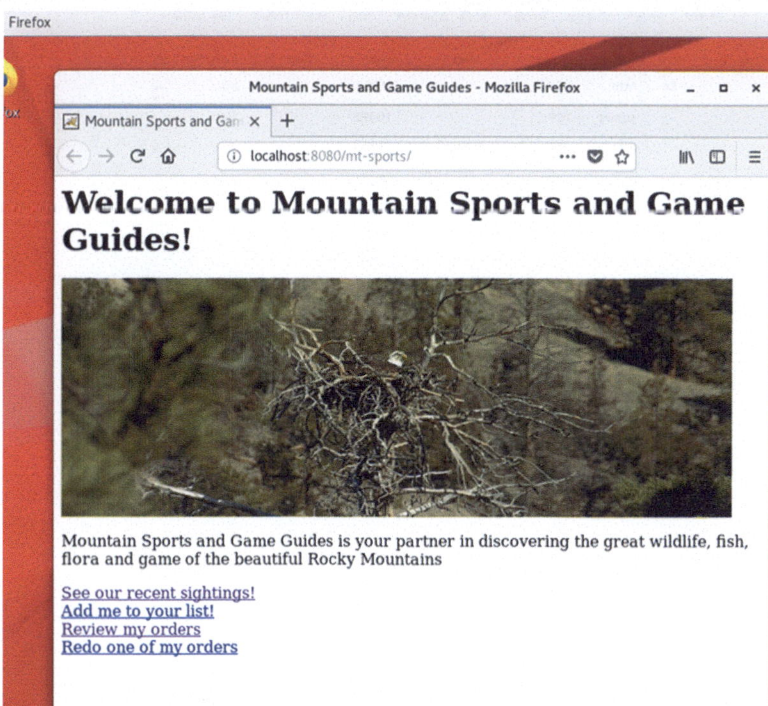

Fig. 5 The "welcome" page for the Mountain Sports scenario site

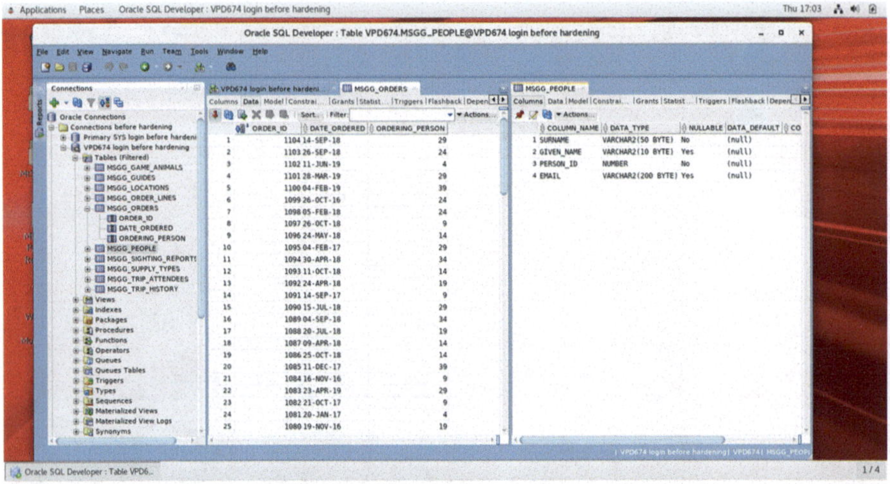

Fig. 6 Using SQL developer within the virtual machine to access scenario database

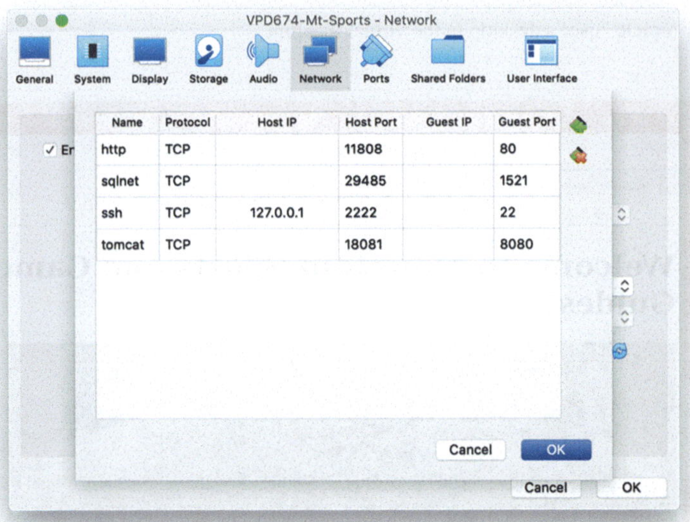

Fig. 7 Port forwarding from the host into the virtual machine

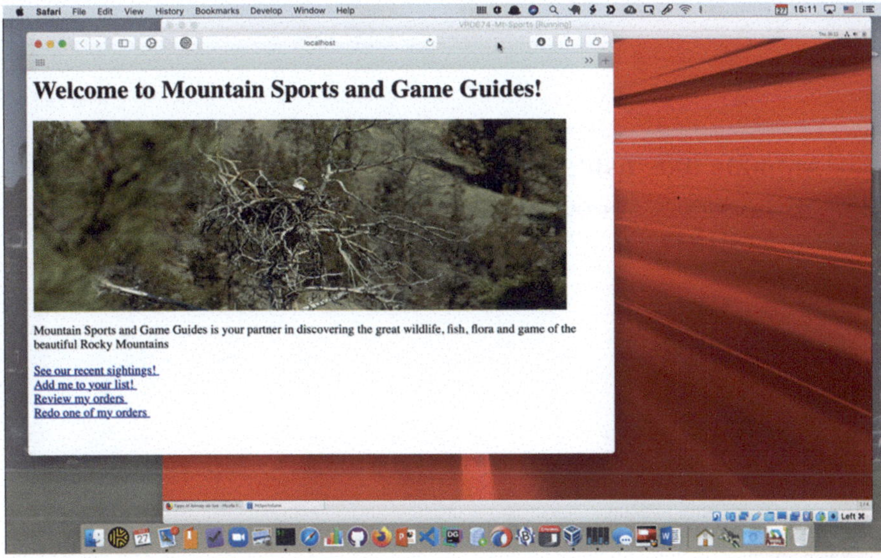

Fig. 8 The Mountain Sports site from the "host" machine

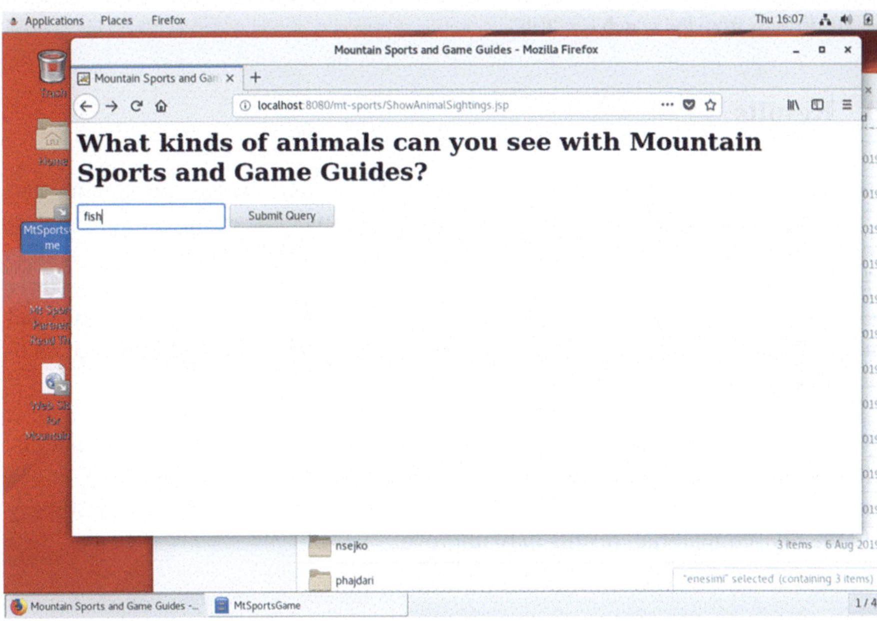

Fig. 9 Mountain Sports page to search for animals

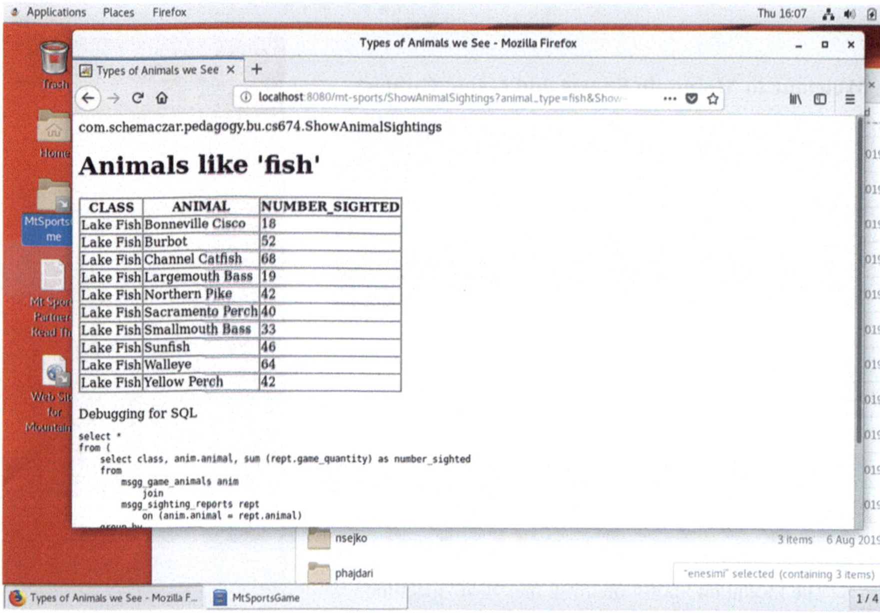

Fig. 10 Results from the search for "fish"

13 Results

Students responded very positively to the scenario as I began introducing and expanding it in the on-campus course around 2014. Students' mastery of the critical virtual private database capability improved significantly, both in testing and as expressed in classroom discussion. Moreover, by covering information security facilities in terms of the meta-mechanisms of Authentication, Authorization, and Audit, the students evidenced a better ability to apply information security basics to non-database topics in cybersecurity.

However, there were areas mitigating the success of the exercise. Most notably, when the revised course was again presented to *online* students, evaluations suffered. While the overall reasons are unclear, shorter timeframe of the online course may be a substantial factor. As the online teaching team gains familiarity with the revised material, results may improve.

14 Conclusion

Cybersecurity is a topic that defies neat delineations and modularization. Information systems security and assurance face many problems in hostile activity, human error, software problems, and natural occurrences like flooding. It remains a regrettable fact that this course conflates substantial specialized knowledge and practice in database security with the initial presentation of cybersecurity principles to students not previously exposed to it. In this circumstance, the choice of following the security meta-mechanisms as the underlying structure for the course allowed substantial progress in covering those database security topics in parallel with the basics. I hope to understand the challenges with this approach in the setting of an online accelerated session. While I am now teaching at a different institution and will no longer be teaching this same course, I will be applying the approaches I worked out and, hopefully, perfecting them.

References

1. D.D. Clark, D.R. Wilson, *A Comparison of Commercial and Military Computer Security Policies*, 1987
2. B.W. Lampson, Computer security in the real world. IEEE Comput. **37**(6), 37–46 (2004)
3. N. Basit, A. Hendawi, J. Chen, A. Sun, A learning platform for SQL injection, in *50th ACM Technical Symposium on Computer Science Education (SIGCSE'19)*, Minneapolis, MN, 2019
4. P.P. Griffiths, B.W. Wade, An authorization mechanism for a relational database system. ACM Trans. Database Syst. **1**(3), 242–255 (1976)
5. S. Rizvi, A. Mendelzon, S. Sudarshan, P. Roy, Extending query rewriting techniques for fine-grained access control, in *Proceedings of the 2004 ACM SIGMOD International Conference of Management of Data*, 2004
6. Oracle Corporation, Using oracle virtual private database to control data access, in *Oracle Database Security Guide*, Oracle Corporation
7. S. McLeod, *Kolb's Learning Styles and Experiential Learning Cycle*, 2017. https://www.simplypsychology.org/learning-kolb.html. Accessed 9 Apr 2020
8. P. Honey, A. Mumford, University of Leicester—Honey and Mumford. https://www2.le.ac.uk/departments/doctoralcollege/training/eresources/teaching/theories/honey-mumford. Accessed 9 Apr 2020
9. D.A. Kolb, *Experiential Learning: Experience as the Source of Learning and Development* (Prentice-Hall, Englewood Cliffs, NJ, 1984)
10. T. Lowe, C. Rackley, Cybersecurity education employing experiential learning, in *Proceedings of the 2018 KSU Conference on Cybersecurity Education, Research and Practice*, Kennesaw, GA, 2018
11. Bell, David E., Leonard J. La Padula, *Secure Computing Systems: Mathematical Foundations*, MITRE Technical Report MTR-2547, (The MITRE Corporation, Bedford, MA, 1973)

Part III
Faculty and Professional Development

Cyber Security Assessment Education for E-Governance Systems

Rajan Gupta, Saibal K. Pal, and Sunil K. Muttoo

1 Background

E-Governance is increasingly being deployed around the world, especially in developing nation for providing quality citizen services at low costs. As per the United Nations report on E-Governance development [1], more numbers of countries are now adopting E-Governance practices in order to provide better governance to their respective citizens. But a major issue is the threat of attacks and security concerns [2–5]. These concerns are related to the E-Government infrastructure and the services that are being offered through public channels to citizens and other institutions. The loss of government data and breach of privacy/confidentiality of the citizen's data are major challenges for stakeholders of E-Governance. Officials are finding it difficult to manage efficient services with rising intrusions and frequent security lapses. This has led governments to think about security loopholes seriously and take necessary actions against any kind of weakness found in the system [6]. Being a type of information system, the electronic governance system and setup must be secured against the threats that are common to information systems in public domain [2, 7].

Confidentiality, integrity, and availability are the three pillars of security (Fig. 1) for any system [8, 9]. Considering the principles of the information security, confidentiality can be defined as the restriction from unauthorized data/information reading and accessing it; integrity can be defined as restriction from unauthorized data/information change; and availability can be defined as restriction from unauthorized data/information access at client and server end, respectively, i.e., information

R. Gupta (✉) · S. K. Muttoo
Department of Computer Science, University of Delhi, Delhi, India

S. K. Pal
Defense Research & Development Organization, Delhi, India

K. Daimi, G. Francia III (eds.), *Innovations in Cybersecurity Education*,
https://doi.org/10.1007/978-3-030-50244-7_10

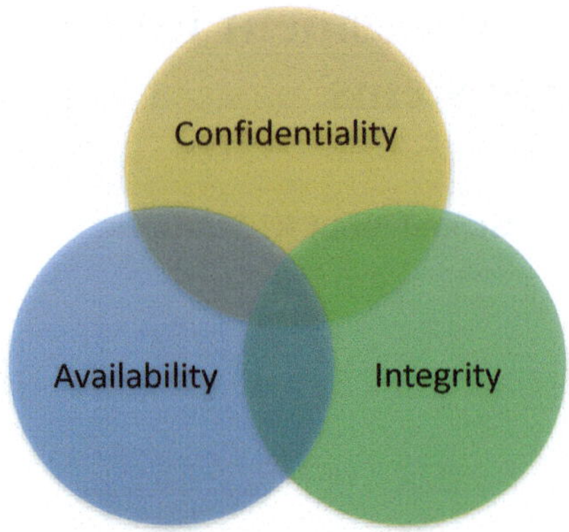

Fig. 1 The three pillars of information systems security

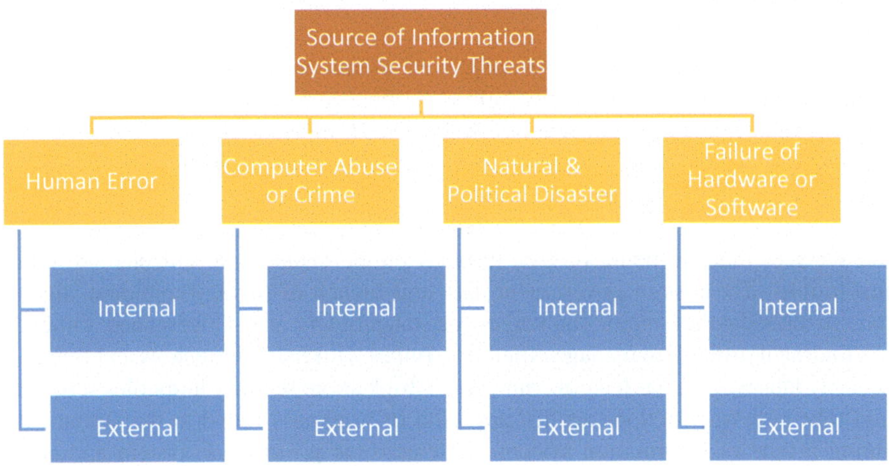

Fig. 2 Sources of information systems security threats

provider and information seeker. Authentication, authorization, and access controls are the other concerns that must be addressed within the information systems security.

As per the different layers of information system [9, 10], the threats depend on the safety of its critical assets. Once the critical assets are protected from the threats, the chances of system lapse become low and financially less impacting. The threat to the critical asset can be through network access, which can be carried out by any internal or external agents (Fig. 2). Internal threats could be due to human or non-human activities. Human activities include poor administrative procedures and deliberate ill act by employees due to grudges or any other negative feeling or due to over-dependence on the unreliable employees. Non-human activities include system failures, power failure, and problem with existing programs, outdated software, and the likes. External threats to the information systems can also be due to the human and non-human activities. In this case, involvement of the humans could be by the hackers or competitors in the market. These malicious users can negatively impact the network, database, or software of the existing system through external means. Non-human activities could be due to natural disasters, unplanned viruses, or accidents due to uncontrolled elements in the surrounding environment. These internal and external agents can pose an accidental or a deliberate attempt to harm the critical assets in form of disclosure, modification, loss, destruction, or interruption [8, 11]. For these reasons, Whitman [12] argued that organizations must invest on appropriate policy formation for security, executing awareness programs among the employees and other stakeholders, educating the staff regarding security policy and practices, and preparing contingency plan communication in case of security breach. Thus, there is a need to assess both the internal and external environment of critical assets involved in information systems in public domain (i.e., E-Governance setup) and evaluate them for the safety of the various organizations.

There are different frameworks for the assessment of security of information systems and E-Governance, but many of these lack depth and scale of the domain. Various security aspects covered up in these frameworks are not exhaustive, and a much simpler scoring system is required to make it understandable by the users at different levels. Some of the existing frameworks are discussed as follows.

Da Veiga and Eloff [13] proposed and discussed an instrument which had the potential to become important for the culture of information security. They focused on the internal threats of the organization's information systems which covered the threats by the employees working in that system. The authors considered internal threats as a major threat in the organizations and proposed that the employees should be advised and trained to work on security principles of information systems. The authors also argued that there must be a culture developed within the organization where security practices are followed seriously on a regular basis.

Rees et al. [14] also proposed a security assessment framework named PFIRES. It included various policies that addresses the information security assessment for any organization. The PFIRES lifecycle had various stages like assessment, planning, delivery, and operation. Assessment, as the first stage, involves risk in any system according to which policies must be developed and requirements must be defined. Delivery and implementation of controls is followed by the monitoring phase. So, the scoring assessment framework that is proposed in this paper can be used at the planning stage, and accordingly the security policies can be implemented and monitored.

Sun et al. [15] also developed a risk assessment model based on the theory of belief functions as suggested by Dempster-Shafer for information systems. Risk-oriented belief function and evidential reasoning were the basis for the risk assessment. It also took the role of cost benefit analysis for risk assessment into consideration. Both theoretical and operational aspects were considered while assessing the risk for the information systems in the organization and were validated through external assurance providers.

Al Hogail [16] discussed the cultural assessment of the various risks associated with the information systems in their assessment framework. The study considered various options related to the security culture like environment, people, technology, strategy, and organization. Different human domain aspects were considered like management, preparedness, responsibility, and society regulations. Various principles related to change management were also taken into consideration, and experts were surveyed for the validation of the framework. It was an overall exhaustive framework covering different dimension which provided base to the current study to consider diverse aspects for E-Governance system.

Sillaber and Breu [17] stressed on the importance of the stakeholder's involvement in risk assessment and management of information systems within the organizations. The various stakeholders were involved in question-answer session, where the various business processes were discussed, and questions related to the security standards of the information system within the organization were stressed upon. Business process matrix was created against stakeholder's awareness for risks for the information systems. Based on the awareness, implications were drawn toward the policy of risk awareness and control mechanisms toward information systems security.

Joshi and Tewari [18], in their study, discussed about the security of E-Governance in India. They discussed about the various cyber threats and security attacks on the E-Governance system and proposed recommendations based on the attacks. But, the study lacked a concrete framework for the security assessment within the E-Governance system in India. Similarly, Roy and Karforma [19] also presented a review of E-Governance security in India. They came up with a protection layout of authentication, integrity, privacy, and non-repudiation which must be there within the E-Governance system.

The other studies [20–23] in the domain of E-Governance either discuss about the various frameworks for development assessment of the different E-Governance projects or are related to specific cloud security assessment of the whole system. There is lack of research studies which can provide a complete and exhaustive information systems security framework for the E-Governance so that any new service setup can take the different aspects of security into consideration. Also, with the help of a new framework, the existing services can be safeguarded against the possible internal and external attacks. The current study proposes such framework for E-Governance risk assessment. But, before defining the security framework, one needs to understand the difference between developed and developing nations. Also, there is a need to define the status of existing frameworks implemented in the developing nations for E-Governance/information systems security.

As suggested by Chen et al. [24], there is a need for a separate framework for the developed and developing nations for E-Governance and related activities. There are lot of differences quoted for the two types of nations. Developing nations are well advanced with respect to their basic technology infrastructure; they have availability of highly skilled labor or have the capacity to outsource their work based on ample amount of funding reserved for them; their citizens have good understanding of working of the computers; economy has been stabilized and growing at a constant rate; and the government is more stable with better utilization of democracy and good amount of transparency being practiced at the top level of the government. On the other side, developing nations have different types of structure and challenges existing within their regions. The basic infrastructure is not in good position and not available to majority of the organization and citizens; the illiteracy is high among the citizens resulting in low availability of skilled labor for the work; since literacy is low, basic computer understanding is very low among the officials and users; economy is still growing with signs of stabilizations; government is still not stable so policies keep on changing; and corruption free working, transparency, as well as efficiency are not present within the system [24].

With respect to developed nations, the security standards are already established. In addition to CIA model, RITE was also proposed to be established for the systems in developed nations [25] in the new millennium. However, the whole concept of digitization came very late in the developing nations. For example, the National E-Governance Plan (NeGP) was launched in 2006 in India [26] and even late in other developing nations. Since digitization happened late in the developing nations, therefore, security concerns were not present earlier. So, the pre-existing models of the developed nations could not be adopted earlier. Now, since the technology penetration happened rapidly in past one decade in the developing nations, the security concerns also increased rapidly. The standard security frameworks were utilized for the establishment of the basic infrastructure, but more customized solutions are not readily available in the market. For example, most of the developed nations have precautions being taken for physical safety of the infrastructure from natural calamities, but developing nations are yet to take these kinds of concerns more seriously in their assessment plans. Similarly, network access, network learning, network economy, and network policy need to be considered for developing nations [24]. Therefore, the developing nation's needs should be assessed separately, and security assessment framework should be developed accordingly, which is easy to use and understand by the users in the developing nations. Few authors [27–32] attempted to work specifically for developing nations, and their analysis summary is shown in Table 1, but more exhaustive framework is required, so that the beginning-level and advanced-level aspects can be considered for threats from both internal and external security.

The main aim of the current chapter is to educate about the cyber security assessment to the personnel working in E-Governance systems through a framework useful for assessment of information security concern under E-Governance. The framework should have all the internal and external aspects covered so that the risk is minimized, at least in the accidental chances of damage to the system. The next section will discuss the various security assessment frameworks with respect to information systems security and E-Governance followed by the proposed framework and its discussion. The case study on an Indian organization is presented afterward followed by the conclusion of the study.

Table 1 Summary of strengths and weaknesses of few prominent existing frameworks

S. no.	Framework	Strengths	Weaknesses
1	PFIRES [14]	Strong feedback-based policy framework; strong assessment of the situation	More focus on implementation of security based on feedback rather than covering exhaustive security areas
2	Sun risk assessment model [15]	Cost benefit analysis of security policy; exhaustive assessment	More focus on the planning of security policy rather than checking the actual implementation
3	PICABUE [27]	Security assessment as part of sustainable goals of an organization has been discussed in detail	Technological aspects have not been covered in depth for the organizations
4	RITE [25]	Good coverage over responsibility, integrity, trust, and ethicality of security policies	Technical controls within the organizations w.r.t. security assessment have not been discussed
5	Stakeholder theory [17]	Very good coverage of business process models and stakeholder participation	Very theoretical model based on the information system risk management
6	Risk assessment [30]	Security policy aspect has been covered from data collected from experts	It's a qualitative model covering general aspect of the security of organizations
7	Information system risk [16]	Very good coverage of management principles for adopting security policy framework	More focused on security culture and human behavior in cyber security aspect

2 Need for Cyber Security Education

The threats of cyber-attacks on computers are increasing at an enormous rate around the world. Various new patterns of such attacks are emerging each passing day. Computer network is the most vulnerable component by an attacker's point of view. The malicious activities occurring on network sites result into loss of computer security that is covered up under confidentiality, integrity, and availability [33]. To deal with this problem, James Anderson first introduced the intrusion detection concept in 1980 [34]. And the first model on this concept was developed by Denning in 1987 [35]. Intrusion detection system (IDS) monitors all the inbound and outbound system activities in order to detect any kind of unwanted patterns [36]. IDS work inside the network and protect the system by any intruders, by generating an alarm anytime a fraudulent activity is observed. The conventional attack detection systems such as firewall, data encryption, or authentication process are not capable to protect network intrusion completely. This may happen due to changing patterns of attack and malware. IDS can be divided into two broad categories, viz., anomaly detection systems and misuse detection systems [37–40].

In the context of developing nation like India, the 2016 ORF report on the Indian National Security Architecture identified that the year 2015 witnessed cyber-attacks on 72% of the Indian firms with attempts made for corporate espionage [41]. India was the topmost targeted nation by the cyber-criminal through the seven social media in the year 2014. The Computer Emergency Response Team (CERT) on 8th January 2015 reported as many as 8311 incidences of security breaches which increased from the previous recoded figure of 5987 in November, 2014, while the reported incidences of disfiguring of the websites increased from 1256 to 2224 in the same period. The CERT in 2015 ranked India as the third most vulnerable nation in Asia susceptible to the most of the continents ransom-ware cyber-attacks. A figure replicated in the Kaspersky Security firm's 2016s third quarterly report which reveals that India faces one third of the ransom-ware incidents in APAC region [42].

Figure 3 shows the security attacks being encountered through the network monitoring on quarterly basis by CERT-IN (Indian Government Web Monitoring Team). Malicious websites and malicious codes are the most encountered anomalies within the Internet traffic surveillance in India in the past few years. Figure 4 shows the activities undertaken by CERT-IN in the past few years. Bot-infected systems

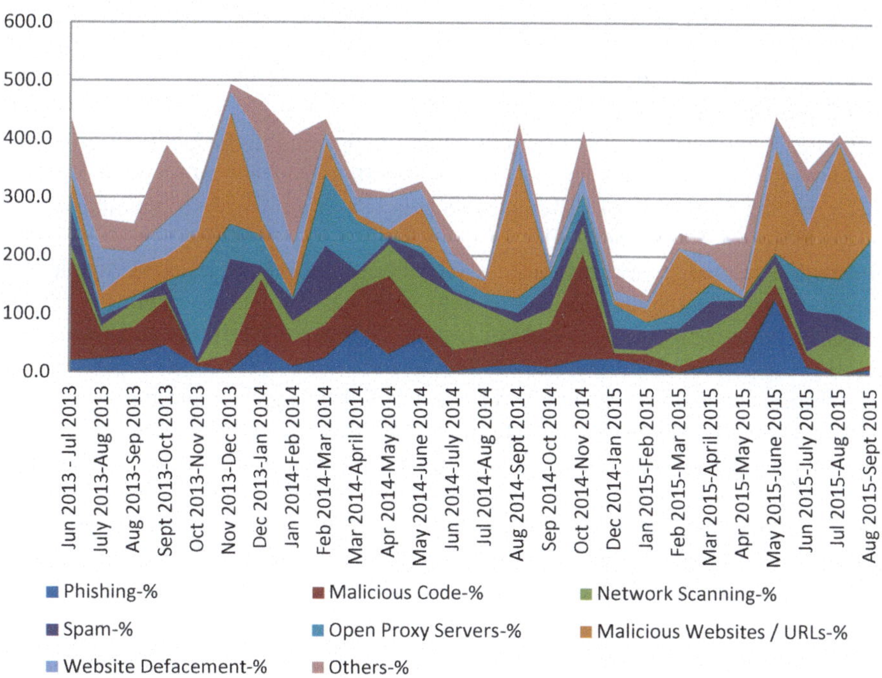

Fig. 3 Natural log of number of attacks detected by CERT-IN in the past few years (Data Source: http://www.cert-in.org.in/)

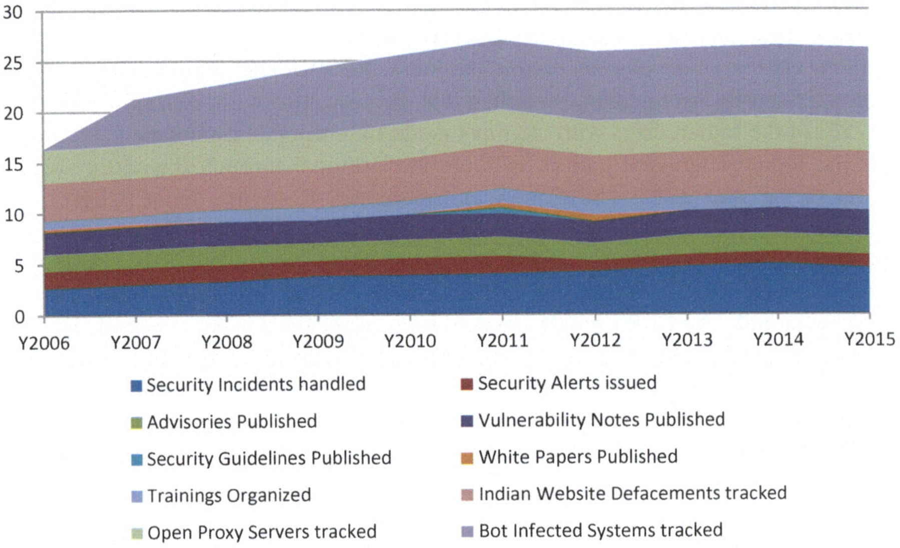

Fig. 4 Natural log of number of activities conducted by CERT-IN in the past few years (Data Source: http://www.cert-in.org.in/)

were tracked highest number of times in the last few years followed by handling of security incidents. All these activities undertaken by CERT-IN can be supported by a good surveillance system.

Figures 5 and 6 suggest the types of attacks happening in the last few years and also show the analysis of the attacks. This is important from education point of view. There are large numbers of different data sets which are currently maintained by the world net data warehouse. One of them is router configuration which includes information related to security, access list, topology, and the likes. A "triple A" system covering authentication along with authorization and accounting systems makes up as the components of the registration records. Similarly, the data under call record comprises of summary related to customer's dial-up session on per-session basis. Email server logs' include SMTP and POP3 transaction summaries. Router statistics is obtained by SNMP polling. It includes link/router utilization, access, and gateway routers [43]. Apart from them, numerous packet scopes exist for IP packet headers' collection, and they are designed for high-performance systems. They act as passive link access in which the modification of the device driver was done for all the "read" commands but not "write" commands for the network interface which was under monitoring. This monitoring can be T3 which was, for a case, terminated at router modeled 7505 by Cisco and is designed for the forwarding of the packets toward monitor for the capture. Now, these captured packets are utilized for the collection of header which contains vital information. Similarly, data apart from textual form like multimedia data is monitored by passing the traffic through various

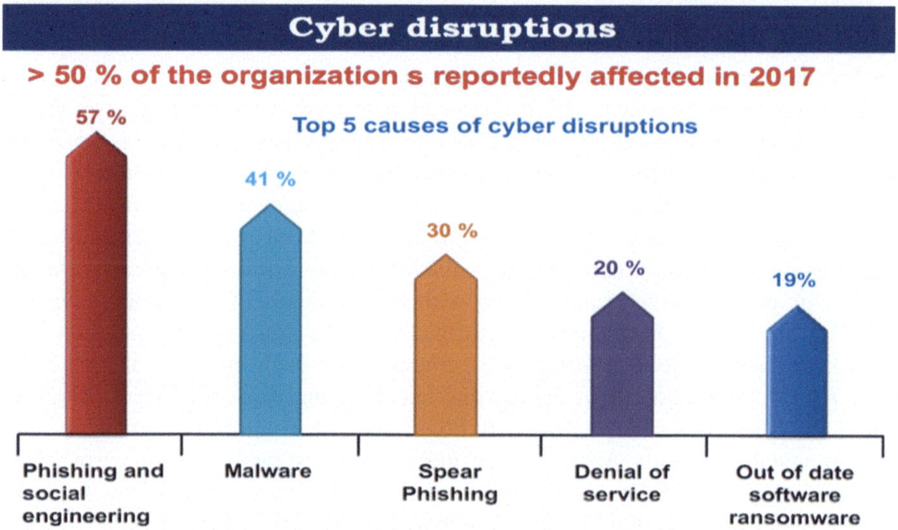

Fig. 5 Cyber disruptions recorded by CERT-IN (Source: https://niti.gov.in/sites/default/files/2019-07/CyberSecurityConclaveAtVigyanBhavanDelhi_1.pdf)

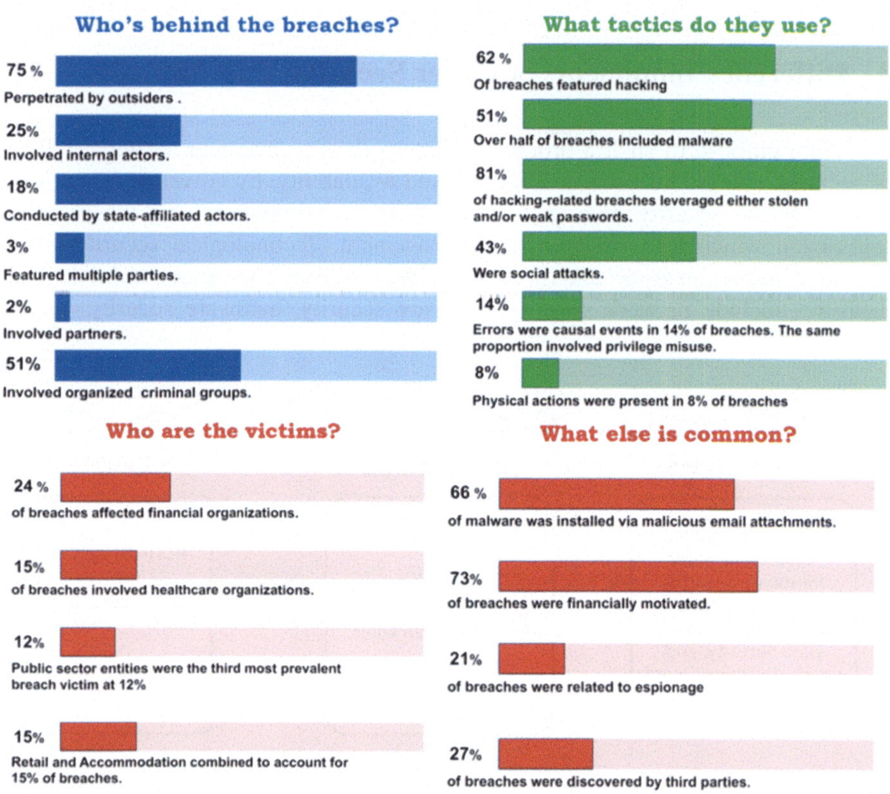

Fig. 6 Analysis of attacks recorded by CERT-IN (Source: https://niti.gov.in/sites/default/files/2019-07/CyberSecurityConclaveAtVigyanBhavanDelhi_1.pdf)

protocols like RTSP, SIP, and other protocols related to session control for setup and packet filter teardown for capturing multimedia sessions [43].

However, there is a lack of basic education toward cyber security assessment which leads to ill preparedness for attacks on various systems. So, there is a need to have an assessment framework through which cyber security can be monitored.

Aiming at strengthening the cyber security ecosystem in India—in line with the government's vision for a "Digital India," the Ministry of Electronics and Information Technology (MeitY) has launched Cyber Surakshit Bharat initiative [44]. This program was in association with the National E-Governance Division (NeGD). Digitization has rapidly transformed the governance system, and therefore the requirement of good governance is crucial. With such initiative, there would be a rise of awareness about cybercrime and building capacity for securing the Chief Information Security Officers (CISOs) and the frontline IT staff across all government departments. Apart from awareness, this first public-private partnership also includes a series of workshops to make people cognizant about the best practices and help the officials with cyber security health tool kits to tackle cyber threats. Similarly all private organizations are also working toward improving cyber security concerns in their respective workplaces.

3 Different Components of Cyber Security

The main purpose of such a proposed framework is to maintain the three pillars of security, i.e., confidentiality, integrity, and availability, by covering the various aspects of the security. Figure 7 shows different components for the security assessment which leads to areas of improvement. Technological security is an extremely important aspect and is covered in the top layer. The various technological measures include network security, software security, hardware security, server

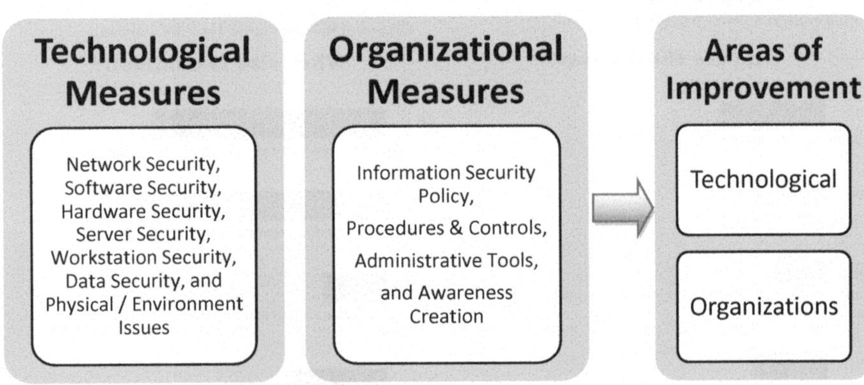

Fig. 7 Components of the basic assessment framework

Table 2 Scoring levels for the statements used for information systems security assessment in E-Governance system

Score level	Status	Description
1	Not at all implemented	There is absolutely no established security measure
2	Partially implemented	Few components of security are implemented in E-Governance system
3	Implemented and not reviewed regularly	Security components are implemented but not mapped with all the components of E-Governance system
4	Almost implemented and regularly reviewed	Most of the security components are implemented and mapped with all the components of E-Governance system
5	Fully implemented and a role model for other E-Governance systems	All the security components are fully implemented and completely mapped to E-Governance system with continuous review and upgradation

security, workstation security, data security, and physical/environment issues. The other important aspect is organizational measures security and is covered at the bottom layer. The various organizational measures are policy and controls for information security, procedures, administrative tools, and awareness creation. To assess all these components, the different aspects should be recorded and noted for an E-Governance system in an organization. Scoring levels of assessment for security are shown in Table 2.

Based on the scoring levels, the various statements of different components of the assessment framework can be rated for the E-Governance systems. The statements have been re-formulated for the assessment and are rated from scale of 1–5 based on Table 1. The statements, for the various components shown in Fig. 7, are shown in Tables 4, 5, 6, 7, 8, 9, 10, 11, 12, 13, and 14 in Appendix. The rating of these statements can be done by any neutral agency in coordination with the director of the organization or by the personnel involved at the management level in the E-Governance systems. The maximum point allocated to any statement is 5, and the minimum is 1 point based on the full availability or non-availability of the specific security aspect, respectively. The tables in Appendix represent the statements used for scoring the various assessment parameters under technological and organizational measures. It has been adopted from studies done on other information systems [45, 46].

Once the statements in Tables 4, 5, 6, 7, 8, 9, 10, 11, 12, 13, and 14 in Appendix related to all the five components are rated as per the scoring assessment from Table 1, the total component score is derived by taking the sum of all the statements divided by the maximum score that can be obtained for the respective component. Maximum component score is obtained by multiplying total number of statements with 5 for every component, as the maximum score for each statement can be rated as 5 only. The various technological and organizational measures are summed up to assess the final situation of the E-Governance service system related to information systems security. Equations (1)–(11) show the score summation of the various statements under different components.

$$\text{Hardware security} = \sum_{i=1}^{4} \frac{[\text{Statement}(i)\ \text{score}]_{\text{Table 4}}}{\text{Maximum component score}} \tag{1}$$

$$\text{Software security} = \sum_{i=1}^{16} \frac{[\text{Statement}(i)\ \text{score}]_{\text{Table 5}}}{\text{Maximum component score}} \tag{2}$$

$$\text{Workstation security} = \sum_{i=1}^{5} \frac{[\text{Statement}(i)\ \text{score}]_{\text{Table 6}}}{\text{Maximum component score}} \tag{3}$$

$$\text{Network security} = \sum_{i=1}^{9} \frac{[\text{Statement}(i)\ \text{score}]_{\text{Table 7}}}{\text{Maximum component score}} \tag{4}$$

$$\text{Server security} = \sum_{i=1}^{10} \frac{[\text{Statement}(i)\ \text{score}]_{\text{Table 8}}}{\text{Maximum component score}} \tag{5}$$

$$\text{Data security} = \sum_{i=1}^{13} \frac{[\text{Statement}(i)\ \text{score}]_{\text{Table 9}}}{\text{Maximum component score}} \tag{6}$$

$$\text{Physical \& environment security} = \sum_{i=1}^{9} \frac{[\text{Statement}(i)\ \text{score}]_{\text{Table 10}}}{\text{Maximum component score}} \tag{7}$$

$$\text{Information security policy} = \sum_{i=1}^{12} \frac{[\text{Statement}(i)\ \text{score}]_{\text{Table 11}}}{\text{Maximum component score}} \tag{8}$$

$$\text{Procedures \& controls policy} = \sum_{i=1}^{6} \frac{[\text{Statement}(i)\ \text{score}]_{\text{Table 12}}}{\text{Maximum component score}} \tag{9}$$

$$\text{Administrative tools \& methods} = \sum_{i=1}^{5} \frac{[\text{Statement}(i)\ \text{score}]_{\text{Table 13}}}{\text{Maximum component score}} \tag{10}$$

Table 3 Overall security assessment in E-Governance system can be assessed through the shown grid in the form of table (as calculated using different equations)

	Organizational measures		
Technological measures	Good	Needs improvement	Bad
Very high	Zone A		Zone B
High			
Medium		Zone C	
Low			
Very low	Zone D		Zone E

$$\text{Awareness creation} = \sum_{i=1}^{9} \frac{[\text{Statement}(i)\ \text{score}]_{\text{Table 14}}}{\text{Maximum component score}} \tag{11}$$

$$\begin{aligned}\text{Technological measures} = \text{Average (}&\text{Hardware security, software security,}\\ &\text{workstation security, network security, server security,}\\ &\text{data security, physical \& environment security)}\end{aligned} \tag{12}$$

$$\begin{aligned}\text{Organizational measures} = \text{Average (}&\text{Informationsecurity policy,}\\ &\text{procedures \& controlspolicy, administrative}\\ &\text{tools \& methods, awareness creation)}\end{aligned} \tag{13}$$

Total scores for technological and organizational measures are calculated using Eq. (12) and Eq. (13), respectively. The assessment evaluation, which has been summarized in Table 3, categorizes the position of any E-Governance system setup into strong or weak zone based on the scores of the technological and organizational measures. The total score of both measures is summated and divided into three equal-sized zones for organizational measures (bad, 0–33%; needs improvement, 34–66%; good, 67–100%) and in five equal-sized zones for technological measures (very low, 0–20%; low, 21–40%; medium, 41–60%; high, 61–80%; very high, 81–100%). The division has been done based on minimum and maximum scores that can be achieved in any system. The scores are divided equally into 3 and 5 classes, respectively, for organizational and technological measures.

Zone A is the best zone where E-Governance system is having very good resistance from the various threats and risks, as it is fully prepared with both types of measures. Zone B is a tricky situation for E-Governance system as even being strong on technological measures, the setup is prone to organizational threats which may lead to inefficient working of the whole system. It is a bad situation where a lot of human interaction are required within the service setup as they may destroy the system setup through various organizational measure risks. Zone C is probably a safe state where less critical E-Governance systems will be present which are semi-prepared for the risks on both the fronts. However, this zone may also signify that the security measures are being implemented regularly and will be made strong in upcoming time period. Zone D seems to be valid for a technologically weak and standalone kind of E-Governance setup where all organizational measures are kept strong. But this is a very dangerous situation for E-Governance setup as a weak technological measure against risks and threats can lead to the complete system failure. It is a very risky situation and even riskier in Zone E which denotes a complete failure of a setup. The system in Zone E would be weak with no risk measures being covered up at either technological or organizational level.

Based on the scores obtained for both the technological and organizational measures for the E-Government system's security, appropriate actions can be taken. The given model fulfills the gap for monitoring both internal and external security loopholes as suggested by information systems security theories [9]. Therefore, this model (as adopted from other information systems) can be customized for the E-Governance system setup, and a comprehensive monitoring can be done for analyzing security of such systems.

Moreover, the proposed framework is specific to the developing nations in the context that developed nations have mostly adopted the CMM level 5 in their organizations [47], involving large-scale information systems; therefore, the relevance for the proposed security assessment framework gets diminished. The developed nation's public information systems are already following the highest level of security standards, so the need for another security assessment framework is not majorly required. On the other hand, the developing nations are weak on the security infrastructure and require less complicated (as compared to CMMI level assessment) assessment criterions to measure their system's security. So, the proposed framework would be applicable majorly for the developing nations.

4 Use of Security Assessment in E-Governance Systems

The proposed framework for the security assessment has been formulated with the help of secondary research through existing literature. The proposed assessment framework for E-Governance has been derived from the information systems security assessment for various other information systems like in the field of higher education [45] and library information management [46]. This has been done because the components used by these information systems were found having the standard security framework requirements as per the theoretical models for information system [9]. Even the literature suggests that various technological- and organizational-level security threats must be covered up while forming the assessment framework for any type of information system [48–50]. All the statements and variables were modified in accordance to their relevance to the E-Governance system. Customization has helped the framework to be more relevant for the E-Governance security assessment, and most of the theoretical aspects have been covered up.

The framework validation has been done through case study analysis using the assessment of an organization. The case data covers up hundreds of technical points as per the information systems security components. Audit team was formed involving the principal researcher, director of the organization, and third-party technical analyst. Various analysts from the organization under study were involved as a part of the audit team for the data collection. A total of 15 members were involved for recording the scores and formulating the reports for the analysis. The purchase reports for the infrastructure in the organization and annual reports for the protocol were also studied by the audit team for the extraction of various technical measures. The survey for organizational measure was conducted within the organization through director's interview. The survey of the infrastructure was also conducted within the organization's premise followed by a face-to-face interview. The survey was conducted at various technical areas as per the availability of the relevant infrastructure. The scores were recorded on real-time basis using a scoring sheet and were analyzed using the scoring pattern given in the next section. Due to security concerns, the organization's identity has been kept anonymous.

The framework, suggested in above sections, has been tested for an organization whose identity has been kept anonymous due to security reasons. However, the organization has an E-Governance system (public information system) in place which is exhaustive in nature and has sensitivity toward security breaches. Initially, the case analysis will describe the nature/details of information system so that its magnitude can be understood and then the scores obtained through the suggested framework will be assessed.

The organization, being assessed in this case, is the central unit for information processing in the education industry and acts as a big information system for various branch nodes for information storage, retrieval, and processing. From infrastructure point of view, the organization has got hundreds of servers, processing the information continuously. These servers act as high storage data centers to various nodes, monitoring systems, media management, core operations management, and technology support for various licensed software. From network point of view, the organization is spread across approximately 50 km in radius having connectivity through optical fibers and other mediums of wireless/wired connections. This organization supports more than 100 sub-organizations with over 10,000 humans interacting through this setup on a regular basis. The organization supports both public access and private access to its servers for information processing. The complete setup is managed by more than 200 staff members on daily basis. The Internet speed varies up to 1 GBPS with adequate networking through cloud as well as other infrastructure. With almost all the data points being present in this organization, it can be considered for the security analysis under E-Governance system as this organization deals with citizens directly along with other entities like private institutions, public institutions, and foreign entities.

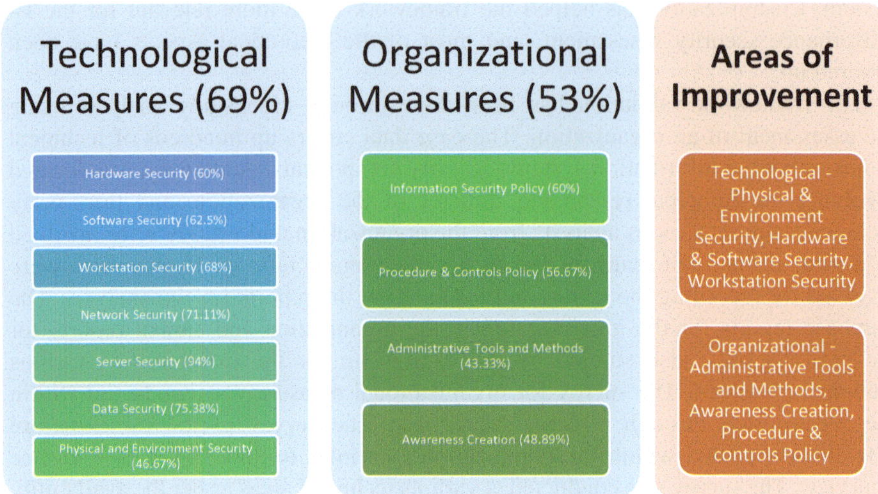

Fig. 8 Scores obtained for each component of the security assessment framework

The scores obtained for the various parameters (as per Eqs. 12 and 13) have been processed and presented in this section. Figure 8 provides the componentwise results for the organization. An organization undertakes the deployment of numerous aspects to ensure a safer and hassle-free environment to its employees. Furthermore, the survey conducted in the current study assesses the different factors such as the implementation of the required hardware and software, an adequate data and network infrastructure, and the server security and administrative tools along with an overall physical and environmental security. Figure 8 depicts that server security is of prime importance and has witnessed the maximum percentage of 94, followed by data security which registered 75.38% and network security which accounted for 71.1%.

Considering the first aspect, i.e., server security, which accounted for the highest percentage, entails different characteristics such as prevention of unauthorized access, implementation of firewalls to protect the systems, detecting the intrusion of malware, and installation of the anti-virus coupled with the maintenance of regular backups for the files. Server security has been rated with the maximum percentage in the organization under the current study. Server security authentication systems are rated excellently, which prevents unauthorized access to information's system server and its implementation for false tolerance, firewalls, intrusion detection, and host auditing software. Organization is also satisfied with the regular updates in the security system of the servers.

The second aspect is data security, witnessing the second highest percentage of 75.38. This aspect considered different features such as keeping a record of the various day-to-day events so that they are easily retraceable. Another aspect considered is the provision of using different forms of external memory devices such as USB, CDs, and DVDs. The features under data security, which were rated excellent on the scoring scale, are password protection for user account and the usage of an anti-virus software. Audit team has given high scores to some of the features which includes proper maintenance of unusual and sensitive media, use of event logging and event management software, fraud detection and prevention measures, use of cryptography techniques, and hardware and software tokens. The overall performance of data security lags behind the top aspect because usage of RFID tags for managing and securing information system has not been properly implemented.

The third aspect is the network security, which exhibited 71.1% and includes different features such as the implementation of the firewall and anti-viruses to prevent any malicious activity and use of digital signatures for ensuring the authenticity of electronic documents. It also ensures security by putting a limit on connection time to restrict the access of high-end applications and databases and separates the cables for public and staff local area networks. The features rated high on the survey questions were related to anti-virus security and desktop security software, use of firewall for various purposes, separation of public and staff LANs, and use of wireless security products. Segmentation of network with a router, limiting the connection time, and use of digital signatures were given average scores.

Workstation security is at the fourth position with 68%. The features entailed are application firewall for connecting mobile laptops to information system's LAN. This parameter also facilitated virus protection and security software programs on web browsers and email programs. Each information system's workstations, laptop screensavers, authentication of networks, office software and browsers of workstations, and configuration of laptops to receive updates in a timely manner were also rated average under this aspect. The features contributing for an overall high percentage of workstation security are use of office productivity software and browsers of the workstations and laptops, application firewall and installation of virus protection programs, configuration settings, and security software programs. The use of user identification and authentication before logging into the information system's workstation, laptop screensavers, or network has not been rated satisfactorily and, hence, needs to be improved.

Software security exhibits the fifth position with a score of 62.5%. Software for security purposes are anti-spy, anti-phishing, desktop security, ID management, rollback web filtering, and spam filtering. Furthermore, software for other uses, such as menu replacement software for replacing the standard windows desktop interface and providing a control over timeouts, logging and browsing and timer software to control the amount of time a patron, can use a workstation. The software elements that scored high are anti-spy software, software related to desktop security, and web filtering software. The software elements which were given average scores are cleanup software management, ID management software, menu replacement software, and unavailability of timer software on a regular basis.

Hardware security witnessed the sixth position with a percentage of 60. The features involved under this aspect are installation of CCTV and visual camera, magnetic detection and electronic theft system at strategic places, and public computers and server areas in information system. For emergency purposes, does it have power sources and alternate communication line? To backup drives within the information system, does the system have a periodic remote monitoring and file monitoring facilities? High ratings were given to the way locks, security cables, locked cable trays, metal cages, emergency power sources, and alternate commination lines are implemented. The performance of CCTV camera and visual camera were given average scores. The performance of hardware security lags because of unavailability of magnetic detection system and electronic theft system in the organization in an advanced form.

Information security policy also holds the sixth position. Various policies are listed under this parameter. Data-related policies for information system are backup and offsite storage policies; data classification, retention, and destruction policies; and secured disposal policies. Furthermore, there are policies for other uses such as policies for sharing, storing, and transmitting of information system data and policies for managing privacy and confidential issues. The policies for protecting individual's hardware and software and secured disposal policies were scored high. Audit team was also satisfied with job responsibility policy; identity management policy; policies on accepted usage of laptops, hand phones, and other wireless devices in the setup environment; policies on accepted usage of workstations, emails, databases, intranet, and Internet in information system; and policies related to sharing, storing, and transmitting of information system data by ISP. The performance of information security policy lags behind because of weak implementation of data backup and offsite storage policies; data classification, retention, and destruction policies; and policies related to reporting, notification, and response of information systems security events.

Procedures and controls policy holds the seventh position with 56.67%. It contains various procedures, such as control and disciplinary procedures applicable if an information system staff or patron breaches the IS security policies or rules and procedures for non-disclosure or confidential agreement for protecting confidential and propriety information. Procedures for updating and reviewing existing information security policies are also listed under this parameter. The policy related to non-disclosure or confidentially agreement gained high scores. Audit team was satisfied with the procedures related to control and discipline, handling sensitive and personal data of patrons and intellectual property, and copyrights to control and protect digital works. The performance of the parameter lags behind due to the low scores being given to the procedures listing all the requirements of outsourcing an information system service or activity and for updating and reviewing existing information systems security policies. The minimum percentages are witnessed by awareness creation (48.89%), physical and environment security (46.67%), and administration tool and methods (43.33%).

The various features for physical and environment security are air conditioners, earthquake warning system, flood detectors, lightning and surge protectors, and security guards, automatic sprinkler system, smoke detectors, fire extinguishers, and fireproof installations in information system building. Furthermore, it includes magnetic stripe, swipe cards, electronic lock, proximity cards, and barcode cards to have a secure and control access to restricted information system areas. The initial level of security is served by warning signs, fencing, vehicle height restrictors, site lightning, and trenches around the information system area. Audit team gave high scores to the way security guard monitored people entering and leaving the organization's building. Also, automatic sprinkler systems, smoke detectors, fire extinguishers, and fireproof installations were well implemented. The performance of physics and environment security lags behind majorly because of unavailability of earthquake early warning system, flood detectors, and lightning and surge detectors. Furthermore, majority of the user identification and authentication forms at the entrance, exit, and other public access areas in the organization were not well implemented.

The various features of awareness creation are related to the awareness in the staff to protect information systems security and training of the staff members for reporting security breach incidences. All the staff and patrons are provided with appropriate security training and education, receive regular updates on information system's policies and procedures, and are trained to monitor and handle information system on their own. KPI provides real insights to have effective security awareness programs. Audit team was satisfied with the way people were made aware of their responsibilities to protect information systems security and were trained to report security breach incidences and monitoring and handling information security on their own. Physical environment and security has various shortcomings such as staff and patrons don't receive approximate information security training and education and don't receive regular updates on information system's policies and procedures, risk assessment approach is not well implemented, and KPI doesn't provide real insights resulting in ineffective security awareness programs.

Administration tool and methods have various procedures too. Some of them are procedures for owner accountability to ensure appropriate protection in information system, procedures to protect information system from all types of threats, and procedures listing all the requirements regarding information system service or activity outsourcing. The procedure for owner accountability was averagely rated. The low percentage of this parameter is due to weak implementation of procedures for the development and implementation of risk analysis; procedures for handling, reporting, notification, and response of IS security events to affected parties; and procedures related to asset classification. Furthermore, internal and external audit programs were not well implemented.

From the scoring assessment, it was found that the overall score for technological measure is 69% and organizational measure is 53%. As per the grid developed in Table 3, it is derived that the current organization most closely falls under Zone B, as shown in Fig. 10. The various components of technological measure are strongly implemented, which can be further improved if physical and environment- and hardware-related security standards can be strengthened. The various components of organizational measure, which are falling in the average range, can be further improved. Administrative tools and awareness campaigns can be developed within the organization.

The total score obtained for various components and the measures (Fig. 9) were verified by the director of the organization and third-party analyst. The mentioned

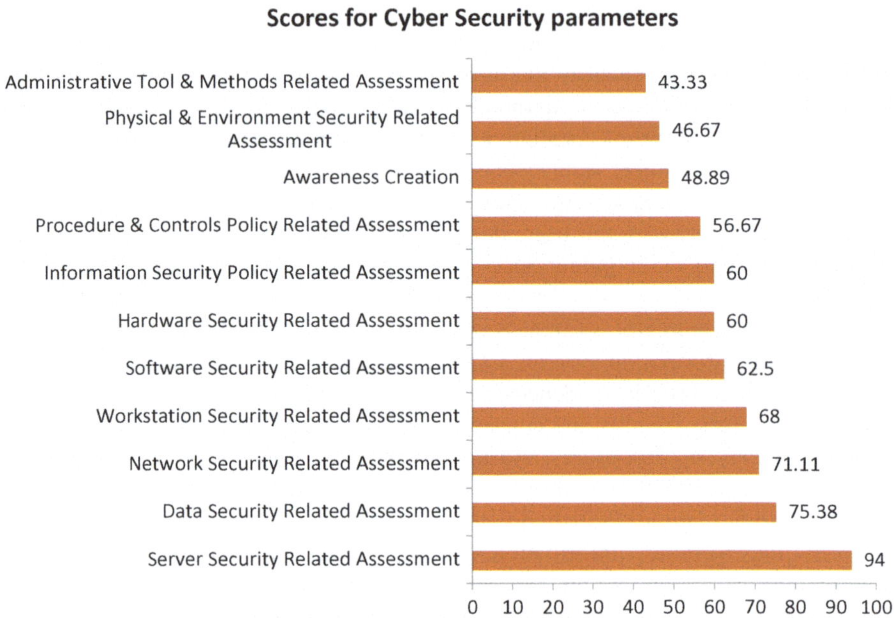

Fig. 9 Scores for various parameters of the security assessment

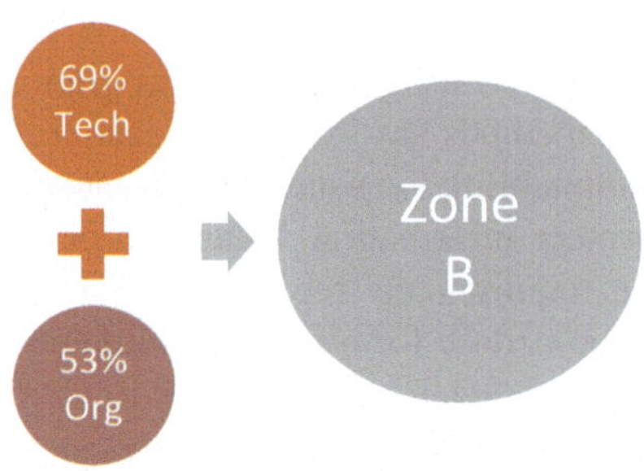

Fig. 10 Zone identification of the organization

two persons were not aware of the scoring pattern and were only told the results for the security components. The consensus was achieved by the team of analysts with respect to the various scores achieved during the analysis. Therefore, the results obtained from the framework of the researcher were validated by the members of the organization and third-party technical analyst.

Given the stature of the organization, the technological measure is nicely placed, but organizational measure can be improved on an immediate basis. However, since the role of the organization is more toward the backend support than frontend processing, low score of organizational measure can be justified in the current case. Better organizational planning will lead to better security coverage of the information system in the public domain.

5 Concluding Remarks

The main aim of the current study was to prepare a framework useful for cyber security education and assessment of information systems security concerns in E-Governance. The framework suggested in the above sections has covered various internal and external aspects of the organization's security to minimize the risk. The components are important from theoretical point of view and are in sync with the frameworks available in the literature. Therefore, the aim of the study has been fulfilled using case study analysis on the proposed framework, and exhaustive coverage has been given to the E-Governance system's security.

With rising concerns of security in the E-Governance environment, it becomes important to audit the system regularly and keep strict check over the information systems security standards. The proposed framework for security assessment seems to be advantageous as it covers the various aspects of security. Moreover, the components are in sync with other assessment models as proposed by various researchers in the past [13, 17]. The current model is exhaustive in the sense that a lot of aspects related to information systems security are covered. This will give the security audit team a ready-to-use questionnaire to assess the level of technological and organizational measures. These measures will protect the system from both internal and external threats. In case of emergency and threats, the overall system will become strong to combat such risks. Currently, one large organization (covering up many small organizations) under the E-Governance system security has been presented. However, the next stage of the research is to collect relevant data from other organizations on the measures mentioned in the proposed framework and check the status of various organizations and institutions involved in the setup of E-Governance. This will help in analyzing the security status of the various E-Governance systems, and hence it would be easier for the stakeholders to take necessary actions.

This model is valid and applicable for all the developing nations as most of these nations are still in the process of setting up their E-Governance system. And referring to such exhaustive framework will provide them confidence in combatting the various types of information systems security threats. The framework can be used by various government officers for E-Governance system, researchers, and industry practitioners to prepare security-related products in E-Governance and public information systems. The framework can be developed into an automated tool, covering self-answered questions in a survey form, for which data can be analyzed using the scoring pattern automatically. The simplicity of the questions will help the users to fill the form themselves and assess the pros and cons of their E-Governance system's security. This will make the institutions self-dependent and self-aware of the situation. The framework can also be used by a third-party security agency to provide security certificates to the various E-Governance systems. Therefore, this study can be a useful contribution under information systems security in E-Governance.

Appendix

Table 4 Hardware security component in E-Governance system

	Hardware security-related assessment	Scoring	Area of improvement
A	CCTV, visual camera, magnetic detection system, and electronic anti-theft system at strategic places, public computer areas, and server areas in E-Governance system		
B	Emergency power sources and alternative communication lines within E-Governance system		
C	Locks, security cables, locked cable trays, metal cages, or anchoring devices to improve the security of hardware equipments in E-Governance system		
D	Periodical remote mirroring or file mirroring to back up disk drives within E-Governance system		

Table 5 Software security component in E-Governance system

	Software security-related assessment	Scoring	Area of improvement
A	Anti-spyware software to detect and remove any threats		
B	Anti-phishing solutions to prevent phishing attacks		
C	Cleanup software to erase files left behind by a user		
D	Desktop security software at application and operating level to monitor, restrict usage, or disable certain features of the workstations with E-Governance system		
E	Automate the process of installing an application or updates to workstations on a network of E-Governance		
F	ID management software to automate administrative tasks such as resetting user passwords and enabling users to reset their own passwords in E-Governance system		
G	Menu replacement software to replace the standard windows desktop interfaces and provide control on timeouts and logging and browsing activities		
H	Multi-user operating systems and application software to allow concurrent access by multiple users of a computer		
I	Automatic debugging and tests to remove any defects from new software or hardware components		
J	Rollback software to keep track and record any changes made to the computers and allow the system to be restored to its original state from any chosen point in time		
K	Single sign-on system for user authentication and authorization to access all computers and systems without the need to enter multiple passwords		

(continued)

Table 5 (continued)

	Software security-related assessment	Scoring	Area of improvement
L	Spam filtering software to detect unwanted spam emails from getting into users' inboxes		
M	Systems recovery to rebuild and repair the library computer systems after disaster or crash		
N	Timer software to control the amount of time a patron can use a workstation		
O	User entrance log to record and monitor user logs which are regularly analyzed		
P	Web filtering software to prevent access to inappropriate materials or sites		

Table 6 Workstation security component in E-Governance system

	Workstation security-related assessment	Scoring	Area of improvement
A	All office productivity software and browsers for the workstations/laptops are configured to receive updates in a timely manner		
B	An application firewall is used for mobile laptops that connect to the E-Governance system's LAN		
C	The computer's BIOS are secured in order to create a secure public access computer		
D	User identification and authentication are required before logging into the E-Governance system's workstations, laptops screensavers, or network		
E	Virus protection programs, configuration settings, and security software programs are installed for web browsers and email programs		

Table 7 Network security component in E-Governance system

	Network security-related assessment	Scoring	Area of improvement
A	Anti-virus software and desktop security software to receive regular updates to protect the internal network from any security breaches		
B	Digital signatures are used to assure the authenticity of any electronic documents sent via the E-Governance system's network (use of passwords, private and public key encryption, or digital certificates)		
C	Firewall to protect internal network from external threats		
D	Firewall with virtual private network is installed for remote and wireless access connections		
E	Limitation of connection time is performed via configuration routines to control and restrict access for the E-Governance system's high-risk applications or databases		

(continued)

Table 7 (continued)

	Network security-related assessment	Scoring	Area of improvement
F	Public and staff's local area networks (LANs) are physically separated by means of separate cabling for each network to provide alternative circuit		
G	Server segregation/perimeter network (DMZ) by using firewalls and some other network access control devices to separate systems that are at a relatively high risk from unsecured network		
H	The network is segmented with a router to increase the bandwidth available to each user and reduce congestions/collisions of the E-Governance system's network		
I	Wireless security products to secure the E-Governance system's wireless network (use of default passwords on wireless access points, network ID, wireless intrusion detection systems, wired equivalency protocol (WEP) encryption, MAC address filtering, or VPN)		

Table 8 Server security component in E-Governance system

	Server security-related assessment	Scoring	Area of improvement
A	Anti-virus software on servers and anti-virus definition files are kept up-to-date		
B	Authentication systems to prevent unauthorized access to the E-Governance system's server		
C	Implement fault tolerance to ensure if one system fails, then there is a backup system that immediately takes over		
D	Firewalls to protect the E-Governance system's network from unwarranted intrusion		
E	Intrusion detection software and host auditing software are installed to monitor servers for signs of intrusion		
F	Regular backups for the data, hard copy of server hardware specifications, installation information, installation software, and passwords are regularly performed and stored at an offsite location		
G	Server logs are reviewed periodically by using a log file monitor utility to monitor any signs of intrusion or security violations		
H	Restrict access to the file system in a server to the directory structure using file or directory permissions		
I	E-Governance system servers' operating systems and applications are hardened to protect from any vulnerabilities		
J	The server is placed in a secure location, such as in a lockable cage, a locked room, and a place with environmental controls		

Table 9 Data security component in E-Governance system

	Data security-related assessment	Scoring	Area of improvement
A	Attributes for each removable media applications in E-Governance system are properly recorded, and the media are kept from any unauthorized devices from accessing, running, or transferring data to E-Governance system workstations and network (USB thumb drives, tapes, CDs, DVDs, disks, drives, etc.)		
B	Combination of authentication systems to restrict access to E-Governance system's data and resources based on a variety of access rights (user identification, passwords, or biometrics system)		
C	Dispose of unused media and sensitive media is properly managed to maintain an audit trail		
D	Enforced path is created between a user terminal and other E-Governance system services to reduce the risk of unauthorized access		
E	Event logging or log management software to ensure the E-Governance system computer security records are stored in sufficient detail for an appropriate period of time (records for security incidents, policy violations, fraudulent activities, and operational problems)		
F	Fraud detection and prevention measures to control fraudulent activity and disclosure of information (use of address verification system, proprietary encryption, internal intrusion detection system, multiple login monitoring, password verification on transactions, or data access controls)		
G	Public key infrastructure (PKI) to secure the exchange of personal data via the E-Governance system network and internet (use of public and private cryptography key pair)		
H	RFID tags to manage and secure the E-Governance system collection as well as to track attendance and prevent unauthorized access into the system premise		
I	Systematic approaches conducted in-house or outsourced to a service provider to address the E-Governance system's vulnerabilities (vulnerability discovery, prioritization, remediation, dynamic protection, verification, and customized reporting)		
J	Use of cryptography techniques [20], hardware and software tokens, and single sign on systems to control data access to E-Governance system's internal and remote computer systems		

(continued)

Table 9 (continued)

	Data security-related assessment	Scoring	Area of improvement
K	Use of password protection of user accounts, anti-virus software, firewalls, wireless network protections, intrusion detection systems, and Internet Protocol Virtual Private Networks/IP VPNs to ensure data insert and sent from one end of a transaction arrives unaltered at the other end		
L	Vital E-Governance system's information or records are regularly backed up		
M	Web access management systems to manage and validate user access to devices, applications, and E-Governance systems (authentication management, single sign-on convenience, audit or reporting systems)		

Table 10 Physical and environment security component in E-Governance system

	Physical and environmental security-related assessment	Scoring	Area of improvement
A	Air conditionings to stabilize the temperature and humidity within the E-Governance system building		
B	Earthquake early warning system to alert E-Governance system staff and patrons prior to damaging ground shaking		
C	Flood detector to provide an early warning of developing floods in an E-Governance system		
D	Lightning protectors and surge protectors to protect any valuable machines or equipments from lighting strikes, voltage spikes, and surges		
E	Security guards to monitor people entering and leaving the E-Governance system buildings and sites		
F	Use of automatic sprinkler systems, smoke detectors, fire extinguishers, and fireproof installations in the E-Governance system buildings and areas adjacent to system's key assets to detect and prevent fires, toxic chemical spills, and explosions		
G	Use of magnetic stripe swipe cards, electronic lock, proximity cards, barcode card, or biometrics to secure and control access to restricted E-Governance system areas		
H	Warning signs, fencing, vehicle height restrictors, site lightings, and trenches around the E-Governance system areas to provide initial layer of security for a system premise		
I	Wireless gates, biometrics, or other user identifications and authentication forms at the system premise main entrances, exits, and public access areas in the E-Governance system		

Table 11 Information security policy component in E-Governance system

	Information security policy-related assessment	Scoring	Area of improvement
A	Backups and offsite storage policies for E-Governance system data, media, or materials that contain sensitive information		
B	Data classification, retention, and destruction policies for E-Governance system data, media, or materials that contain sensitive information		
C	Identity management policies for E-Governance system user registration and password management		
D	Job responsibility policy for individual employee responsibilities related to the E-Governance system IS security practices		
E	Policies on access control, authentication, and authorization practices for using the E-Governance systems		
F	Policies on protection of E-Governance system assets to protect your system's hardware, software, data, and people		
G	Secure disposal policies for E-Governance system data, media, or materials that contain sensitive information		
H	Polices on reporting, notification, and response of information systems security events to affected parties such as individuals, law enforcement, campus, or parent organizations		
I	Policies on acceptable use of wireless devices in E-Governance system such as laptops and hand phones		
J	Policies on acceptable use of workstations, e-mails, databases, intranet, and Internet in E-Governance system		
K	Policies on managing privacy and confidentiality issues, including breaches of personal information		
L	Policies on sharing, storing, and transmitting of E-Governance system data via ISPs, external networks, or contractors' systems		

Table 12 Procedures and controls policy component in E-Governance system

	Procedures and controls policy-related assessment	Scoring	Area of improvement
A	Controls and disciplinary procedures if an E-Governance system staff or patrons breach the IS security policies or rules (verbal warning, written warning, suspension, and dismissal)		
B	Procedures for handling sensitive data and personal data of patrons to prevent errors, unauthorized disclosure, or misuse by those who handle it		
C	Procedures for non-disclosure agreement or confidentiality agreement to all staff and patrons to protect any type of confidential and proprietary information		
D	Procedures for update and review existing information security policies		
E	Procedures on the intellectual property rights and copyrights in controlling and protecting any digital works		
F	Procedures which list all requirements with regard to outsourcing any E-Governance system service or activities		

Table 13 Administrative tool and methods in E-Governance system

	Administrative tool and methods-related assessment	Scoring	Area of improvement
A	Procedure for owner accountability to ensure appropriate protection is maintained for each library E-Governance system asset (e.g., information assets, software assets, physical assets, and library services)		
B	Procedures for the development and implementation of risk analysis to protect E-Governance system from all types of threats (performance of assets analysis, threat analysis, annual loss expectancy analysis, identification and evaluation of security measures)		
C	Procedures on handling, reporting, notification, and response of IS security events to affected parties such as individuals, law enforcement, campus, or parent organization		
D	Procedures related to asset classification in order to organize it according to its importance and sensitivity to loss (unclassified, confidential, secret, and top secret)		
E	Regular internal and external audits programs appropriate for E-Governance system size, complexity of activities, scope of operations, risk profile, and compliance with the relevant standards		

Table 14 Awareness creation component in E-Governance system

	Awareness creation-related assessment	Scoring	Area of improvement
A	All staff and patrons at various levels are made aware of their responsibilities with regard to protecting the E-Governance system's security and trained to report any security breach incidences		
B	All staff and patrons at various levels receive appropriate information security trainings and education		
C	All staff and patrons at various levels receive regular updates on E-Governance system's policies and procedures		
D	Information security awareness trainings have become mandatory to all staff and patrons at various levels		
E	Risk assessment approach exists and follows a defined process that is documented		
F	Staff and patrons at various levels are trained to monitor and handle the E-Governance system on their own		
G	There are balanced set of key performance indicators (KPIs) and metrics used to provide the real insight into the effectiveness of security awareness programs		
H	There are positive supports and commitments from the top management to coordinate the implementation of E-Governance system's security controls (e.g., via allocation of budget, strong interest, and active involvements)		
I	Threats that could harm and adversely affect critical operations of E-Governance system's security are identified and updated regularly		

References

1. United Nations, *E-Governance survey*, 2014. https://publicadministration.un.org/egovkb/Portals/egovkb/Documents/un/2014-Survey/E-Gov_Complete_Survey-2014.pdf. Accessed 15 Mar 2016
2. S. Singh, D.S. Karaulia, E-governance: information security issues, in *Proceedings of the International Conference on Computer Science and Information Technology, India*, 2011, pp. 120–124. http://www.academia.edu/download/38526006/77_1211468.pdf.
3. R. Gupta, S.K. Pal, S.K. Muttoo, Analysis of information systems security for E-governance in India, in *National Workshop on Cryptology*, (DESIDOC, DRDO & CRSI, Delhi, 2013), pp. 17–25
4. R. Gupta, S.K. Pal, S.K. Muttoo, Review based security framework for E-governance services. Chakravyuh DRDO **11**(1), 42–50 (2015)
5. R. Gupta, S.K. Pal, S.K. Muttoo, Network monitoring and internet traffic surveillance system: issues and challenges in India, in *Intelligent Systems Technologies and Applications*, (Springer International Publishing, New York, 2016), pp. 57–65. https://doi.org/10.1007/978-3-319-23258-4_6

6. A. Miller, R. Horne, C. Potter, *Information Security Breach Survey* (Pricewaterhouse Coopers, London, 2016)
7. S.K. Muttoo, R. Gupta, S.K. Pal, Analysing security checkpoints for an integrated utility-based information system, in *Emerging Research in Computing, Information, Communication and Applications*, (Springer, Singapore, 2016), pp. 569–587. https://doi.org/10.1007/978-981-10-0287-8_53
8. M. Stamp, *Information Security: Principles and Practice* (Wiley, Hoboken, 2011)
9. N. Godbole, *Information Systems Security: Security Management, Metrics, Frameworks and Best Practices (With CD)* (Wiley, Hoboken, 2008)
10. R. Gupta, S.K. Muttoo, S.K. Pal, Proposal for integrated system architecture in utilities, in *Proceedings of the Advances in Computing, Communications and Informatics (ICACCI) IEEE*, 2014, pp. 1995–1998. doi: https://doi.org/10.1109/ICACCI.2014.6968652
11. K.D. Loch, H.H. Carr, M.E. Warkentin, Threats to information systems: today's reality, yesterday's understanding. MIS Q. **16**, 173–186 (1992). https://doi.org/10.2307/249574
12. M.E. Whitman, In defense of the realm: understanding the threats to information security. Int. J. Inf. Manag. **24**(1), 43–57 (2004). https://doi.org/10.1016/j.ijinfomgt.2003.12.003
13. A. Da Veiga, J.H. Eloff, A framework and assessment instrument for information security culture. Comput. Secur. **29**(2), 196–207 (2010). https://doi.org/10.1016/j.cose.2009.09.002
14. J. Rees, S. Bandyopadhyay, E.H. Spafford, PFIRES: a policy framework for information security. Commun. ACM **46**(7), 101–106 (2003). https://doi.org/10.1145/792704.792706
15. L. Sun, R.P. Srivastava, T.J. Mock, An information systems security risk assessment model under the Dempster-Shafer theory of belief functions. J. Manag. Inf. Syst. **22**(4), 109–142 (2006). https://doi.org/10.2753/MIS0742-1222220405
16. A. AlHogail, Design and validation of information security culture framework. Comput. Hum. Behav. **49**(1), 567–575 (2015). https://doi.org/10.1016/j.chb.2015.03.054
17. C. Sillaber, R. Breu, Using business process model awareness to improve stakeholder participation in information systems security risk management processes, in *Wirtschafts Informatik*, 2015, pp. 1177–1190. http://www.wi2015.uni-osnabrueck.de/Files/WI2015-D-14-00044.pdf.
18. A. Joshi, H. Tiwari, Security for E-governance. J. Inf. Oper. Manag. **3**(1), 254 (2012). http://furooshgah.ir/wp-content/uploads/2016/12/SECURITY-FOR-E-GOVERNANCE.pdf
19. A. Roy, S. Karforma, A survey on E-governance security. Int. J. Comp. Eng. Comp. Appl. **8**(2), 50–62 (2011)
20. H. Singh, A.K. Kar, P.V. Ilavarasan, Assessment of e-governance projects: an integrated framework and its validation, in *Proceedings of the Special Collection on eGovernment Innovations in India*, (ACM, New York, 2017), pp. 124–133. https://doi.org/10.1145/3055219.3055228
21. V. Singh, G. Singh, Citizen centric assessment framework for e-governance services quality. Int. J. Business Informat. Syst. **27**(1), 1–20 (2018). https://doi.org/10.1504/IJBIS.2018.088568
22. A. Mateen, S. Sabir, K. Ullah, A development of hybrid framework for E-Government. arXiv preprint arXiv:1702.02442 (2017). https://arxiv.org/ftp/arxiv/papers/1702/1702.02442.pdf.
23. S.L. Kim, T.S. Teo, A. Bhattacherjee, K. Nam, IS auditor characteristics, audit process variables, and IS audit satisfaction: an empirical study in South Korea. Inf. Syst. Front. **19**(3), 577–591 (2017). https://doi.org/10.1007/s10796-015-9612-z
24. Y.N. Chen, H.M. Chen, W. Huang, R.K. Ching, E-government strategies in developed and developing countries: an implementation framework and case study. J. Glob. Inf. Manag. **14**(1), 23–46 (2006). https://doi.org/10.4018/jgim.2006010102
25. G. Dhillon, J. Backhouse, Technical opinion: information system security management in the new millennium. Commun. ACM **43**(7), 125–128 (2000). https://doi.org/10.1145/341852.341877
26. K. Prasad, E-governance policy for modernizing government through digital democracy in India. J. Inf. Policy **2**, 183–203 (2007). https://doi.org/10.5325/jinfopoli.2.2012.0183

27. G. Mitchell, A. May, A. McDonald, PICABUE: a methodological framework for the development of indicators of sustainable development. Int. J. Sustain. Dev. World Ecol. **2**(2), 104–123 (1995). https://doi.org/10.1080/13504509509469893
28. D.H. Meadows, Indicators and information systems for sustainable development, in *A Report to the Balaton Group, The Sustainability Institute*, 1998. https://pdfs.semanticscholar.org/3372/06350e14a75581b88550fadfd0b39d144d87.pdf. Accessed 25 Jan 2017
29. R.T. Watson, G.G. Kelly, R.D. Galliers, J.C. Brancheau, Key issues in information systems management: an international perspective. J. Manag. Inf. Syst. **13**(4), 91–115 (1997). https://doi.org/10.1080/07421222.1997.11518144
30. C. Harland, L. Knight, R. Lamming, H. Walker, Outsourcing: assessing the risks and benefits for organisations, sectors and nations. Int. J. Oper. Prod. Manag. **25**(9), 831–850 (2005). https://doi.org/10.1108/01443570510613929
31. S. Basu, E-government and developing countries: an overview. Int. Rev. Law Comput. Technol. **18**(1), 109–132 (2004). https://doi.org/10.1080/13600860410001674779
32. T. Almarabeh, A. AbuAli, A general framework for e-government: definition maturity challenges, opportunities, and success. Eur. J. Sci. Res. **39**(1), 29–42 (2010). http://unpan1.un.org/intradoc/groups/public/documents/apcity/unpan045348.pdf
33. H.J. Liao, C.H.R. Lin, Y.C. Lin, K.Y. Tung, Intrusion detection system: a comprehensive review. J. Netw. Comput. Appl. **36**(1), 16–24 (2013)
34. J.P. Anderson, *Computer Security Threat Monitoring and Surveillance (Vol. 17), Technical Report* (James P. Anderson Company, Fort Washington, PA, 1980)
35. D.E. Denning, An intrusion-detection model. IEEE Trans. Softw. Eng. **13**(2), 222–232 (1987)
36. A.M. Chandrashekhar, K. Raghuveer, Performance evaluation of data clustering techniques using KDD Cup-99 intrusion detection data set. Int. J. Inform. Netw. Secur. **1**(4), 294–305 (2012)
37. C.F. Endorf, E. Schultz, J. Mellander, *Intrusion detection & prevention* (McGraw-Hill Osborne Media, Osborne, 2004)
38. X. Wang, S. Chen, S. Jajodia, Tracking anonymous peer-to-peer VoIP calls on the internet, in *Proceedings of the 12th ACM conference on computer and communications security*, (ACM, New York, 2005), pp. 81–91
39. S. Kaplantzis, N. Mani, M. Palaniswanmi, G. Egan, Security Models for Wireless Sensor Networks, PhD Conversion Report, Monash University, Australia, 2006
40. S. Rathore, A. Saxena, M. Manoria, Intrusion detection system on KDDCup99 dataset: a survey. Int. J. Comp. Sci. Informat. Technol. **6**(4), 3345–3348 (2015)
41. R. Bhattacharya, Indian companies faced cyber-attack in 2015: KPMG survey, *The Economic Times* (2015). http://articles.economictimes.indiatimes.com/2015-12-01/news/68688315_1_cyber-risks-cyber-forensicskpmg-survey. Accessed 15 Jan 2016
42. A.M. Sukumar, C. R. Sharma, *The Cyber Command: Upgrading India's National Security Architecture* (2016). http://www.orfonline.org/wp-content/uploads/2016/03/SR_9_Arun-Mohan-Sukumar-and-RK-sharma.pdf. Accessed 15 Sep 2016
43. R. Caceres, N. Duffield, A. Feldmann, J.D. Friedmann, A. Greenberg, R. Greer, J.E. van der Memle, Measurement and analysis of IP network usage and behavior. Commun. Mag. IEEE **38**(5), 144–151 (2000)
44. S. Das, 9 Cybersecurity policies & initiatives by Indian Govt in 2019 (2019). https://analyticsindiamag.com/9-cybersecurity-policies-initiatives-by-indian-govt-in-2019/. Accessed 15 Mar 2020
45. ISG-IHE, Information security governance assessment tools for higher education, 2005. https://net.educause.edu/ir/library/pdf/SEC0421.pdf. Accessed 12 Mar 2016
46. R. Ismail, A.N. Zainab, Information systems security in special and public libraries: an assessment of status, 2013. https://arxiv.org/ftp/arxiv/papers/1301/1301.5386.pdf.
47. P. Brudenall, *Technology and Offshore Outsourcing Strategies* (Palgrave Macmillan, Basingstoke, 2005). https://doi.org/10.1057/9780230518568

48. H. Berghel, The two sides of RoI: return on investment vs. risk of incarceration. Commun. ACM **48**(4), 15–20 (2005). https://doi.org/10.1145/1053291.1053305
49. C. Sundt, Information security and the law. Inform. Secur. Technol. Represent. **1**(1), 2–9 (2006). https://doi.org/10.1016/j.istr.2005.11.003
50. B. Von Solms, Information security—the third wave. Comput. Secur. **19**(7), 615–620 (2000). https://doi.org/10.1016/S0167-4048(00)07021-8

A Survey on the Effectiveness of the Secure Software Development Life Cycle Models

Jing-Chiou Liou and Saniora R. Duclervil

Introduction

Cybersecurity has become prevalent and important in today's society. Some recent cybersecurity incidents have made many people victims of cybercrimes. For an example, Equifax underwent a data breach in 2017 [1], and over 143 million Americans became victims of the data breach. Information stolen included credit card numbers and social security numbers. In 2019, 2 billion records were exposed in a massive smart home device breach with the Chinese-based company, Orvibo [2]. In December 2019, a home security camera company, Ring, suffered from a security breach that leaks sensitive information belonging to over 3600 customers [3]. There are also reports that the Ring security cameras were hacked and controlled by cyber predators [3].

Two primary factors contribute to the modern cybersecurity problems. One is the connectivity, and another is the software. In the Internet era, almost every computing device connects to some sort of networks and eventually connects to the Internet. These Internet-connected devices provide convenient applications and services and have become an integrated part of our lives. Disconnecting from the Internet to reduce the cybersecurity risks is not a realistic option for many.

Therefore, to tackle the cybersecurity issues, we have to focus on software security. Software security is about understanding the software-induced security risks and how to manage them. To manage software security well, we need to comprehend the process of designing, building, and testing software for security.

J.-C. Liou (✉) · S. R. Duclervil
School of Computer Science and Technology, Kean University, Union, NJ, USA
e-mail: jliou@kean.edu; duclersa@kean.edu

K. Daimi, G. Francia III (eds.), *Innovations in Cybersecurity Education*,
https://doi.org/10.1007/978-3-030-50244-7_11

Software Development Life Cycle

Information technology project management uses a process to develop a software application by planning, designing, executing, and monitoring that application in order to meet organizational needs. This type of process or model is called the System Development Life Cycle or Software Development Life Cycle (SDLC). The Software Development Life Cycle is a framework that defines the process used by organizations to build an application from its inception to decommission. The SDLC can take months or years to complete.

There are quite a few different models of SDLC proposed and used in the IT industry. The two most popular ones are Predictive and Iterative. Based on their characteristics and the technology advancements, each is used in industry sectors in the past and current time. Regardless of the type of SDLC models, most divide the development process into phases or stages through which software is developed. The objectives of each phase are to:

- Define the work items and steps that will be performed.
- Produce deliverables.
- Ensure the deliverables meet the quality criteria.

At the end of each phase, a quality gate or milestone is used to verify the completion of all required works, and deliverables are produced as expected. For an example, at the end of Requirement Analysis phase, before the project can move into System Design phase, a Software Requirements Document (SRD) will be reviewed and baselined. Project stakeholders, including project team and end users or customers, will then sign off a quality gate document that indicates officially the end of such phase. A successful review of the quality gate marks the time to move into the next phase.

Predictive Model

The predictive life cycle is used when the scope of the project can be clearly articulated in advance of the project beginning, and the schedule and cost can be predicted. In the early era of computing, this was a feasible and preferable process to develop a software: As this model focuses on concept planning and the design of software before its implementation, the predictive process model provides more confidence to the software development team on the success of software project.

One of the most widely used models in this category is the Waterfall. The Waterfall process was initially proposed by Benington [4] in 1956 and later refined by W. W. Royce in 1970 [5]. In his original proposal, there are seven phases in the process:

- Systems Requirements.
- Software Requirements.

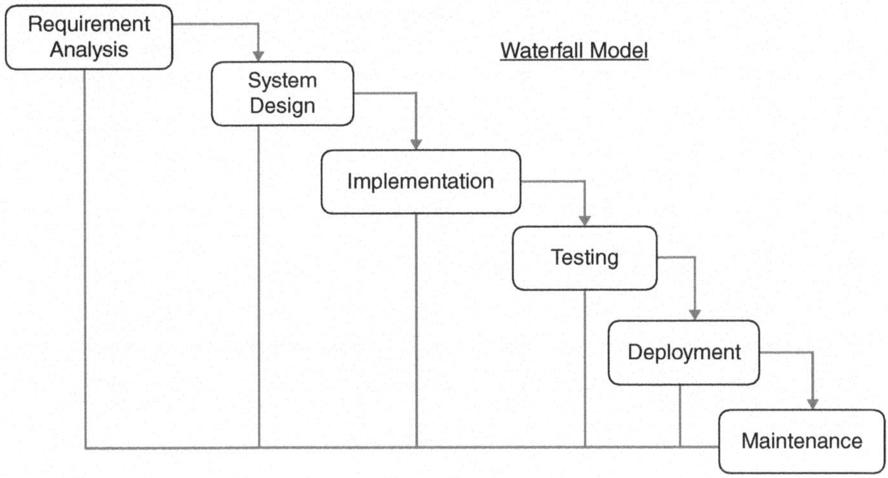

Fig. 1 Waterfall model

- Analysis.
- Program Design.
- Coding.
- Testing.
- Operations.

However, through the evolution in software engineering, variations in the phases were developed in the past. As shown in Fig. 1, the most recent accepted version combines both "Systems Requirements" and "Software Requirements" into a single "Requirement phase" phase. The "Analysis" and "Program Design" were also integrated into just a "Design" phase. Some change the "Coding" phase into an "Implementation" or "Development" phase. People also use "Verification" in place of "Testing." Finally, the "Deployment" phase is also added into the process to distinguish the efforts taken to install the software into the production environment.

The Waterfall model was used heavily in the late twentieth century in the software industry. There are other variations of Waterfall models. This rigid, linear, and inflexible process has declined in its popularity to the more adaptive process models in the early twenty-first century. Nonetheless, it is still used by some of the software projects that are not web applications today.

Iterative Models

There are many process models proposed under the category of iterative SDLC. One that is getting attention is the Spiral model proposed by Boehm in 1988 [6]. The Spiral model, as shown in Fig. 2, is a meta-life cycle model because many iterative

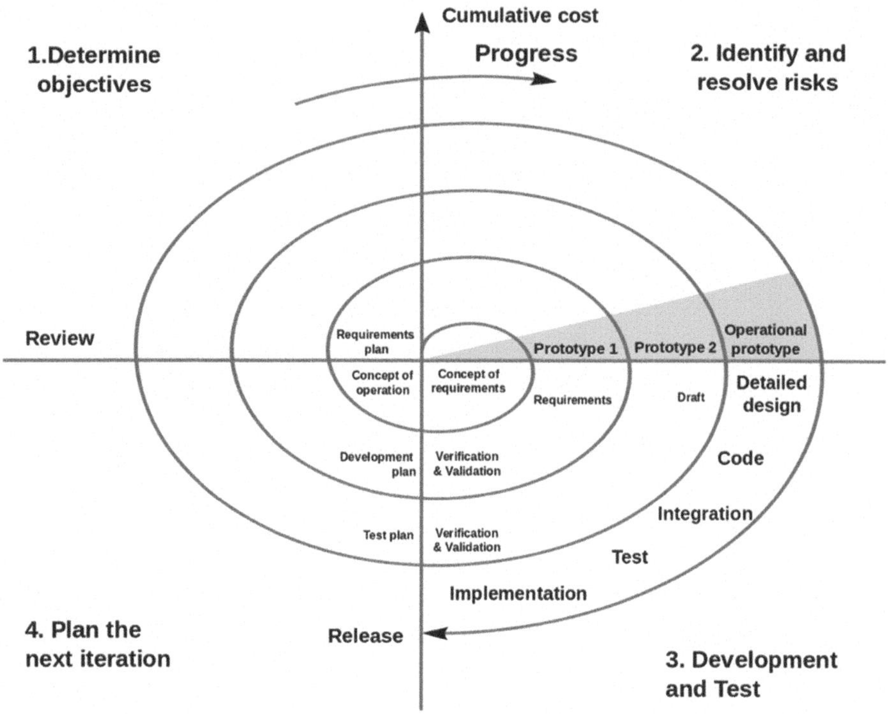

Fig. 2 A Spiral model

SDLC models proposed later are based on the same concept as the Spiral model. A metamodel is a model of a model, and metamodeling is the process of generating such metamodels. In each iteration, a prototype is developed and reviewed using a Waterfall-like phases. The review results are used to improve on the prototype in the next iteration. The iteration continues until the software meets user expectations.

Although the Spiral is considered to be a very good process model, it is never used by the software industry: During the time of the client-server architecture, using the Spiral requires contract flexibility and the developers being able to assess any project risks to avoid failure caused by constantly changing tasks. As the phases in each iteration are similar to those in the Waterfall, the Spiral seems just like using Waterfall to develop multi-releases of software.

One iterative process model that later became the dominate SDLC in the software industry is the Agile model. J. Sutherland and K. Schwaber first conceived the Scrum as a set of processes at OOPSLA '95 [7]. Both, along with 15 others, later initiated with the same concepts of Scrum in 2001 the term "Agile" in the "Manifesto for Agile Software Development [8].

The Agile model uses an iterative spiral SDLC but extends the advantages of an iterative model by rapid delivering of software. Agile divides the software into small builds and encourages developing these builds in iterations, called Sprints. As each build consists of only a small set of functionalities, a Sprint is usually timeboxed to take 2 to 4 weeks to complete.

In addition to Scrum, there are different subtypes of the Agile process models. These subtypes of the Agile process model include Rapid Application Development (RAD) [9], Prototyping [10], Rational Unified Process (RUP) [11], and Extreme Programming (XP) [12]. However, we will simply use the umbrella term Agile as a type of SDLC in the comparison.

In the Agile model, especially in Scrum, it is noted that the software requirements are presented in the form of "user stories." User stories describe the use scenarios or the user behaviors which need to be supported by the software. A user story describes the type of user, what they want, and why. Hence, a user story is intended to capture a description of a software function from an end user perspective.

After carefully planning and designing the user stories at the Initiate stage, those user stories are allocated into different Sprints. During a Sprint, the allocated set of user stories are developed, tested, and released into the production system. Once the Sprint is complete, the life cycle proceeds into next Sprint with another set of user stories. Each user story forms a software module, and the development of the module follows a streamlined Waterfall process.

Secure Software Development Life Cycle

Regardless of the specific approach, whether Waterfall or Agile, the SDLC is the process that IT project managers and teams use to develop a software. However, security components are conventionally not incorporated into the SDLC process. It is when the software is completed and operational that security is considered. Hence, this leaves room for vulnerabilities and for hackers to attack the system.

To enhance security features, in the past it was common practice to perform security-related activities only as part of testing. However, most testing is performed based on the functionalities or features described in the software requirements or user stories. Therefore, there is a need to add security-related functionalities or features into requirements or user stories. However, challenges present while using this practice.

The first challenge is that how many security requirements are actually needed for the software to be considered secure. There could be a false sense of software security when many security requirements are implemented. Moreover, most of the security requirements, such as prevention of SQL injection and session hijacking, are not user-requested functions. Should the software company implement a software with such security features at its own cost? In a 2017 EU public opinion survey, most respondents (61%) say they consider security and privacy features when choosing an information technology product, and only more than one quarter (27%) say they are ready to pay more for better security and privacy features [13]. In another PwC US Protect.me survey, it shows that 85% will not engage with a company if they have concerns about its security practices [14]. Therefore, a

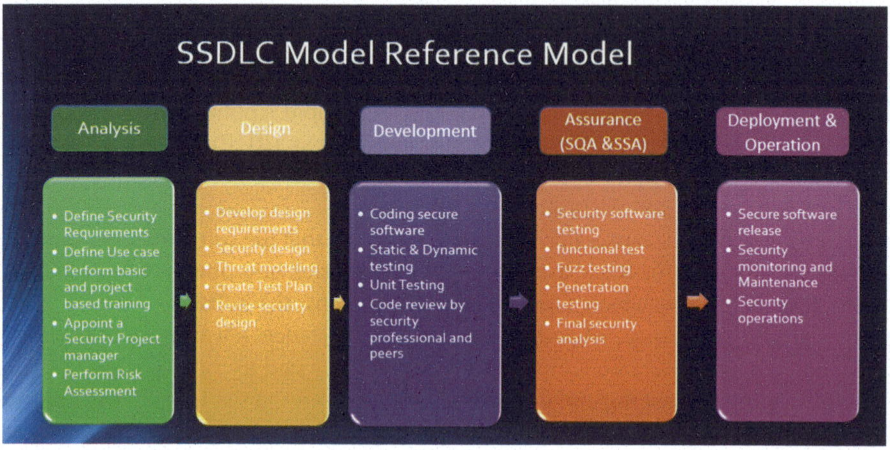

Fig. 3 SSDLC reference model

software company needs to balance the number of security features for its software and the cost to developing such features at its own expense.

That is when the Secure Software Development Life Cycle (SSDLC) comes in to fill the gap in recent secure software development practices. The Secure Software Development Life Cycle usually follows the same process as SDLC that the organization adopted, and it also has the same phases. However, in this case, security is incorporated in each phase of the SSDLC. The only problem is that, just like the SDLC, the SSDLC is not one size fits all approach. There are many new SSDLC models which have been proposed or modified from existing SDLC models, and not all models can work for all types of IT projects because of the different needs and specifications.

Most of the proposed SSDLC models are developed with additional activities or components inserted in the corresponding SDLC. Some SSDLC are proposed primary based on the Waterfall modes [15–20]; others target the Agile model [21, 22]. To find out how effective a SSDLC will work for most or all IT projects, we have to understand the fundamental characteristics of many proposed SSDLC models. A basic Waterfall-based Secure Software Development Life Cycle model, depicted in Fig. 3, is provided below as a reference for discussion.

Under each phase, we have included some of the common and necessary components for an effective SSDLC:

- Analysis: Among the activities in the phase, performing risk assessment is considered a more effective and leading technique in the phase.
- Design: Threat modeling will help for secure software design.
- Development: Adopt best practice in secure coding, such as using static code analyzer and code review will be more effective.

- Assurance: For SSDLC, adopting a software assurance approach, instead of just testing, provides more benefits to the process. It consists of activities for penetration test and risk-based test.
- deployment and operation: Secure operation practice is usually ignored in the SSDLC.

Analysis of SSDLC Models

There are several comparisons of different SSDLC models from other researchers [23–25]. However, these articles do not perform actual comparison based on any measurements or use criteria derived from the characteristics embedded in the SSDLC models that they compare, such as if additional training is required for the SSDLC. They simply list the SSDLC models and described the security-related activities/components inserted into the original SDLC.

To perform the comparison of SSDLC models in more technical depth, we need to look into the characteristics of those popular models and select the common characteristics across the models for comparison. To the end, we have to firstly determine the SSDLC models to compare.

There are many SSDLC models proposed in the last two decades: Some are used by certain software organizations, and some later became a recommended process and were partially or completely adopted into standards. For the comparison, we have carefully selected some of the popular models from them. The first model considered is Microsoft's Trustworthy Computing Security Development Life Cycle model [17, 19, 26]. The Microsoft's Trustworthy Computing Security Development Life Cycle model, or the SDL (see Fig. 4) as Microsoft calls it, was developed and adapted by Microsoft Corporation in 2004 [17] for internal use and as a way to prevent any vulnerabilities in the development of applications. This model has a total of six phases: Requirements, Design, Implementation, Verification (which is the Testing phase), Release, and Support and Servicing. Most security components are the same as the basic SSDLC, but Microsoft has a security kick-off meeting in the first phase and the need to register with Microsoft's SWI (Secure Windows Initiative).

The second model is the Seven Touchpoints model [18]. This model was developed by Gary McGraw in 2004, and the highlights involve software security touch points, which are also known as "best practices." As shown in Fig. 5 [27], it is later called "Building Security In" and is adopted by some organizations, such as the Software Engineering Institute (SEI). This model has six phases: Requirements and Use Case, Architecture and Design, Test Plans, Code, Tests and Test Results, and Feedback from the Field. The security components are more or less the same, but the difference is that the Touchpoints model is a Waterfall and an Agile development model combined to make one model.

The Software Engineering Institute (SEI) Team Software Process (TSP) for Secure Software Development [28, 29] is chosen for the third process model. This

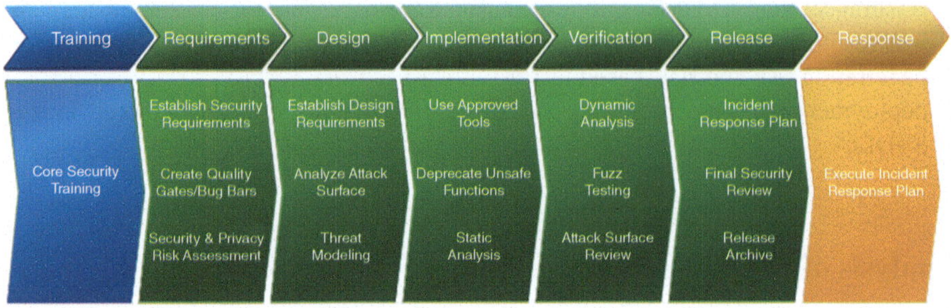

Fig. 4 Secure Software Development Process Model at Microsoft©

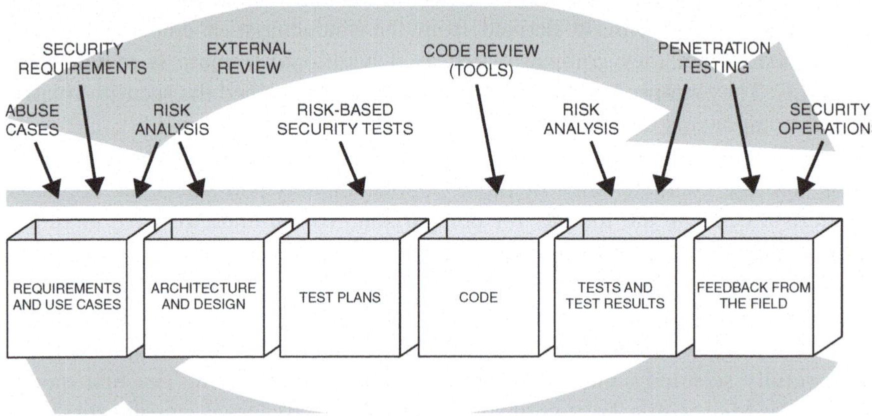

Fig. 5 Building Security In Seven Touchpoints model

SSDLC was developed by SEI. The Software Engineering Institute developed this SDLC model that ties with the Team Software Process (TSP) and Personal Software Process (PSP). It aims to decrease the likelihood of appearances of vulnerabilities in software developed. Shown in Fig. 6, this model has only four phases which is not the same for the basic SSDLC reference model shown in Fig. 3. It has a Requirement, Design, Implementation, and Testing phase but no Operation and Deployment phase. Organizational policies, management oversight, resources and training, project planning, project tracking, risk management, measurement, and feedback are incorporated into all of the phases. It is a more thoughtful but also a very complicated model.

The last model identified is the SSDLC model developed by Jones and Rastogi [15]. This model, as we can see in Fig. 7, has a total of five phases: Design, Development, Testing, Operation and Maintenance, and Disposal phase.

Compared with the other three, there are few differences in this model. This model has no separate Requirement phase. The requirement gathering and workshops are done in the Design phase. Training is not provided to the development team but to the end user in the Operation and Maintenance phase and is ongoing

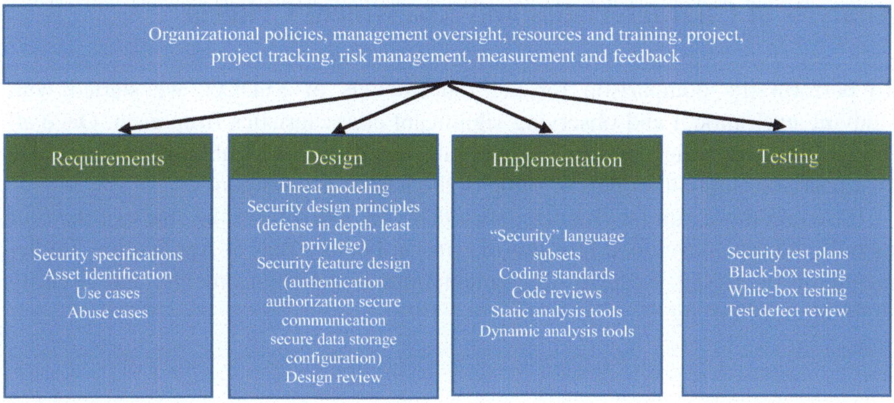

Fig. 6 SEI TSP for Secure Software Development

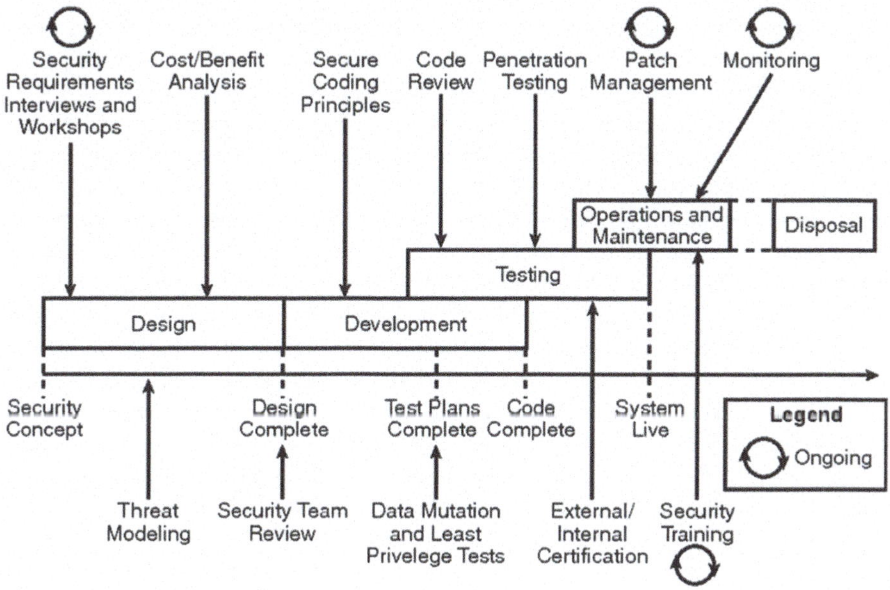

Fig. 7 Jones and Rastogi SSDLC

even after the process. Certification is required at the end of the testing. There is a Disposal phase that focuses on if and when the program is being disposed of, moved, remodeled, or archived. This last phase is a plus because there are security protocols and components that need to be added when disposing of, redeveloping, or moving the software.

Analysis of Characteristics and Assessment Criteria

To perform the comparison for the four models of SSDLC, we started with studying each model and observing significant characteristics from each. Once all information was collected after studying all models, we were able to identify some characteristics that all models or some of them have in common.

Based on those similar characteristics that concluded from the study of the four models, we have developed four sets of criteria: Focus Areas of Application, Implementation of Model, Security Implementations and Enhancements, and Security Training and Staffing.

Focus Areas of Application

This group focuses on the process area of the application or the software, as to whether it fits well for Agile, Waterfall, or both. Moreover, its relevance to a particular industry sector is also examined and if it contains all phases.

The first criterion identifies if the model is either Waterfall or Agile rooted. When a model is based on Waterfall, this is an indication that the project is going to last few months to few years. When a SSDLC model is originally targeted on an Agile model, this is an indication that each Sprint in the project will take the most a month to complete. Because the scope is much clear in Waterfall, while Agile is more of risk-driven, a SSDLC or a SDLC model for that matter is more effective when it is applied in either Waterfall or Agile, but not both. The measure of this criterion uses a binary measurement which is yes or no. In this criterion, it determines if the SSDLC models would work well in a Waterfall approach, Agile approach, or both.

The second criterion in this group is that the model has to be universal for most organizations. For this criterion a 1–5-scale measurement is applied: 1 being specific and 5 being general for public use. The purpose of this criterion is to find a model that is most effective for the use of most or all information technology projects.

The third criterion in this group is that the particular SSDLC model has to contain all the phases that are required in the SSDLC reference model. The phases include Analysis/Requirement phase, Design phase, Development phase, Assurance (SQA, SSA) phase, and deployment and operation phase. This criterion has a scaling measurement that is either yes or no. All phases must be included according to the reference model that is provided earlier in this study.

Implementation of Model

This set of criteria focuses on the implementation of the model and how the implementation will impact on the application when it is being developed.

The first criterion is that the SSDLC model has to be compatible with the development of the software. It means that the model has to incorporate certain components that will enhance and benefit the development of the software. To that end, a percentage measurement for this criterion is chosen, 100% being that the model is completely compatible with the development of the software. This means that regular as well as security components are added into the process and the process model is able to bring forth the functionality and the security components in an effective manner. If the model is at 75% compatibility, that means that the model satisfies only part of the regular function and security function requirements. This could happen if the model does not include all phases of the SSDLC or if one or more phases are lacking security components. If the model is at 50% compatibility, that means that only the regular components are added, but not the security components. If the model is at 49% or below, that means that it does not perform neither the regular nor security requirements well.

The Implementation of Model criterion is necessary but can be optional. It relates to if the SSDLC is good for a new or a continued development. This criterion focuses on disposing of the software or developing a new software entirely or moving the software to a new location. The data involved with the software need to be protected to avoid data breaches or vulnerabilities while disposing of, changing the location of, or developing a new software from the existing software.

The third criterion of the group is that the Preliminary/Initiation phases have to be effective for the development of the software. In the Preliminary phase which is mostly the Requirement phase in SSDLC, the requirements analysis for the software is performed. Project managers, security project managers, and stakeholders are selected as well. Security training takes place for the team involved with the project as well. In a Requirement phase, there can also be user stories developed in Agile development. User stories are tasks that can be manipulated at a later time. Even though it is already completed, it gives the team the opportunity to go back and change the user story or fix a problem that occurred. This is usually true in Agile development, because a Sprint in Agile development lasts less than 1 month. This criterion tests if the team is able to redo a Sprint, add to it, or correct it.

The measurements provided are highly effective, moderately effective, and not effective. A model's Requirement phase is highly effective if it has both a requirement analysis and user requirements in a Waterfall model or if there are user stories in an Agile model and if it has training or workshops (security training or a certification or coding lessons) for the project team. Security requirements are usually described in one or more user stories in an Agile model. If a model's Requirement phase is moderately effective, it has a requirement analysis and a user story but no training or workshop for the team involved with the project. If a SSDLC model's Requirement phase is considered ineffective, that means it has a Security Requirement phase, but provides no training.

The last criterion requires a SSDLC model to have a deployment and operation phase. At the end of the software development, security components and protocols should be taken into account in order to release the software into production environment and to monitor and maintain the software. The measurement provided for this criterion is yes or no. The measure is yes if it has a Deployment/Operation phase or no if it does not have a deployment and operation phase.

Security Implementations and Enhancements

Criteria in this group focus on the security components of the SDLC which provides the ground for SSDLC.

The first criterion for this group is that security testing has to be performed in at least two or more phases of the SSDLC. In a SSDLC, there are security components in all of the SSDLC phases. Testing is usually conducted mainly in the Implementation (UT/IT), Testing (ST), and Deployment (UAT) phases. This criterion makes a SSDLC more effective with integrated security testing components, so that the software will be less vulnerable to attack. The measurements provided are: Yes, it has two or more phases that does testing, or no if it does not.

The last criterion for this group is that all phases have to have a security component. That is the whole idea for the Secure Software Development Life Cycle model. Security has to be emphasized throughput all phases to ensure in everything that is being done in SSDLC will ensure the safety of the software. The model is 100% effective if all phases have a security component. And it is 50% effective if it does not.

Security Training and Staffing

This group focuses on whether there is security staff and how difficult training is.

Security training is very essential in a process model for a secure software development. Security training involves educating the team on vulnerabilities, secure coding, threat modeling, security protocols, etc. This criterion is being measured by its difficulty in a 1–10 scale. 1–3 is easy, 4–6 is moderate, and 7–10 is difficult. Security training has to be easy and understandable. If a type of security training requires a software team to get security certification, they wouldn't be able to concentrate on the development of the software. Security training has to be specific and attainable.

The last criterion is if there requires security staff to oversee all security aspect of a SSDLC. Security staff can make sure all security components in the SSDLC are carried out successfully. Implementing a security staff is the same thing as implementing a staff/team for a regular System Development Life Cycle. In a SSDLC, a security project manager, stakeholder, and software team (such as a tester and a software engineer who is skilled in the cybersecurity field) have to be present. A yes or no measurement is provided for this criterion.

Comparison of SSDLC Models

Comparison from Criteria

Based on information collected, as shown in Table 1, for the Focus Areas of Application group of criteria, no model as of now works well for both Waterfall and Agile development. In the second criterion, the more universal model is the Jones and Rastogi model and Microsoft's. In the third criterion, the only model that does not have all the phases required in the SSDLC is the SEI model. The more effective model is the Jones and Rastogi model; the second effective model is Microsoft's model.

For the Implementation of Model group, SEI is at 75% compatibility because it does not have a deployment and operation, phase that has security components and protocols to protect the software when it is released and being monitored. The Gary McGraw Touchpoints model is at 0% compatibility because as of now, Agile and Waterfall development cannot be combined. In the third criterion, two models are leading: the Jones and Rastogi model which has a phase that has new or continued development and the Touchpoints model because of the fact that it can work with Agile development which allows for continued development. SEI is the least effective in this criterion because it is the only one that does not have a deployment and operation phase. The more effective models in this group of criteria are the Jones and Rastogi model and Microsoft's model.

For the Security Implementations and Enhancements group, as depicted in Table 2, the more effective models are the Jones and Rastogi model and Microsoft's model. Security testing is done in at least two of the phases for all models. All models have security components in each phase except the SEI model because it

Table 1 Comparison results from criteria 1 to 5

SSDLC model	Criterion 1: agile, waterfall, or both	Criterion 2: Universal (1–5)	Criterion 3: model has all phases	Criterion 4: compatibility	Criterion 5: For new or continued development
Microsoft SDL	Waterfall: Yes Agile: Yes Both: No	Scale of 2	Yes	100%	No
BSI Touchpoints	Waterfall: Yes Agile: No Both: No	Scale of 1	Yes	0%	Yes
SEI TSP	Waterfall: Yes Agile: Yes Both: No	Scale of 2	No	75%	Yes
Jones and Rastogi	Waterfall: Yes Agile: No Both: No	Scale of 3	Yes	100%	No

Table 2 Comparison results from criteria 6 to 11

SSDLC model	Criterion 6: Initiation phase effectiveness	Criterion 7: has operation and maintenance phase	Criterion 8: is security testing done in two or more phases	Criterion 9: security is integrated in all phases	Criterion 10: difficulty of implementing security training	Criterion 11: has security staff
Microsoft SDL	Highly effective	Yes	Yes	100%	Scale of 2	Yes
BSI Touchpoints	Ineffective	Yes	Yes	100%	Scale of 10	No
SEI TSP	Highly effective	No	Yes	75%	Scale of 5	Yes
Jones and Rastogi	Highly effective	Yes	Yes	100%	Scale of 8	Yes

does not have a deployment and operation phase which has security components and protocols releasing and monitoring the software.

For the Security Training and Staff group of criteria, all models have a security staff except for the Touchpoints model, and the models that implemented an easy training are Microsoft's model and the SEI model. The Touchpoints model has no security training. In the Jones and Rastogi model, what is involved in the security training is not specified, and it is only implemented in the Operations phase and continues on after that. The most effective models in this group are Microsoft's model and the SEI model.

Summary of the Comparison

Microsoft's model seems an effective model; however, it is customized by Microsoft Corporation and is not meant to be used by the general public. In other terms, this model is less effective for most or all IT projects.

The Gary McGraw Touchpoints is a model that has the potential to be a Waterfall model but not both Agile. As of now we have not found a model that could work well for both Waterfall and Agile development. If this model were to be used, considering the model leans more toward Agile, there would not be enough time to complete all the phases that is crucial to the successful development of the software because Agile development only takes at most a month in a Sprint, while Waterfall could take months to years. As of now, a model like this is not effective for SSDLC development yet alone being used by most organizations for software development.

The SEI team process for Secure Software Development has a great training process for the team members in the Requirement phase. However, this model does not have the deployment and operation phase that contains security components and protocols for releasing the software to the public and monitoring it to prevent any attacks. As of now, this model is not as effective.

The model that is most effective in the comparison is the SSDLC model developed by Jones and Rastogi. It contains all of the phases and meets most of the important criteria. It has an Operation and Maintenance phase, it is universal enough for public use, and the process model is compatible with the all types of software development organization. Although a Requirement phase is not added, the requirement gathering and workshops are added into the Design phase. Security training is not added until the Maintenance phase, but it is ongoing even after the development of the software. There is a Disposal phase which is not in the universal model, but it does prove useful if the software needs to be moved, redeveloped, or disposed of.

Conclusion

Since cyber-attacks are becoming more common and more dangerous and unavoidable, it is important that our software is protected, yet alone the process used to develop our software. There are many Secure Software Development Life Cycle models available to the IT industry. Some of them are very similar to each other. However, since most IT projects are different in terms of sizes, operation objectives, as well as the sectors of industry, the SSDLC model has to be general enough to meet at least some of the criteria for an IT organization's software development.

This study is only meant to compare existing popular SSDLC models and to identify which one is most effective for being used for most or many IT projects and the one that would be most effective in protecting a software from various attacks and vulnerabilities.

Overall, the comparison results demonstrate that the Jones and Rastogi model is considered to be an effective one for many IT projects, especially for Agile projects. However, it is worthy to mention that, because of the various types of IT projects, one specific model cannot be applied for use in all types of IT projects. For an IT project operated in Waterfall, BSI Seven Touchpoints model can be an excellent alternative.

References

1. *Identity Theft*. https://www.usa.gov/identity-theft. Accessed 24 Feb 2020
2. *Massive Smart Home Breach Leads To Consumer Security Concerns*. https://www.idtheftcenter.org/massive-smart-home-breach-leads-to-consumer-security-concerns/. Accessed 24 Feb 2020
3. *Ring Throws Customers Under The Bus After Data Breach*. https://www.eff.org/deeplinks/2019/12/ring-throws-customers-under-bus-after-data-breach. Accessed 24 Feb 2020
4. H.D. Benington, Production of large computer programs, in *Proceedings, ONR Symposium on Advanced Programming Methods for Digital Computers*, (1956), pp. 15–27

5. W.W. Royce, Manage the development of large software systems, proceedings. IEEE WESCON **26**, 1–9 (1970). http://www-scf.usc.edu/~csci201/lectures/Lecture11/royce1970.pdf. Accessed 24 Feb 2020
6. B.W. Boehm, *A Spiral Model of Software Development and Enhancement, Computer* (1988), pp. 61–72
7. The Scrum Guide, pp. 17. https://www.scrumguides.org/docs/scrumguide/v2016/2016-Scrum-Guide-US.pdf. Accessed 24 Mar 2020
8. K. Beck et al. *Manifesto for Agile Software Development.* https://Agilemanifesto.org/iso/en/manifesto.html. Accessed 24 Feb 2020
9. J.M. Kerr, R. Hunter, *Inside RAD: How to Build a Fully Functional System in 90 Days or Less* (McGraw-Hill, 1994)
10. Software Prototyping. https://www.ingsoftware.com/software-prototyping. Accessed 24 Mar 2020
11. I. Jacobson, G. Booch, J. Rumbaugh, *The Unified Software Development Process* (Addison-Wesley Professional, 1999)
12. K. Beck, *Extreme Programming Explained: Embrace Change* (Addison-Wesley, 2000)
13. European Commission, *Special Eurobarometer 460, Attitudes towards the impact of digitisation and automation on daily life* (2017). https://ec.europa.eu/commfrontoffice/publicopinion/index.cfm/Survey/getSurveyDetail/instruments/SPECIAL/surveyKy/2160. Accessed 24 Feb 2020
14. PwC Consumer Intelligence Series: Protect.me https://www.pwc.com/us/en/services/consulting/library/consumer-intelligence-series/cybersecurity-protect-me.html. Accessed 24 Feb 2020
15. R.L. Jones, A. Rastogi, Secure coding: Building security into the software development life cycle. 29-39. Inf. Syst. Secur. **13**(5) (2004)
16. Keary, E., & Manico, J. (n.d.). Secure Development LifeCycle. https://www.owasp.org/images/7/76/Jim_Manico_(Hamburg)_-_Securiing_the_SDLC.pdf. Accessed 24 Feb 2020
17. S. Lipner, The trustworthy computing security development LifeCycle, in *Proc. 20thAnnual Computer Security Applications Conference, Pp 2-15, Tucson, AZ*, (2004)
18. G. McGraw, Software security. IEEE Secur. Priv. **2**(2), 80–83 (2004)
19. Microsoft. (2012). Microsoft Security Development Lifecycle (SDL) – Version 5.2. https://docs.microsoft.com/en-us/previous-versions/windows/desktop/cc307748%28v%3dmsdn.10%29. Accessed 24 Feb 2020
20. M. Morana. *Building Security into the Software Life Cycle*, a Business Case (n.d.). https://www.blackhat.com/presentations/bh-usa-06/bh-us-06-Morana-R3.0.pdf. Accessed 24 Feb 2020
21. T. Ayalew, T. Kidane, B. Carlsson, Identification and evaluation of security activities in agile projects, in *2013 Nordic Conference on Secure IT Systems*, (Ilulissat, Greenland, 2013), pp. 139–153
22. M.I. Daud, Secure software development model: A guide for secure software life cycle, in *Proc. the International MultiConference of Engineerings and Computer Scientist 2010, Vol. I, Hongkong*, (2010)
23. J. Gregoire, K. Buyens, B.D. Win, R. Scandariato, W. Joosen, On the secure software development process: CLASP and SDL compared, in *Proc. 29thInternational Conference on Software Engineering Workshops, 2007.* https://www.researchgate.net/publication/4261954_On_the_Secure_Software_Development_Process_CLASP_and_SDL_Compared. Accessed 24 Feb 2020
24. K. Tiirik, *Comparison of SDLC and Touchpoints*. https://courses.cs.ut.ee/MTAT.03.246/2013_spring/uploads/Main/essay09.pdf. Accessed 24 Feb 2020
25. B.D. Win, R. Scandariato, K. Buyens, J. Gregoire, W. Joosen, On the secure software development process: CLASP, SDL and Touchpoints compared, information and software technology archive. **51**(7), 1152–1117 (2009)
26. Microsoft, Security development Lifecycle for agile development, in *Microsoft Security Development Lifecycle*, (2009). https://www.blackhat.com/presentations/bh-dc-10/Sullivan_Bryan/BlackHat-DC-2010-Sullivan-SDL-Agile-wp.pdf. Accessed 24 Feb 2020

27. G. McGraw, *Software Security, Building Security* In. http://www.swsec.com/resources/touchpoints/. Accessed 3/24/2020
28. D. NooPur, Developing secure software, in secure software engineering. The DoD software Tech News **8**(2), 3–7 (2005). http://www.sis.pitt.edu/jjoshi/devsec/securesoftware.pdf. Accessed 24 Feb 2020
29. J. W. Over. *Team Software Software Process for Secure Software Development* (2002) . https://resources.sei.cmu.edu/asset_files/Presentation/2002_017_001_24393.pdf. Accessed 24 Feb 2020

Flexible Access Control over Privacy-Preserving Cloud Data Processing

Wenxiu Ding, Xinren Qian, Rui Hu, Zheng Yan, and Robert H. Deng

Introduction

Cloud computing has been widely adopted in various application domains owing to its specific advantages, which enables cloud users to store their data and perform various computations on the data without incurring a high cost. It even becomes the "lifeline" of many institutes or organizations. With the advent of Internet of Things, enormous amounts of data are produced and outsourced to the cloud for storage and analysis. Data analysis helps to gain insights on related entities in a physical world, which can provide tremendous value for various applications in multifarious domains, e.g., medical [1], cybersecurity education [2], and business [3]. However, the cloud may not be fully trusted by cloud users since it may reveal or disclose the data outsourced by the cloud users or the processed results of these data, which may seriously undermine user privacy. For example, to train the future generation or employees with cybersecurity skills, the customized cybersecurity exercises will be more suitable if more related information are collected and analyzed. But they may be reluctant to offer too much data (such as work culture, associated threats) due to privacy concern. Therefore, it has great significance to protect sensitive data and data processing results from being leaked to any unauthorized parties. A

W. Ding · X. Qian · R. Hu
School of Cyber Engineering, Xidian University, Xi'an, China

Z. Yan (✉)
School of Cyber Engineering, Xidian University, Xi'an, China

Department of Communications and Networking, Aalto University, Espoo, Finland
e-mail: zheng.yan@aalto.fi

R. H. Deng
School of Information Systems, Singapore Management University, Singapore, Singapore
e-mail: robertdeng@smu.edu.sg

K. Daimi, G. Francia III (eds.), *Innovations in Cybersecurity Education*,
https://doi.org/10.1007/978-3-030-50244-7_12

standard solution is to encrypt the data before uploading. However, data encryption introduces several challenges as described below.

First, encryption seriously restricts the computations/analysis over the outsourced data in the cloud. With traditional encryption algorithms (such as AES), it is impossible for the cloud to process the encrypted data directly. Some existing efforts adopted partially homomorphic encryption (PHE) to solve the problem, but they are limited only to multiplication and addition operations on encrypted data [4, 5], which are not sufficient to satisfy user demands in many applications. More operations, such as comparison and equality test, are required in practical applications [6, 7]. This requests further study on privacy-preserving computations. The basic operations can be widely applied to realize complex and useful applications, e.g., privacy-preserving classifications in machine learning [8], trust evaluation in Internet of Things [9], and medical analysis in e-health [10]. Obviously, the more basic computations available over encrypted data, the more support on complete and complex functions and algorithms. To realize arbitrary computations over ciphertext, schemes based on fully homomorphic encryption (FHE) were designed [11–13]; however, most FHE-based schemes suffer from huge computation overhead and high storage cost, which make them impractical for real-world deployment and wide usage.

Second, secure multi-user access control over processed results also needs to be supported [14]. Both existing PHE and FHE are single-user systems, which inherently lack support for multi-user access to the processing results of encrypted data. The scheme based on PHE in [15] supports distribution of addition operation results through an interactive protocol between two servers, but the protocol must be executed for each data request, thus is inefficient. Attribute-based encryption (ABE) is an effective tool to support fine-grained access control and multi-user access and has been applied in many application scenarios [16–18]. However, to our knowledge, there is no effort in the literature on fine-grained access control over the results of encrypted data computation. Previous work [19] aims to solve this problem by combining homomorphic encryption and proxy re-encryption, but it only supports one requester access at one time. In case multiple users want to access the same result, it needs to execute the designed scheme for each requester, which obviously incurs high communication and computation costs.

In this chapter, we propose a novel system in order to overcome the challenges as described above. It supports multiple basic computations over encrypted data and realizes flexible access control over the processing results by employing PHE and ABE, which can be easily extended and implemented to cybersecurity education. We present a family of protocols to efficiently realize several basic computations over encrypted data. Then, we extend the system with maximum, minimum, and division computations over integers. We propose to combine the ciphertext of ABE with homomorphism to realize a fine-grained access control of the processing results.

Related Work

With the development of cloud computing, cybersecurity education becomes critical because the traditional cybersecurity cannot guarantee the security of organizations due to the sophisticated networks, while it also becomes flexible by taking advantage of educational testbeds and framework. The cloud users (such as students and engineers) can be greatly benefited through cybersecurity education and get trained with enough technical skills. However, the risk of revealing personal data makes it urgent to enhance data security and user privacy.

Secure Data Processing Based on SMC

Secure multi-party computation (SMC) enables computations over multi-user outsourced data without revealing each input. It lays a technical foundation for many problems, such as database query, intrusion detection, and data mining with privacy preservation [9]. Several Schemes [20, 21] based on the popular SMC construction Sharemind [22] were proposed to achieve various secure computations. But the product of N pieces of data needs about 3^N multiplications of 32-bit numbers under the cooperation of three involved servers in Sharemind, which obviously cannot adapt to big data processing.

Secure Data Processing Based on Homomorphic Encryption

FHE Schemes [11–13] are designed to realize arbitrary computations over encrypted data. Due to their high computation overhead, some extended Schemes [23, 24] are proposed to improve efficiency. However, their computation and storage costs are still not satisfactory for practical applications [25, 26]. PHE can only support limited computations, but it is more efficient and practical than FHE and has been widely used in various applications. Some Schemes [4, 5] can only support addition and multiplication over a limited number of data inputs. In [4], decryption requires solving the problem of discrete logarithm, which seriously restricts the length and the number of data inputs. The multi-party computation framework proposed in [5] achieves addition and multiplication by following the idea of secret sharing. Similar to the SMC-based scheme in [21], it is unable to support the multiplication of a large number of data inputs. Liu et al. [6] proposed a framework for efficient outsourced data calculations with privacy preservation.

Secure Data Access Control

Cloud storage enables cloud users to upload their data to the cloud for storage and further sharing. However, it leads to a new problem that the cloud users lose full control over their data. Proxy re-encryption can also be adopted to manage data sharing in cloud [27, 28]. But it cannot support fine-grained access control on homomorphic computation result. Role-based access control (RBAC) can provide partial flexibility based on one level policy, which ensures that only the user with specified role can access the data. But, these constructions [29, 30] based on RBAC cannot support flexible access policies with various attribute structures.

ABE [31, 32] has been widely applied in cloud storage management for achieving fine-grained access control [33–35]. Furthermore, trust-based Schemes [16–18] simplify the attributes involved in ABE and take into consideration only trust levels. These schemes highly reduce the computation cost. But, only one entity is in charge of the access control, which makes this entity obviously knows the results.

Secure Division Based on Arithmetic Transformations

Katzenbeisser et al. [36] chose a tuple $(\rho_x, \sigma_x, \tau_x)$ to represent a value $x \in D_l$, which belongs to a certain interval $D_l = [-l; +l]$ with $l > 0$, where $\rho_x = 1$, σ_x encodes the sign of the value x and τ_x indicates the absolute of the value. The division result can be computed by basic operations on corresponding element through function $\mathrm{LDIV}([\overline{x}], [\overline{y}]) = \left([\rho_x], [\sigma_x][\sigma_x], [\tau_x][\tau_y]^{-1}[\tau_{C^2}]\right)$. Though the representation of numbers can support secure computations on non-integers, its division result is an approximation with bounded relative error, and encoding increases the overhead of data preprocessing.

To overcome this issue and get an accurate result, Dahl et al. [37] performed a Taylor expansion on the reciprocal of a denominator to transform the division computation over encrypted data into multiplication and addition over encrypted data. The implementation of several sub-protocols brings high computational overhead. Also, the frequent interactions bring high communication overhead.

Veugen [38] presented three protocols based on a client-server model where the client has encrypted data $[x]$ and the server has the corresponding decryption key K. In order to improve the precision of data analysis, Catrina and Saxena [39] attempted to approximately get a division result over two floating point numbers by applying the Goldschmidt method [40]. But this scheme cannot support division computations over encrypted input data. To overcome this issue, Ugwuoke et al. [41] proposed a division protocol to support encrypted floating point numbers based on homomorphic encryption. However, both of the above schemes use fixed rounds of iterative computations to guarantee fixed precise of results, which results in high computational overhead.

Secure Division Based on Secure Bit Decomposition Protocol (SBD)

The modulo value operation limits the length of the data in division computation. In order to protect the confidentiality of both the divisor and the dividend, some studies use the secure bit decomposition protocol [42] to realize secure division [6, 20]. After data providers upload their encrypted data, the cloud first decomposes encrypted data as binary string and then executes division to get a quotient and a remainder by operating secure bit shift. But the bit decomposition protocol is generally very complicated, thus hard to be deployed.

Cybersecurity Education

With the development of cloud computing, the information system of organizations or schools becomes large-scale and complicated, which makes it difficult to deploy defensive mechanisms and suffers from undetected cyber-attacks [43]. Cybersecurity education aims to train IT-related employee or the future generation with technical skills. To customize specified cybersecurity exercises and enhance entities' security knowledge [44], a lot of data (such as work environment and threats) should be provided, which may breach privacy.

However, currently most researches [45–47] focus on the design and implementation of frameworks for cybersecurity exercises and testbeds. The STEAM framework [46] inserts the Arts into cybersecurity education, while the EDU Range framework [45] eliminates the dependence on virtual machine or private cloud replaced by public cloud-based framework. Frank et al. [47] introduced life cycle into testbed design. Abir et al. [48] pointed out that universities and industries lack communications about training courses and curriculum, which leads to that students do not gain adequate knowledges required by the specific workplace. To solve this issue, the cooperative education was designed to enhance the involvement of students and industry and enrich their work skills [49]. Rakesh et al. [50] discussed about the significance of security analytics and shared their educational experiences. But all work above ignored the privacy issues and do not provide a secure and privacy-preserving way to share data and customize courses.

System Model

Our proposed system mainly comprises five types of entities as shown in Fig. 1:

Data service provider (DSP) is served by the cloud, which stores user data, provides some computation service, and controls user access.

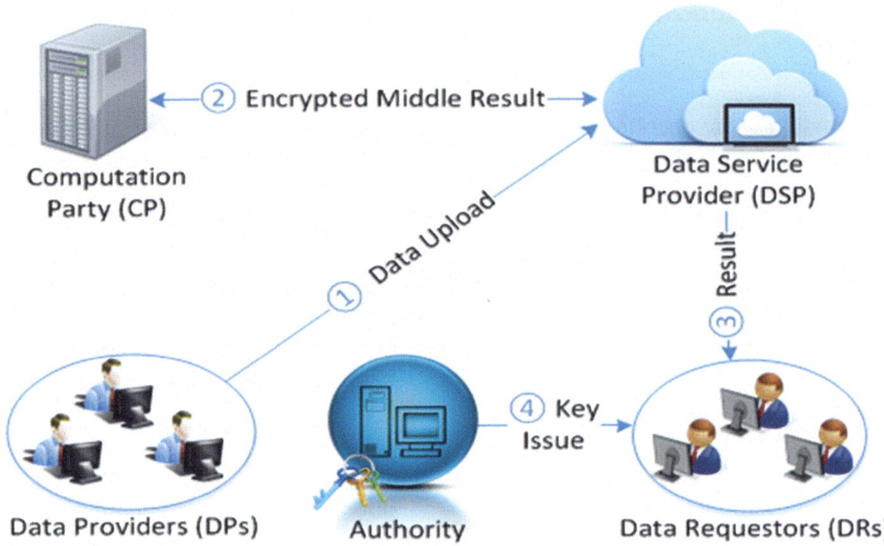

Fig. 1 A system model

Computation party (CP) fulfills partial computations and control access. It can be any party (a company or a university) who wants to train its employee or students with cybersecurity skills. There may exist multiple CPs for different applications. Herein, we simplify our design by considering only one CP in this chapter.

Data providers (DPs) are the data collectors or producers that encrypt data and store them in the DSP.

Data requesters (DRs) are the data consumers that acquire the result of data processing in a specific context.

Authority is fully trusted, which is responsible for system parameter generation and ABE key issue.

Preliminaries

For a better understanding of the scheme designs, please refer to previous work [51–53] for the detailed notation tables.

Additive Homomorphic Encryption

Paillier's cryptosystem [54] is one of the most important additive homomorphic encryption. Suppose we have N pieces of encrypted data under same key pk, which can be presented as $[m_i]_{pk}$ $(i = 1, 2, \ldots, N)$. Additive homomorphic encryption satisfies this equation,$D_{sk}\left(\prod_{i=1}^{N}[m_i]_{pk}\right) = \sum_{i=1}^{N} m_i$, where $D_{sk}()$ is the homomorphic decryption algorithm with secret key sk.

Key-Policy Attribute-Based Encryption (KP-ABE)

In KP-ABE, ciphertexts are generated based on some descriptive attributes, while decryption keys are associated with policies. For more details about KP-ABE, refer to [32]. Notably, ciphertext-policy attribute-based encryption (CP-ABE) [31] can also be applied to implement our scheme.

Homomorphic Re-Encryption Scheme (HRES)

We revise the Scheme [55] (named as EDD) and design the HRES to provide two-level decryption and achieve secure data processing. The complete version of HRES is introduced in work [19].

Data Processing Procedure

Step 1 (System Setup @ All Entities): Authority calls the algorithm ***KeyGen*** and $Setup^{ABE}(\lambda, U)$ to complete the setup of HRES and ABE.

Step 2 (Data Upload @ DPs): DPs encrypt their personal data before uploading it to the DSP. It directly recalls ***EncTK*** to encrypt data m_i (unless otherwise specified, $|m_i| < \mathcal{L}(n)/4$):

$$[m_i] = \left(T_i, T_i'\right) = \left\{(1 + m_i * n) * PK^{r_i}, g^{r_i}\right\} \bmod n^2$$

Step 3 (Data Preparation @ DSP): Upon receiving the data from DPs, the DSP needs to do some analyses over the encrypted data. It provides a data packet and ABE ciphertext for access control to the CP. In addition, CP chooses a random partial key ck_1 for access control, which will be used in Step 5.

Step 4 (Data Process @ CP): Upon receiving the preprocessing results from DSP, CP chooses another random partial key ck_2 to obtain the preprocessing result $\left[\hat{m}\right]_{pk_{ck_2}}$ or $\left[\hat{f}\right]_{pk_{ck_2}}$. Regarding access control, CP encrypts ck_2 using ABE to get $CK_2' = Enc^{ABE}\left(ck_2, \gamma, PK'\right)$ and forwards it to DSP.

Step 5 (Additional Process @ DSP): The DSP needs to remove the mask from ciphertext $[\hat{m}]_{pk_{ck_2}}$ or $\left[\hat{f}\right]_{ck_2}$ to obtain processed ciphertext $[m]_{pk_{ck}}$ or $[f]_{pk_{ck}}$ where $pk_{ck} = g^{ck}$ and $ck = ck_1 * ck_2$.

Regarding access control, the DSP encrypts ck_1using ABE under the same policy to get CK'_1 and further gets CK' through the homomorphism of ABE:$CK' = CK'_1 * CK'_2 = Enc^{ABE}\left(ck_1 * ck_2, \gamma, PK'\right)$. Finally, the DSP keeps $[m]_{pk_{ck}}$ or $[f]_{pk_{ck}}$ and CK' for user access.

Step 6 (Data Access @ DR): If the DR satisfies the access policy, Authority issues a secret key SK' to the DR. Hence, the DR can decrypt CK' to get ck and further obtain m or f.

Detailed Data Processing

System setup and data collection are the same as those in part 4. Thus, we do not introduce the details in this part; we mainly focus on the steps from 3 to 6 in each basic operation.

Addition

This function aims to obtain the sum of all raw data, $m = \sum_{i=1}^{N} m_i$, which can be accomplished by multiplying all ciphertexts. Note that the number of the data in *Addition* affects the length of the provided data. If we want to get the sum result of N pieces of data, it should guarantee that $m_i < n/N$.

Step 3 (Data Preparation @ DSP): Due to additive homomorphism, the DSP can directly multiply encrypted data one by one as follows:$[m] = \left(T, T'\right) = \prod_{i=1}^{N} [m_i] = \left(\prod_{i=1}^{N} T_i, \prod_{i=1}^{N} T_i'\right)$. To realize group access control, it chooses a random number r_1 and the first partial key ck_1 and then computes as follows:

1. Compute $c_1 = {ck_1}^{-1} \, mod \, n^2$.
2. Mask ciphertext: $[c_1 (m + r_1)] = \left(\widetilde{T}, \widetilde{T'}\right) = \left\{(T (1 + r_1 * n))^{c_1}, (T')^{c_1}\right\}$.
3. Call ***PDec*1** to partially decrypt it:$[c_1 (m + r_1)]_{pk_{CP}} = \left(\hat{T}, \hat{T'}\right) = \left\{\widetilde{T}, \left(\widetilde{T'}\right)^a\right\}$.

Then DSP sends $[c_1 (m + r_1)]_{pk_{CP}}$ to the CP.

Step 4 (Data Process @ CP): The CP calls the algorithm ***PDec*2** with sk_{CP} to decrypt the encrypted data and obtain $c_1(m + r_1)$. Then the CP chooses the second partial key ck_2 and a random number r to encrypt data as follows, $[c_1 (m + r_1)]_{pk_{ck_2}} = \left\{(1 + c_1 (m + r_1) n) g^{ck_2 * r}, g^r\right\}$,where $pk_{ck_2} = g^{ck_2}$. The CP encrypts ck_2 to obtain CK'_2 and then forwards $[\hat{m}]_{pk_{ck}}$ and CK'_2 back to the DSP.

Step 5 (Additional Process @ DSP): The DSP computes to obtain the final processed data with ck_1 and r_1, $[m]_{pk_{ck}} = \left(\overline{T}^{ck_1}(1-r_1 n), \overline{T}'\right) = \left\{(1+m*n)\,g^{ck_1*ck_2*r}, g^r\right\}$, where $pk_{ck} = g^{ck_1*ck_2}$ and $ck = ck_1 * ck_2$. It encrypts ck_1 using ABE and gets $CK' = CK'_1 * CK'_2 = Enc^{ABE}\left(ck_1 * ck_2, \gamma, PK'\right)$.

Subtraction

This function aims to obtain the subtraction of some data $\left(m=\sum_{i=1}^{W} m_i - \sum_{i=W}^{N} m_i\right)$ with encrypted data $[m_i]$ ($i = 1, \ldots, N$). It can be accomplished by negating the subtracted terms (by raising to the power of ($n - 1$)) and then following the procedure of *Addition*.

Step 3 (Data Preparation @ DSP): The DSP first computes $\left[\sum_{i=1}^{W} m_i\right] = \prod_{i=1}^{W}[m_i]$ and $\left[\sum_{i=W+1}^{N} m_i\right] = \prod_{i=W+1}^{N}[m_i]$. It further calculates $\left[-\sum_{i=W+1}^{N} m_i\right] = \left(\left[\sum_{i=W+1}^{N} m_i\right]\right)^{n-1}$ and multiplies them to obtain: $[m] = \left[\left(\sum_{i=1}^{W} m_i - \sum_{i=W+1}^{N} m_i\right)\right] = \left[\sum_{i=1}^{W} m_i\right] * \left[-\sum_{i=W+1}^{N} m_i\right]$. Then the subsequent process is the same to that in *Addition*. Due to length and simplicity reasons, we skip its details.

Multiplication

This function aims to obtain the product of all raw data $\left(m = \prod_{i=1}^{N} m_i\right)$ For ease of presentation, we describe the details with two pieces of data ($[m_1]$, $[m_2]$). Note that if we need to get the product of N pieces of data, it must be guaranteed that $\mathcal{L}(m_i) < \mathcal{L}(n)/(2N)$.

Step 3 (Data Preparation @ DSP): First, the DSP chooses a random partial key ck_1 and a random number c_1 and sets another one as $c_2 = (ck_1 * c_1)^{-1}\ mod\ n$.

To conceal each raw data from the CP, the DSP does one exponentiation and one decryption with its own secret key by calling ***PDec1***:

1. $[c_1 * m_1] = \left\{T_1{}^{c_1}, (T'_1)^{c_1}\right\}$; $[c_1 * m_1]_{pk_{CP}} = \left(T_1{}^{(1)}, T_1{}'^{(1)}\right) = \left\{T_1{}^{c_1}, (T'_1)^{c_1*a}\right\}$
2. $[c_2 * m_2] = \left\{T_2{}^{c_2}, (T_2{}')^{c_2}\right\}$; $[c_2 * m_2]_{pk_{CP}} = \left(T_2{}^{(1)}, T_2{}'^{(1)}\right) = \left\{T_2{}^{c_2}, (T_2{}')^{c_2*a}\right\}$.

The data packet sent to the CP is $\{[c_1 * m_1]_{pk_{CP}}, [c_2 * m_2]_{pk_{CP}}$.

Step 4 (Data Process @ CP): Upon receiving the data packet from the DSP, the CP uses the algorithm ***PDec2*** to decrypt the data:$c_1 * m_1 = T_1{}^{(1)}/(T_1{}'^{(1)})^b$, $c_2 * m_2 = T_2{}^{(1)}/(T_2{}'^{(1)})^b$.

It then chooses ck_2 and a random number r and encrypts $c_1 * m_1 * c_2 * m_2$ and ck_2 as follows:

$\left[\hat{m}\right]_{pk_{ck_2}} = [c_1c_2m]_{pk_{ck_2}} = \left\{(1 + c_1m_1c_2m_2 * n)\, g^{ck_2 * r}, g^r\right\}$; $CK_2' = Enc^{ABE}\left(ck_2, \gamma, PK'\right)$.

Finally, the CP forwards $\left[\hat{m}\right]_{pk_{ck_2}}$ and CK_2' to the DSP.

Step 5 (Additional Process @ DSP): The DSP further processes the data packet with ck_1 and gets ciphertext as follows:$[m]_{pk_{ck}} = \left\{\overline{T}^{ck_1}, \overline{T}'\right\}$;$CK' = CK_2' * Enc^{ABE}\left(ck_1, \gamma, PK'\right)$.

Sign Acquisition

We assume that $\mathcal{L}(m) < \mathcal{L}(n)/4$ and *BIG* is the largest raw data of m. Then the raw data is in the scope $[-BIG, BIG]$. Sign Acquisition can be achieved by masking the original ciphertext with random numbers of limited length and then checking the length of the masked data to further determine the real length of original data. Here, the DR targets to obtain the final sign indicator f from $[m_1]$.

Step 3 (Data Preparation @ DSP): The DSP chooses three random numbers R $\left(\mathcal{L}(R) < \mathcal{L}(n)/4\right)$, c_1, and ck_1. It first encrypts "1" and then computes as follows:

1. $[1] = \{(1 + n) * PK^{r'}, g^{r'}\}$;$[2 * m_1 + 1] = (T, T') = [m_1]^2 * [1]$.
2. Then it flips a coin s. If $s = -1$, it computes:

$$\left(T_1{}^{(1)}, T_1{}'^{(1)}\right) = \left\{T^{n-R}, (T')^{a*(n-R)}\right\} = [-R * (2 * m_1 + 1)]$$

Otherwise, if $(s = 1)$, it calls ***PDec1*** and computes:

$$\left(T_1{}^{(1)}, T_1{}'^{(1)}\right) = \left\{T^R, T'^{a*R}\right\} = [R * (2 * m_1 + 1)]$$

3. The DSP Computes $c_2 = (ck_1)^{-1} \bmod n$ and $s' = c_1 * c_2 * s \bmod n$.

The data packet sent to the CP is $\{(T_1{}^{(1)}, T_1{}'^{(1)}), s'\}$.

Step 4 (Data Process @ CP): Upon receiving the data packet from the DSP, the CP decrypts $(T_1{}^{(1)}, T_1{}'^{(1)})$ with ***PDec2*** to obtain raw data $m' = R * (2 * m_1 + 1) \bmod n$ if $s = 1$ or $m' = R * (2 * m_1 + 1) \bmod n$ if $s = -1$. The CP compares $\mathcal{L}\left(m'\right)$ with $\mathcal{L}(n)/2$. If $\mathcal{L}\left(m'\right) < \mathcal{L}(n)/2$, it sets $u = 1$; otherwise, $u = -1$.

The CP chooses a random number r and a second partial key ck_2 and further computes as follows:$\left[\hat{f}\right]_{pk_{ck_2}} = \left(\overline{T}, \overline{T}'\right) = \left\{\left(1 + s'u * n\right) g^{ck_2 * r}, g^r\right\}$. Encrypt ck_2 using ABE: $CK_2' = Enc^{ABE}\left(ck_2, \gamma, PK'\right)$.

Finally, the CP forwards $\left[\hat{f}\right]_{pk_{ck_2}}$ to DSP.

Step 5 (Additional Process @ DSP): The DSP further processes the data packet as follows:

Compute $c_3 = c_1{}^{-1} \bmod n$;$[f]_{pk_{ck}} = \left\{\overline{T}^{ck_1 * c_3}, \left(\overline{T}'\right)^{c_3}\right\}$;$CK' = Enc^{ABE}(ck_1, \gamma, PK') * CK'_2$.

Step 6 (Data Access @ DR): The DR satisfying the access policy in ABE can decrypt CK' to obtain ck and further decrypts $[f]_{pk_{ck}}$ to obtain f. Note: if $f = 1$, $m_1 \geq 0$; otherwise, $m_1 < 0$.

Absolute

We assume that $\mathcal{L}(m) < \mathcal{L}(n)/4$ and that *BIG* is the largest raw data of m. Then the raw data is in the scope $[-BIG, BIG]$. Here, given ciphertext $[m_1]$, DR wants to get the absolute value $|m_1|$.

Step 3 (Data Preparation @ DSP): The DSP chooses three random numbers R where $\mathcal{L}(R) < \mathcal{L}(n)/4$, c_1, and c_2 and chooses the first partial key ck_1. It first encrypts "1" and computes as follows:

1. $[1] = \{(1+n) * PK^{r'}, g^{r'}\}$;$[2 * m_1 + 1] = (T, T') = [m_1]^2 * [1]$.
2. Then it flips a coin s. If $s = -1$, $(T_1{}^{(1)}, T_1{}'^{(1)}) = [-R * (2 * m_1 + 1)]$.
   Otherwise, it calls ***PDec1*** and computes $(T_1{}^{(1)}, T_1{}'^{(1)}) = [R * (2 * m_1 + 1)]$.
3. Compute $[c_1 m_1] = [m_1]^{c_1}$, and call ***PDec1*** to obtain $[c_1 m_1]_{pk_{CP}}$.
4. The DSP sets $c_3 = (ck_1)^{-1} \bmod n$ and $s' = c_2 * c_3 * s \bmod n$.

The data packet sent to the CP is $\{(T_1{}^{(1)}, T_1{}'^{(1)}), s', [c_1 m_1]_{pk_{CP}}\}$.

Step 4 (Data Process @ CP): Upon receiving the data packet from DSP, the CP decrypts $(T_1{}^{(1)}, T_1{}'^{(1)})$ and $[c_1 m_1]_{pk_{CP}}$ with ***PDec2*** to obtain raw data: $m' = (-1)^{s+1} * R * (2 * m_1 + 1) \ mod\ n$ and $c_1 m_1$, respectively. CP compares $\mathcal{L}(m')$ with $\mathcal{L}(n)/2$. If $\mathcal{L}(m') < \mathcal{L}(n)/2$, it sets $u = 1$; otherwise, $u = -1$. Then CP chooses r and the second partial key ck_2 and further computes as follows:$[c_1 m_1 s' u]_{pk_{ck_2}} = (\overline{T}, \overline{T}')$. Encrypt $ck_2$with ABE: $CK'_2 = Enc^{ABE}(ck_2, \gamma, PK')$. Finally, the CP forwards $[c_1 m_1 s' u]_{pk_{ck_2}}$ and CK'_2 to DSP.

Step 5 (Additional Process @ DSP): The DSP further processes the data packet as follows:

1. Set $c_4 = (c_1)^{-1}\ mod\ n$ and $c_5 = (c_2)^{-1} \bmod n$.

$$[su * m_1]_{pk_{ck}} = \left\{\overline{T}^{ck_1 * c_4 * c_5}, \overline{T}'^{c_4 * c_5}\right\}; CK' = Enc^{ABE}(ck_1, \gamma, PK') * CK'_2$$

Step 6 (Data Access @ DR): The DR that satisfies the access policy in ABE can decrypt CK' to obtain ck. The DSP sends the data packet $[su * m_1]_{pk_{ck}}$ to the DR in a secure way. Then the DR can decrypt it to obtain $su * m_1$. Note: if $m_1 \geq 0$, $su = 1$; otherwise, $su = -1$. Hence, $su * m$ is the absolute of data m.

Comparison

Comparison can be simply accomplished by checking the sign of the difference value of two data by calling *Sign Acquisition*. For ease of presentation, $m_1 - m_2$ is denoted as m_{1-2}.

$[m_1] = \left(T_1, T_1'\right) = \{(1 + m_1 * n) * PK^{r_1}, g^{r_1}\}; [m_2] = \left(T_2, T_2'\right) = \{(1 + m_2 * n) * PK^{r_2}, g^{r_2}\}$

Step 3 (Data Preparation @ DSP): DSP first computes to get the subtraction of encrypted data:

$$\left(T, T'\right) = \left\{T_1 * (T_2)^{n-1}, T_1' * (T_2')^{n-1}\right\} = [(m_1 - m_2)].$$

The following steps are the same as those in *Sign Acquisition*, which are skipped for the reason of chapter length limitation. Through the cooperation of the DSP and the CP, the DR finally gets the sign of $m_{1-2} = m_1 - m_2$. DR can obtain the comparison result. If $m_{1-2} \geq 0$, $m_1 \geq m_2$; otherwise, $m_1 < m_2$.

Equality Test

Equality test needs to check the signs of both difference value and negative difference value of original two data by calling *Comparison* twice. DR wants to know whether m_1 is equal to m_2 or not from encrypted data ($[m_1]$, $[m_2]$). The DSP and CP directly interact with each other in two parallel computations of *Comparison*.

They compare m_1 and m_2 in two forms: 1) $m_{1-2} = m_1 - m_2$ and 2) $m_{2-1} = m_2 - m_1$. Through the operations in *Comparison*, DSP can get two results $[f_1]_{pk_{ck}}$ and $[f_2]_{pk_{ck}}$, respectively. Then the DSP can obtain $[f]_{pk_{ck}} = [f_1 + f_2]_{pk_{ck}} = [f_1]_{pk_{ck}} * [f_2]_{pk_{ck}}$. Finally, DR that satisfies the access policy in ABE can decrypt CK' to obtain *ck*. DSP sends the data packet $[f]_{pk_{ck}}$ to the DR in a secure way. Then the DR can further decrypt $[f]_{pk_{ck}}$ to obtain *f*. Note: if $f = 2$, $m_1 = m_2$; otherwise, $m_1 \neq m_2$.

Maximum and Minimum

Two-to-One (T2O)

This scheme aims to obtain the max and min values from two encrypted data for a data requester.

Step 3 (@ DSP): First, the DSP randomly selects some numbers R_1, R_2, and $R_3$where $\mathcal{L}(R_1) < \mathcal{L}(n)/4$ and then executes the following operations: here, $m_- = m_1 - m_2$ and $m_+ = m_1 + m_2$.

1. $[1] = \{(1+n) * PK^{r'}, g^{r'}\};[m_-] = [m_1 - m_2] = [m_1] * [m_2]^{n-1}$
2. $[R_2 * m_+ + R_3] = ([m_1 + m_2])^{R_2} * [R_3];[R_2 m_-] = (T_-, T_-') = [m_1 - m_2]^{R_2}$

$$[2 * m_- + 1] = (T, T') = \left\{(1 + (2 * m_- + 1) * n) * PK^{r'+2*r_1}, g^{r'+2*r_1}\right\}$$

Then it flips a coin s. If $s = -1$, then compute$(T_1{}^{(1)}, T_1{}'^{(1)}) = \left\{T^{n-R_1}, (T')^{a*(n-R_1)}\right\} = [-R_1 * (2 * m_- + 1)]_{pk_{CP}}$ and$(T_2, T_2') = [-R_2 m_-]_{pk_{CP}} = \left\{T_-{}^{n-R_1}, (T_-')^{a*(n-R_1)}\right\}$. Otherwise, if ($s = 1$), it calls $PDec1$ and computes $(T_1{}^{(1)}, T_1{}'^{(1)}) = \left\{T^{R_1}, T'^{a*R_1}\right\} = [R_1 * (2 * m_- + 1)]_{pk_{CP}}$ and $(T_2, T_2') = [R_2 m_-]_{pk_{CP}} = \left\{T_-, (T_-')^a\right\}$. It further calls $PDec1$ on $[R_2 * m_+ + R_3]$ to get $[R_2 * m_+ + R_3]_{pk_{CP}}$. Finally, it forwards CP the data packet $\{(T_1{}^{(1)}, T_1{}'^{(1)})$, $[R_2 * m_+ + R_3]_{pk_{CP}}, (T_2, T_2')\}$.

Step 4 (@ CP): CP further processes the data packet from the DSP. It first decrypts $(T_1{}^{(1)}, T_1{}'^{(1)})$ and $(T_2, T_2')$with $PDec2$ to obtain raw data $\hat{m} = R_1 * (2 * m_- + 1) \bmod n, \hat{m_-} = (R_2 m_-) \bmod n$ if $s = 1$ or $\hat{m} = -R_1 * (2 * m_- + 1) \bmod n, \hat{m_-} = (-R_2 m_-) \bmod n$ if $s = -1$.

Then CP needs to compare $\mathcal{L}(\hat{m})$ with $\mathcal{L}(n)/2$. If $\mathcal{L}(\hat{m}) < \mathcal{L}(n)/2$, it sets $u = 1$; otherwise, $u = -1$. The CP further encrypts the raw data $u * \hat{m_-}$ with the public key of the targeted DR as $[u * \hat{m_-}]_{pk_{DR}} = (\overline{T}, \overline{T}') = \{(1 + u\hat{m_-} * n)\, pk_{DR}{}^{r}, g^r\}$.

Decrypt $[R_2 * m_+ + R_3]_{pk_{CP}}$ and then encrypt it with pk_{DR} to get $[R_2 * m_+ + R_3]_{pk_{DR}}$. Finally, the CP forwards the data packet to DSP: $\{[u * \hat{m_-}]_{pk_{DR}}, [R_2 * m_+ + R_3]_{pk_{DR}}\}$.

Step 5 (@ DSP): The DSP first removes the mask R_3 by computing $[R_2 m_+]_{pk_{DR}} = [R_2 * m_+ + R_3]_{pk_{DR}} * [-R_3]_{pk_{DR}}$. Then it can get the max and min with $r = (2R_2)^{-1} \bmod n$:

$$[\max]_{pk_{DR}} = \left([u * \hat{m_-}]_{pk_{DR}} * [R_2 m_+]_{pk_{DR}}\right)^r;$$

$$[\min]_{pk_{DR}} = \left([u * \hat{m_-}]_{pk_{DR}}{}^{n-1} * [R_2 m_+]_{pk_{DR}}\right)^r.$$

Step 6 (@ DR): The DR with the corresponding secret key can decrypt the ciphertext ($[\max]_{pk_{DR}}$ and $[\min]_{pk_{DR}}$) to obtain the maximum and minimum values.

Multiple-to-One (M2O)

Given an example of n pieces of ciphertexts ($[m_1]$,$[m_2]$,$\cdots[m_i]$,$\cdots[m_n]$), this scheme can get the maximum and minimum results $[\max]_{pk_{DR}}$ and $[\min]_{pk_{DR}}$ for the targeted data requester DR. Note that the T2O can provide the maximum and minimum values from ciphertext $[m_1]$ and $[m_2]$ for DR. If we use the PK to replace the public key of DR (pk_{DR}) in T2O, we can get the ciphertext $[max]$ and $[min]$ through parallel processing. Herein, we take maximum computation as an example, which has the same procedure as minimum computation.

In order to get the final maximum from more than two ciphertext, we need to execute several rounds of the T2O scheme. The computation follows a tree structure. It divides the data into many groups and each group has two pieces of data. Then T2O is executed over every group with PK to get the ciphertext $[max]$. Until the last two pieces of data in the last layer, DSP and CP execute T2O with pk_{DR} to get the final ciphertext $[\max]_{pk_{DR}}$.

Two-to-Multiple (T2M)

Given two ciphertext $[m_1]$ and $[m_2]$, this scheme can provide the sorting results $[\max]_{pk_{ck}}$ and $[\min]_{pk_{ck}}$, which indicates the ciphertext of max and the min results under the public key pk_{ck}.

Step 3 (@ DSP): DSP randomly selects four numbers, R_1, R_2, R_3, ck_1, which satisfies $R_1 = R_2 * ck_1 \bmod n^2$ and $\mathcal{L}(R_1) < \mathcal{L}(n)/4$ and then preprocesses the data from DPs as follows:

1. $[1] = \{(1+n) * PK^{r'}, g^{r'}\};[m_-] = [m_1 - m_2] = [m_1] * [m_2]^{n-1}$
2. $[R_2 m_+ + R_3] = [m_1 + m_2]^{R_2} * [R_3];[R_2 m_-] = \left(T_-, T'_-\right) = [m_1 - m_2]^{R_2}$

$$[2*m_- + 1] = \left(T, T'\right) = \left\{(1 + (2*m_- + 1)*n) * PK^{r'+2*r_1}, g^{r'+2*r_1}\right\}$$

The DSP calls ***PDec*1** to decrypt $[R_2 * m_+ + R_3]$ to get $[R_2 * m_+ + R_3]_{pk_{CP}}$. Then, it further flips a coin s. If $s = -1$, it computes $\left(T_1{}^{(1)}, T_1{}'^{(1)}\right) = \left\{T^{n-R_1}, \left(T'\right)^{a*(n-R_1)}\right\} = [-R_1 * (2*m_- + 1)]_{pk_{CP}}, \left(T_2, T'_2\right) = [-R_2 m_-]_{pk_{CP}} = \left\{T_-{}^{n-R_1}, \left(T'_-\right)^{a*(n-R_1)}\right\}$. Otherwise if $(s = 1)$, it directly calls ***PDec*1** to compute $\left(T_1{}^{(1)}, T_1{}'^{(1)}\right) = \left\{T^{R_1}, T'^{a*R_1}\right\} = [R_1 * (2*m_- + 1)]_{pk_{CP}}, \left(T_2, T'_2\right) = [R_2 m_-]_{pk_{CP}} = \left\{T_-, \left(T'_-\right)^a\right\}$. Then it sends CP the data packet $\{\left(T_1{}^{(1)}, T_1{}'^{(1)}\right), \left(T_2, T'_2\right), [R_2 * m_+ + R_3]_{pk_{CP}}\}$.

Step 4 (@ CP): The CP calls $PDec2$ to decrypt $(T_1{}^{(1)}, T_1{}'^{(1)})$ and $\left(T_2, T'_2\right)$from DSP to obtain raw data $m' = R_1 * (2 * m_- + 1) \bmod n, \hat{m_-} = (R_2 m_-) \bmod n$ if s $= 1$ or $m' = -R_1 * (2 * m_- + 1) \bmod n, \hat{m_-} = (-R_2 m_-) \bmod n$ if s $= -1$.

The CP checks the sign of m' by comparing $\mathcal{L}\left(m'\right)$ with $\mathcal{L}(n)/2$. If $\mathcal{L}\left(m'\right) < \mathcal{L}(n)/2$, it sets $u = 1$; otherwise, $u = -1$. And it further encrypts the raw data $u * \hat{m_-}$ with a randomly chosen key pair $\left(ck_2, pk_{ck_2} = g^{ck_2}\right)$: $\left[u * \hat{m_-}\right]_{pk_{ck_2}} = \left(\overline{T}, \overline{T}'\right) = \{\left(1 + u\hat{m_-} * n\right) g^{ck_2 * r}, g^r\}$.

Decrypt $[R_2 * m_+ + R_3]_{pk_{CP}}$ to get $R_2 * m_+ + R_3$ and re-encrypt it as $[R_2 * m_+ + R_3]_{pk_{ck_2}}$. Moreover, it needs to encrypt ck_2 with ABE to get

$CK_1' = Enc^{ABE}(ck_2, \gamma, PK')$. Finally, the CP forwards the data packet to DSP: $\{[u * \hat{m_-}]_{pk_{ck_2}}, [R_2 * m_+ + R_3]_{pk_{ck_2}}, CK_1'\}$.

Step 5 (@ DSP): First, the DSP sets $\{T, T'\}$=$[R_2 * m_+ + R_3]_{pk_{ck_2}}$ and then computes $[R_2 * m_+]_{pk_{ck_2}} = \{T * (1 - R_3 * n), T'\}$. The DSP computes $r = (2R_1)^{-1}$ *mod n* and finally obtains the encrypted max and min: $[\max]_{pk_{ck}} = \left(\left([u * \hat{m_-}]_{pk_{ck_2}} * [R_2 * m_+]_{pk_{ck_2}}\right)^{1,ck_1}\right)^r$; $[\min]_{pk_{ck}} = \left(\left(\left([u * \hat{m_-}]_{pk_{ck_2}}\right)^{n-1} * [R_2 * m_+]_{pk_{ck_2}}\right)^{1,ck_1}\right)^r$.

The DSP calls HE^{ABE} to obtain $CK = CK_1' * Enc^{ABE}(ck_2, \gamma, PK') = Enc^{ABE}(ck_1 * ck_2, \gamma, PK')$.

Step 6 (@ DR): The DR can access the computation results if it satisfies the access policy.

Multiple-to-Multiple (M2M)

DSP and CP invoke the T2M rather than the T2O to obtain the final result $[\max_{\lfloor lb(n)\rfloor,1}]_{pk_{ck}}$. Owing to chapter length limitation, we skip the details of above process.

Division

Scheme 1

Scheme 1 can provide the ciphertext of division result $[\lfloor m_1/m_2 \rfloor]_{pk_{DR}}$ as shown in Fig. 2.

Step 3 (Data Preparation @ DSP): DSP first chooses two random numbers r_1, r_2, where $L(r_i) < L(n)/4$. Then, it processes data to conceal each raw data from CP, as described below:

1. $[m_1r_1] = \{T_1{}^{r_1}, (T_1')^{r_1}\} = [m_1]^{r_1}$, $[m_2r_1] = \{T_2{}^{r_1}, (T_2')^{r_1}\} = [m_2]^{r_1}$.
2. $[m_2r_1r_2] = [m_2r_1]^{r_2} = [m_2]^{r_1r_2} = \{T_2{}^{r_1r_2}, (T_2')^{r_1r_2}\}$
   ; $[m_1r_1 + m_2r_1r_2] = [m_1r_1] * [m_2r_1r_2]$.
3. $[m_2r_1]_{pk_{CP}} = \left\{T_2{}^{r_1}, (T_2')^{r_1 * sk_{DSP}}\right\} = \{(1 + r_1 * m_2 * n)\, PK^{r*r_1}, g^{r*a*r_1}\}$.

$$[m_1r_1 + m_2r_1r_2]_{pk_{CP}} = \{T_1{}^{r_1} T_2{}^{r_1r_2}, (T_2')^{r_1} (T_2')^{r_1r_2}\}^a.$$

Next, DSP sends the data packet $([m_2r_1]_{pk_{CP}}, [m_1r_1 + m_2r_1r_2]_{pk_{CP}})$ to CP.

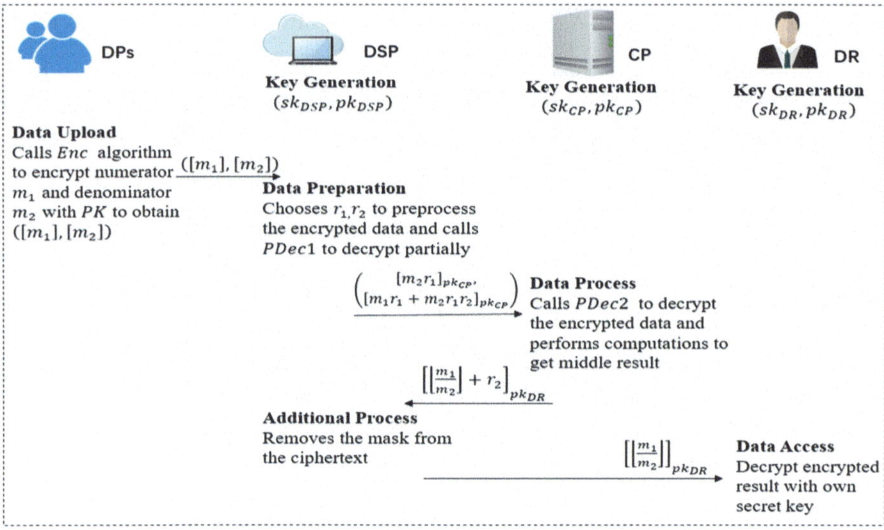

Fig. 2 The procedure of division computation for a targeted data requester

Step 4 (Data Process @ CP): Upon receiving the data packet from DSP, CP calls ***PDec*2** to decrypt the packet. Then, CP performs division operations on plaintexts and encrypts the computational result with pk_{DR}.

1. $m_2 r_1 = T_2{}^{r_1} / \left(\left(T_2'\right)^{r_1 * a}\right)^b \bmod n;\ m_1 r_1 + m_2 r_1 r_2$
$= T_1{}^{r_1} T_2{}^{r_1 r_2} / \left(\left(T_2'\right)^{r_1} \left(T_2'\right)^{r_1 r_2}\right)^{a*b} \bmod n.$
2. $(m_1 r_1 + m_2 r_1 r_2) / m_2 r_1 = \left\lfloor \frac{m_1}{m_2} \right\rfloor + r_2;\ \left[\left\lfloor \frac{m_1}{m_2} \right\rfloor + r_2\right]_{pk_{DR}}$
$= \left\{\left(1 + \left(\left\lfloor \frac{m_1}{m_2} \right\rfloor + r_2\right) * n\right) pk_{DR}^r, g^r\right\}.$

The data sent to DSP is the ciphertext $\left[\left\lfloor \frac{m_1}{m_2} \right\rfloor + r_2\right]_{pk_{DR}}$. We use $\left\lfloor \frac{m_1}{m_2} \right\rfloor$ to represent the quotient.

Step 5 (Additional Process @ DSP): DSP encrypts the random number r_2 as $[r_2]_{pk_{DR}}$ and computes $\left([r_2]_{pk_{DR}}\right)^{n-1}$. Then, DSP removes the mask from the ciphertext as below.

$$\left[\left\lfloor \frac{m_1}{m_2} \right\rfloor + r_2\right]_{pk_{DR}} * \left([r_2]_{pk_{DR}}\right)^{n-1} = \left[\left\lfloor \frac{m_1}{m_2} \right\rfloor + r_2\right]_{pk_{DR}} * \left([-r_2]_{pk_{DR}}\right)$$
$$= \left[\left\lfloor \frac{m_1}{m_2} \right\rfloor\right]_{pk_{DR}}.$$

Step 6 (Data Access @ DR): Upon receiving the final ciphertext from DSP, the targeted DR can call $Dec\left(\left[\left\lfloor \frac{m_1}{m_2} \right\rfloor\right]_{pk_{DR}}, sk_{DR}\right)$ to get the final quotient of the division.

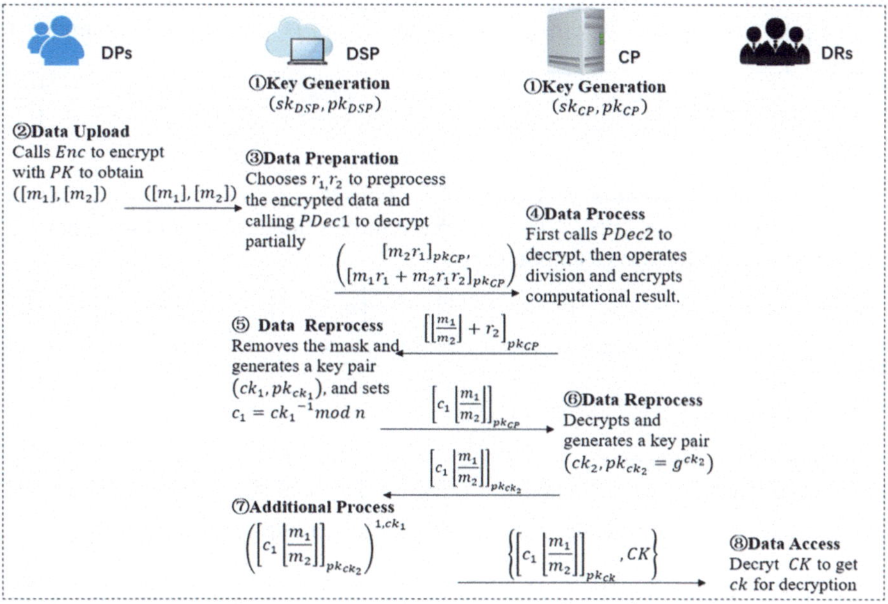

Fig. 3 The procedure of division computation with flexible access control

Scheme 2

We design Scheme 2 to enable flexible access control over computational results as shown in Fig. 3.

Step 3 (Data Preparation @ DSP): DSP chooses two random numbers r_1 and r_2 where $L(r_i) < L(n)/4$ and preprocesses data to mask raw data as follows, which is the same as Scheme 1.

1. $[m_1 r_1] = \left\{T_1^{r_1}, \left(T_1'\right)^{r_1}\right\} = [m_1]^{r_1}$, $[m_2 r_1] = \left\{T_2^{r_1}, \left(T_2'\right)^{r_1}\right\} = [m_2]^{r_1}$.
2. $[m_2 r_1 r_2] = [m_2]^{r_1 r_2} = \left\{T_2^{r_1 r_2}, \left(T_2'\right)^{r_1 r_2}\right\}$;$[m_1 r_1 + m_2 r_1 r_2] = [m_1 r_1] * [m_2 r_1 r_2]$.
3. $[m_2 r_1]_{pk_{CP}} = \left\{T_2^{r_1}, \left(T_2'\right)^{r_1 * sk_{DSP}}\right\} = \left\{(1 + r_1 * m_2 * n)\, PK^{r * r_1}, g^{r * a * r_1}\right\}$.

$$[m_1 r_1 + m_2 r_1 r_2]_{PK_{CP}} = \left\{T_1^{r_1} T_2^{r_1 r_2}, \left(T_2'\right)^{r_1} \left(T_2'\right)^{r_1 r_2}\right\}^{a}.$$

Similarly, DSP sends the data packet $\left([m_2 r_1]_{pk_{CP}}, [m_1 r_1 + m_2 r_1 r_2]_{pk_{CP}}\right)$ to CP.

Step 4 (Data Process @ CP): CP calls ***PDec2***$(_*,\ $***sk***$_{CP})$ to decrypt received data from DSP to get $m_2 r_1$ and $m_1 r_1 + m_2 r_1 r_2$, and then performs division operations on plaintexts with perturbations, as well as encrypts the computational result by calling $Enc(_*, pk_{CP})$.

1. $\left\lfloor \frac{m_1}{m_2} \right\rfloor + r_2 = (m_1 r_1 + m_2 r_1 r_2)/m_2 r_1$, where $\left\lfloor \frac{m_1}{m_2} \right\rfloor$ is quotient and remainder is ignored.
2. CP sends the data $\left[\left\lfloor \frac{m_1}{m_2} \right\rfloor + r_2\right]_{pk_{CP}} = \left\{\left(1 + \left(\left\lfloor \frac{m_1}{m_2} \right\rfloor + r_2\right) * n\right) pk_{CP}^r, g^r\right\}$ to DSP.

Step 5 (Data Reprocess @ DSP): DSP chooses a partial key ck_1 and sets a random number as $c_1 = (ck_1)^{-1}$ *mod n*. DSP removes the mask from the ciphertext and performs the following computations:

$$\left[\left|\frac{m_1}{m_2}\right| + r_2\right]_{pk_{CP}} * \left([r_2]_{pk_{CP}}\right)^{n-1} = \left[\left|\frac{m_1}{m_2}\right|\right]_{pk_{CP}}; \left[c_1 \left|\frac{m_1}{m_2}\right|\right]_{pk_{CP}} = \left(\left[\left|\frac{m_1}{m_2}\right|\right]_{pk_{CP}}\right)^{c_1} = \left\{\tilde{T}, \tilde{T}'\right\}$$

The data sent to CP is $\left[c_1 \left\lfloor \frac{m_1}{m_2} \right\rfloor\right]_{pk_{CP}}$.

Step 6 (Data Reprocess @ CP): CP first calls ***PDec2***$(*,$ $\boldsymbol{sk_{CP}})$ to decrypt the received data. Then, it chooses a partial key ck_2 to generate a key pair $(ck_2, pk_{ck_2} = g^{ck_2})$ and calls $Enc\,(*, pk_{ck_2})$ to encrypt the data:$c_1 \left\lfloor \frac{m_1}{m_2} \right\rfloor = \tilde{T} / \left(\tilde{T}'\right)^b \bmod n$;$\left[c_1 \left\lfloor \frac{m_1}{m_2} \right\rfloor\right]_{pk_{ck_2}} = \left\{\left(1 + \left(c_1 \left\lfloor \frac{m_1}{m_2} \right\rfloor\right) * n\right) pk_{ck_2}^r, g^r\right\} = \left\{\overline{T}, \overline{T}'\right\}$.

In addition, CP calls Enc^{ABE} to encrypt ck_2:$CK_2 = Enc^{ABE}(ck_2, \mathcal{T}, PK')$. Furthermore, the ABE key CK_2 is sent to DSP along with the ciphertext $\left[c_1 \left\lfloor \frac{m_1}{m_2} \right\rfloor\right]_{pk_{ck_2}}$.

Step 7 (Additional Process @ DSP): DSP operates partial modular computation on received ciphertext with its partial key ck_1 and performs ABE algorithms to obtain encrypted access keys.

$$\left[\left\lfloor \frac{m_1}{m_2} \right\rfloor\right]_{pk_{ck}} = \left(\left[c_1 \left\lfloor \frac{m_1}{m_2} \right\rfloor\right]_{pk_{ck_2}}\right)^{1, ck_1} = \left\{\overline{T}^{ck_1}, \overline{T}'\right\}$$

$$= \left\{\left(1 + ck_1 * c_1 * \left\lfloor \frac{m_1}{m_2} \right\rfloor * n\right) g^{ck_1 * ck_2 * r}, g^r\right\} = \left\{\left(1 + \left\lfloor \frac{m_1}{m_2} \right\rfloor * n\right) g^{ck * r}, g^r\right\},$$

where $pk_{ck} = (pk_{ck_2})^{ck_1} = (pk_{ck_1})^{ck_2}$.

1. Calling Enc^{ABE} to encrypt ck_1: $CK_1 = Enc^{ABE}(ck_2, \mathcal{T}, PK')$.
2. ABE homomorphic computation: $CK = CK_1 * CK_2 = Enc^{ABE}(ck_1 * ck_2, \mathcal{T}, PK')$.

Finally, DSP keeps $\left[\left\lfloor \frac{m_1}{m_2} \right\rfloor\right]_{pk_{ck}}$ and *CK* for user access.

Step 8 (Data Access @ DRs): Upon receiving the computational results and *CK* from DSP, the DRs who satisfy the access policy can obtain a secret key *SK*′ from the authority. Thus, the DRs can decrypt CK to get ck by calling Dec^{ABE} and get the final quotient by calling $Dec\left(\left[\left\lfloor \frac{m_1}{m_2} \right\rfloor\right]_{pk_{ck}}, ck\right)$.

Division and Rest

Scheme 3

To support accurate division computation, we design Scheme 3 to further calculate remainder based on Scheme 1. We omit the same first three steps as in Scheme 1 and introduce the additional part as below.

Step 4 (Data Process @ CP): Upon receiving the data packet from DSP, CP first calls $\boldsymbol{PDec2}(_*, \boldsymbol{sk_{CP}})$ to obtain masked plaintext and performs the following computations:

$\left\lfloor \frac{m_1}{m_2} \right\rfloor + r_2 = (m_1 r_1 + m_2 r_1 r_2)/m_2 r_1; R r_1 = (m_1 r_1 + m_2 r_1 r_2) - m_2 r_1 * \left(\left\lfloor \frac{m_1}{m_2} \right\rfloor + r_2\right)$.

Then, CP calls $Enc(_*, pk_{DR})$ to encrypt the above computational result as $\left\{\left[\left\lfloor \frac{m_1}{m_2} \right\rfloor + r_2\right]_{pk_{DR}}, [Rr_1]_{pk_{DR}}\right\}$ and sends the data packet to DSP.

Step 5 (Data Additional Process @ DSP): DSP removes the mask from received ciphertext to get encrypted quotient and remainder as follows:

$\left[\left\lfloor \frac{m_1}{m_2} \right\rfloor\right]_{pk_{DR}} = \left[\left\lfloor \frac{m_1}{m_2} \right\rfloor + r_2\right]_{pk_{DR}} * \left([r_2]_{pk_{DR}}\right)^{n-1}; [R]_{pk_{DR}} = \left([Rr_1]_{pk_{DR}}\right)^{r_1^{-1}}$.

Step 6 (Data Access @ DR): Upon receiving the computational results from DSP, the targeted DR can decrypt two ciphertext $\left[\left\lfloor \frac{m_1}{m_2} \right\rfloor\right]_{pk_{DR}}$ and $[R]_{pk_{DR}}$ to get the final quotient and remainder by calling $Dec(_*, sk_{DR})$.

Scheme 4

Similarly, Scheme 4 is proposed by adding the computations of remainder based on Scheme 2. We introduce its details below by omitting the same first three steps as in Scheme 2.

Step 4 (Data Process @ CP): Upon receiving data packet from DSP, CP first calls $\boldsymbol{PDec2}(_*, sk_{CP})$ to obtain two messages $m_2 r_1$ and $(m_1 r_1 + m_2 r_1 r_2)$. Then, it performs basic computations to get $\left\lfloor \frac{m_1}{m_2} \right\rfloor + r_2$ and Rr_1. Furthermore, CP calls $Enc(_*, pk_{CP})$ to encrypt the computational result and sends the encrypted data packet $\left\{\left[\left\lfloor \frac{m_1}{m_2} \right\rfloor + r_2\right]_{pk_{CP}}, [Rr_1]_{pk_{CP}}\right\}$ to DSP.

Step 5 (Data Reprocess @ DSP): DSP first chooses a partial key ck_1 and sets a random number as $c_1 = (ck_1)^{-1}$ *mod* n. Then, it removes the mask from received ciphertext and conceals the data by performing the following computations:

1. $\left[\left\lfloor\frac{m_1}{m_2}\right\rfloor + r_2\right]_{pk_{CP}} * \left([r_2]_{pk_{CP}}\right)^{n-1} = \left[\left\lfloor\frac{m_1}{m_2}\right\rfloor\right]_{pk_{CP}};\left[c_1\left\lfloor\frac{m_1}{m_2}\right\rfloor\right]_{pk_{CP}} = \left(\left[\left\lfloor\frac{m_1}{m_2}\right\rfloor\right]_{pk_{CP}}\right)^{c_1}$.
2. $\left([Rr_1]_{pk_{CP}}\right)^{r_1^{-1}} = [R]_{pk_{CP}};[c_1R]_{pk_{CP}} = \left([R]_{pk_{CP}}\right)^{c_1} = \left\{\hat{T}, \hat{T}'\right\}$.

Next, the data packet $\{\left[c_1\left\lfloor\frac{m_1}{m_2}\right\rfloor\right]_{pk_{CP}}, [c_1R]_{pk_{CP}}\}$ is sent to CP.

Step 6 (Data Reprocess @ CP): With received data packet, CP first performs ***PDec2***$(*, sk_{CP})$ on encrypted data. Then, it chooses a partial key ck_2 to generate a key pair $\left(ck_2, pk_{ck_2} = g^{ck_2}\right)$ and calls $Enc\left(*, pk_{ck_2}\right)$ to encrypt the masked data. Detailed processes are described below:

$$\left[c_1\left\lfloor\frac{m_1}{m_2}\right\rfloor\right]_{pk_{CP}} \overset{\textbf{PDec2}(*, sk_{CP})}{\rightarrow} c_1\left\lfloor\frac{m_1}{m_2}\right\rfloor \overset{Enc\left(*, pk_{ck_2}\right)}{\rightarrow} \left[c_1\left\lfloor\frac{m_1}{m_2}\right\rfloor\right]_{pk_{ck_2}}.$$

$$[c_1R]_{pk_{CP}} \overset{\textbf{PDec2}(*,\ sk_{CP})}{\rightarrow} c_1R \overset{Enc\left(*, pk_{ck_2}\right)}{\rightarrow} [c_1R]_{pk_{ck_2}}.$$

In addition, CP calls ABE encryption algorithm to encrypt ck_2:$CK_2 = Enc^{ABE}\left(ck_2, \mathcal{T}, PK'\right)$.

The data packet $\{\left[c_1\left\lfloor\frac{m_1}{m_2}\right\rfloor\right]_{pk_{ck_2}}, [c_1R]_{pk_{ck_2}}, CK_2\}$ is sent to DSP.

Step 7 (Additional Process @ DSP): Upon receiving the data packet, DSP performs the following operations:

1. $\left[\left\lfloor\frac{m_1}{m_2}\right\rfloor\right]_{pk_{ck}} = \left(\left[c_1\left\lfloor\frac{m_1}{m_2}\right\rfloor\right]_{pk_{ck_2}}\right)^{1,ck_1};[R]_{pk_{ck}} = \left([c_1R]_{pk_{ck_2}}\right)^{1,ck_1}$.
2. Using Enc^{ABE} to encrypt ck_1: $CK_1 = Enc^{ABE}\left(ck_2, \mathcal{T}, PK'\right)$.
3. Homomorphism of ABE: $CK = CK_1 * CK_2 = Enc^{ABE}\left(ck_1 * ck_2, \mathcal{T}, PK'\right)$.

DSP keeps the encrypted data packet $\left\{\left[\left\lfloor\frac{m_1}{m_2}\right\rfloor\right]_{pk_{ck}}, [R]_{pk_{ck}}\right\}$ and ABE key CK for user access.

Step 8 (Data Access @ DR): The DRs that satisfy the access policy can obtain a secret key SK' from the authority, which can be used to get ck by calling $Dec^{ABE}(PK', SK', CK)$. Then DRs decrypt the received ciphertext $\left[\left\lfloor\frac{m_1}{m_2}\right\rfloor\right]_{pk_{ck}}$ and $[R]_{pk_{ck}}$ obtained from DSP to get the quotient and remainder.

Applications in Cybersecurity Education

Privacy-preserving data processing with ABE guarantees data security and user privacy. In the field of cybersecurity education, privacy-sensitive data are generated and issued, e.g., course feedback, survey inputs, security-related data for intrusion/malware detection provided by different parties for course exercises, multi-party sensitive data processing, etc. By analyzing these data in a privacy-preserving way, we can judge teaching performance, support further course improvement, offer essential course practice to allow students to deeply understand cybersecurity theories and technologies, etc. Herein, our schemes offer an efficient and privacy-preserving measure to conduct data analysis, which provides a good practice for students to understand homophonic encryption and its usage. Some concrete examples are listed below:

Privacy-Preserving Data Analysis

The feedbacks and opinions of all students and faculties are essential to improve course quality. Our schemes can be adopted for data collection and dispel privacy concerns. It can be used in the following two scenarios:

Teaching Performance Evaluation: Our schemes can collect, process, and analyze the student ratings in a privacy-preserving way, especially in online courses. With our schemes, students are encouraged to provide their feedback or survey inputs honestly. Furthermore, the students can select preferred courses by comparing different course evaluation results and personal study expectation.

Preparing and Rating Examination Questions: The design goal of flexible data sharing and access control in our schemes would be a key point for remote cooperation among experts or teachers.

Teachers can prepare examination questions cooperatively in a privacy-preserving and flexible way. Our schemes can protect the content of examination papers and enable the teachers to get the feedback of other teachers to assess the rationality of papers. Moreover, they can also be applied to exchange the statistics of examination results from students and complete remote rating through cooperation. This kind of online cooperation can greatly improve education efficiency.

Cybersecurity Experimental Platform

Apart from the above, our schemes can be integrated to build up an experimental platform for cybersecurity education. It will help students gain a deep insight into privacy and security of outsourced data processing.

Cybersecurity Course Exercises: Our schemes offer a good experimental platform to conduct cybersecurity experiments with regard to secure data analytics for flexible and fine-grained access control over the processing results. For example, a number of students can collect sensitive security-related data from different sources and perform secure processing on those data at an untrusted party, and then different students get the processing results without knowing other inputs. For another example, students can provide their own mobile phone apps' usage data to process in a secure way with our schemes in order to know the trust and popularity of the apps without disclosing their personal app usage information. Through these experimental exercises, the students can get deep insight on encrypted data processing and flexible access control over processing results.

Conclusion

With the development and widely deployment of information systems, cybersecurity education becomes popular and significant. In order to gain customized courses, some private information are offered but may erode their privacy. In this chapter, we proposed an efficient and secure system to achieve privacy-preserving data processing with ABE-based flexible access control. It can support several operations and achieve fine-grained access control without the need of fully trusted cloud servers, which can be deployed in cybersecurity education framework. We also illustrate a number of applications of our system for the purpose of cybersecurity education.

Acknowledgment The work is supported in part by the National Natural Science Foundation of China under Grants 61672410 and 61802293, the National Postdoctoral Program for Innovative Talents under grant BX20180238, the Project funded by China Postdoctoral Science Foundation under grant 2018M633461, the Academy of Finland under Grants 308087, 314203, and 335262, the Shaanxi Innovation Team project under grant 2018TD-007, and the 111 project under grant B16037.

References

1. A. Belle, R. Thiagarajan, S. Soroushmehr, F. Navidi, D.A. Beard, K. Najarian, Big data analytics in healthcare. Biomed. Res. Int. **2015**, 1–16 (2015)
2. G. Javidi, E. Sheybani, K-12 Cybersecurity education, research, and outreach, in *2018 IEEE Frontiers in Education Conference (FIE)*, (San Jose, CA, USA, 2018), pp. 1–5
3. J.J. Stephen, S. Savvides, R. Seidel, P. Eugster, Practical confidentiality preserving big data analysis, in *6th USENIX Workshop on Hot Topics in Cloud Computing (HotCloud 14)*, (Philadelphia, PA, USA, 2014)
4. B. Wang, M. Li, S.S. Chow, H. Li, A tale of two clouds: Computing on data encrypted under multiple keys, in *2014 IEEE Conference on Communications and Network Security (CNS)*, (San Francisco, CA, USA, 2014), pp. 337–345

5. A. Peter, E. Tews, S. Katzenbeisser, Efficiently outsourcing multiparty computation under multiple keys. IEEE Transactions on Information Forensics and Security (TIFS) **8**, 2046–2058 (2013)
6. X. Liu, R. Choo, R. Deng, R. Lu, J. Weng, Efficient and privacy-preserving outsourced calculation of rational numbers. IEEE Transactions on Dependable and Secure Computing (TDSC) **15**, 27–39 (2016)
7. X. Liu, R. Deng, W. Ding, R. Lu, B. Qin, Privacy-preserving outsourced calculation on floating point numbers. IEEE Transactions on Information Forensics and Security **11**, 2513–2527 (2016)
8. R. Bost, R.A. Popa, S. Tu, S. Goldwasser, Machine learning classification over encrypted data, in *NDSS*, (San Diego, California, USA, 2015)
9. Z. Yan, P. Zhang, A.V. Vasilakos, A survey on trust management for internet of things. J. Netw. Comput. Appl. **42**, 120–134 (2014)
10. A. Khedr, G. Gulak, SecureMed: Secure medical computation using GPU-accelerated Homomorphic encryption scheme. IEEE Journal of Biomedical & Health Informatics **22**, 597–606 (2017)
11. Z. Brakerski, C. Gentry, V. Vaikuntanathan, (leveled) fully homomorphic encryption without bootstrapping, in *Proceedings of the 3rd Innovations in Theoretical Computer Science Conference*, (Cambridge, MA, USA, 2012), pp. 309–325
12. C. Gentry, Computing arbitrary functions of encrypted data. Commun. ACM **53**, 97–105 (2010)
13. M. Van Dijk, C. Gentry, S. Halevi, V. Vaikuntanathan, Fully homomorphic encryption over the integers, in *Annual International Conference on the Theory and Applications of Cryptographic Techniques*, (Riviera, French, 2010), pp. 24–43
14. V.C. Hu, T. Grance, D.F. Ferraiolo, D.R. Kuhn, An access control scheme for big data processing, in *10th IEEE International Conference on Collaborative Computing: Networking, Applications and Worksharing*, (Chicago, Illinois, USA, 2014), pp. 1–7
15. Z. Yan, W. Ding, V. Niemi, A.V. Vasilakos, Two schemes of privacy-preserving trust evaluation. Future Generation Computer Systems (FGCS) **62**, 175–189 (2015)
16. C. Huang, Z. Yan, N. Li, M. Wang, Secure pervasive social communications based on Trust in a Distributed way. IEEE Access **4**, 9225–9238 (2016)
17. Z. Yan, X. Li, M. Wang, A. Vasilakos, Flexible data access control based on trust and reputation in cloud computing. IEEE Transactions on Cloud Computing **5**, 485–498 (2015)
18. Z. Yan, X. Li, R. Kantola, Controlling cloud data access based on reputation. Mobile Networks and Applications **20**, 828–839 (2015)
19. W. Ding, Z. Yan, R.H. Deng, Encrypted data processing with Homomorphic re-encryption. Inf. Sci. **409**, 35–55 (2017)
20. J. Feng, L.T. Yang, Q. Zhu, K.-K.R. Choo, Privacy-preserving tensor decomposition over encrypted data in a federated cloud environment. IEEE Transactions on Dependable and Secure Computing (2018). https://doi.org/10.1109/TDSC.2018.2881452
21. L. Kamm, J. Willemson, Secure floating point arithmetic and private satellite collision analysis. Int. J. Inf. Secur. **14**, 531–548 (2015)
22. D. Bogdanov. *Sharemind: Programmable Secure Computations With Practical Applications* (Tartu University, 2013), PhD Thesis
23. J.H. Cheon, J.-S. Coron, J. Kim, M.S. Lee, T. Lepoint, M. Tibouchi, A. Yun, Batch fully homomorphic encryption over the integers, in *Annual International Conference on the Theory and Applications of Cryptographic Techniques*, (Athens, 2013), pp. 315–335
24. W. Wang, Y. Hu, L. Chen, X. Huang, B. Sunar, Exploring the feasibility of fully homomorphic encryption. IEEE Trans. Comput. **64**, 698–706 (2015)
25. L. Morris, Analysis of partially and fully homomorphic encryption. Rochester Institute of Technology, 1–5 (2013)
26. X. Liu, R.H. Deng, Y. Yang, H.N. Tran, S. Zhong, Hybrid privacy-preserving clinical decision support system in fog–cloud computing. Futur. Gener. Comput. Syst. **78**, 825–837 (2017)

27. Z. Yan, W. Ding, H. Zhu, A scheme to manage encrypted data storage with deduplication in cloud, in *International Conference on Algorithms and Architectures for Parallel Processing*, (Zhangjiajie, China, 2015), pp. 547–561
28. C. Dong, G. Russello, N. Dulay, Shared and searchable encrypted data for untrusted servers, in *IFIP Annual Conference on Data and Applications Security and Privacy*, (London, 2008), pp. 127–143
29. W.C. Garrison III, A. Shull, S. Myers, A.J. Lee, On the practicality of cryptographically enforcing dynamic access control policies in the cloud, in *2016 IEEE Symposium on Security and Privacy*, (San Jose, 2016), pp. 819–838
30. Z. Tianyi, L. Weidong, S. Jiaxing, An efficient role based access control system for cloud computing, in *IEEE 11th International Conference on Computer and Information Technology (CIT)*, (Paphos, Cyprus, 2011), pp. 97–102
31. J. Bethencourt, A. Sahai, B. Waters, Ciphertext-policy attribute-based encryption, in *2007 IEEE Symposium on Security and Privacy (SP'07)*, (Oakland, 2007), pp. 321–334
32. V. Goyal, O. Pandey, A. Sahai, B. Waters, Attribute-based encryption for fine-grained access control of encrypted data, in *13th ACM Conference on Computer and Communications Security*, (Alexandria, 2006), pp. 89–98
33. S. Yu, C. Wang, K. Ren, W. Lou, Achieving secure, scalable, and fine-grained data access control in cloud computing, in *2010 Proceedings IEEE INFOCOM*, (San Diego, 2010), pp. 1–9
34. M. Li, S. Yu, Y. Zheng, K. Ren, W. Lou, Scalable and secure sharing of personal health records in cloud computing using attribute-based encryption. IEEE Transactions on Parallel and Distributed Systems **24**, 131–143 (2013)
35. Z. Wan, J.E. Liu, R.H. Deng, HASBE: A hierarchical attribute-based solution for flexible and scalable access control in cloud computing. IEEE Transactions on Information Forensics and Security (TIFS) **7**, 743–754 (2012)
36. M. Franz, B. Deiseroth, K. Hamacher, S. Jha, S. Katzenbeisser, H. Schröder, Secure computations on non-integer values, in *2010 IEEE International Workshop on Information Forensics and Security*, (Seattle, Washington, USA, 2010), pp. 1–6
37. M. Dahl, C. Ning, T. Toft, On secure two-party integer division, in *International Conference on Financial Cryptography and Data Security*, (Bonaire, 2012), pp. 164–178
38. T. Veugen, Encrypted integer division and secure comparison. International Journal of Applied Cryptography **3**, 166–180 (2014)
39. O. Catrina, A. Saxena, Secure computation with fixed-point numbers, in *International Conference on Financial Cryptography and Data Security*, (Canary Islands, Spain, 2010), pp. 35–50
40. R. Bhoyar, P. Palsodkar, S. Kakde, Design and implementation of goldschmidts algorithm for floating point division and square root, in *International Conference on Communications*, (London, 2015), pp. 1588–1592
41. C. Ugwuoke, Z. Erkin, R.L. Lagendijk, Secure fixed-point division for Homomorphically encrypted operands, in *Proceedings of the 13th International Conference on Availability, Reliability and Security*, (Hamburg, Germany, 2018), pp. 1–10
42. B.K. Samanthula, H. Chun, W. Jiang, An efficient and probabilistic secure bit-decomposition, in *Proceedings of the 8th ACM SIGSAC Symposium on Information, Computer and Communications Security*, (Hangzhou, 2013), pp. 541–546
43. R. Gurnani, K. Pandey, S.K. Rai, A scalable model for implementing cyber security exercises, in *2014 International Conference on Computing for Sustainable Global Development (INDIACom)*, (New Delhi, 2014), pp. 680–684
44. E. Amankwa, M. Loock, E. Kritzinger, Enhancing information security education and awareness: Proposed characteristics for a model, in *2015 Second International Conference on Information Security and Cyber Forensics (InfoSec)*, (Cape Town, 2015), pp. 72–77
45. R. Weiss, F. Turbak, J. Mache, M.E. Locasto, Cybersecurity education and assessment in EDURange. IEEE Security & Privacy **15**(3), 90–95 (2017)

46. J. LeClair, K.M. Hollis, D.M. Pheils, Cybersecurity education and training and its reliance on STEAM, in *2014 IEEE Integrated STEM Education Conference*, (Princeton, NJ, 2014), pp. 1–5
47. M. Frank, M. Leitner, T. Pahi, Design considerations for cyber security Testbeds: A case study on a cyber security Testbed for education, in *2017 IEEE 15th Intl Conf on Dependable, Autonomic and Secure Computing, 15th Intl Conf on Pervasive Intelligence and Computing, 3rd Intl Conf on Big Data Intelligence and Computing and Cyber Science and Technology Congress(DASC/PiCom/DataCom/CyberSciTech)*, (Orlando, FL, 2017), pp. 38–46
48. A. M'Baya, J. Laval, N. Moalla, Y. Ouzrout, A. Bouras, Ontology based system to guide internship assignment process, in *2016 12th International Conference on Signal-Image Technology & Internet-Based Systems (SITIS)*, (Naples, 2016), pp. 589–596
49. F. Ghemri, A. Bouras, Innovative education in cyber security field through collaborative education, in *2018 3rd Technology Innovation Management and Engineering Science International Conference (TIMES-iCON)*, (Bangkok, Thailand, 2018), pp. 1–5
50. R. Verma, M. Kantarcioglu, D. Marchette, E. Leiss, T. Solorio, Security analytics: Essential data analytics knowledge for Cybersecurity professionals and students. IEEE Security & Privacy **13**(6), 60–65 (2015)
51. W.X. Ding, R. Hu, Z. Yan, X.R. Qian, R.H. Deng, L.T. Yang, M.X. Dong, An extended framework of privacy-preserving computation with flexible access control. IEEE Trans. Netw. Serv. Manag., 1 (2019). https://doi.org/10.1109/TNSM.2019.2952462
52. W. Ding, Z. Yan, R. Deng, Privacy-preserving data processing with flexible access control. IEEE Transactions on Dependable & Secure Computing **17**, 363–376 (2017)
53. W.X. Ding, Z. Yan, X.R. Qian, R.H. Deng, Computing maximum and minimum with privacy preservation and flexible access control, in *IEEE GLOBECOM 2019*, (Hawaii, USA, 2019), pp. 1–7
54. P. Paillier, Public-key cryptosystems based on composite degree residuosity classes, in *International Conference on the Theory and Applications of Cryptographic Techniques*, (Berlin, Germany, 1999), pp. 223–238
55. E. Bresson, D. Catalano, D. Pointcheval, A simple public-key cryptosystem with a double trapdoor decryption mechanism and its applications, in *International Conference on the Theory and Application of Cryptology and Information Security*, (Berlin, Germany, 2003), pp. 37–54

Part IV
Laboratory Enhancements

Stepping-Stone Intrusion Detection and its Integration into Cybersecurity Curriculum

Jianhua Yang

Introduction to Stepping-Stone Intrusion

The ability to detect stepping-stone intrusion and protect computers from being attacked is essential for the safe and reliable operation of ubiquitous computing, information, and network systems. With the development of advanced computing technologies, intruders have learned to use various techniques to hide their identities in order to avoid detection. Of course, many computing systems have intrusion detection software installed. However, to avoid being detected and captured, most intruders launch their attacks via a long connection chain involving multiple zombie hosts. The zombie hosts are called stepping-stones [1]. The attacks launched via stepping-stones are called stepping-stone intrusion. To the best of our knowledge, stepping-stone intrusion and its detection techniques are rarely taught in cybersecurity curriculum even though the techniques have been widely used by most of the professional hackers. In this chapter, we propose to integrate the cutting-edge techniques of stepping-stone intrusion detection into cybersecurity curriculum not only to educate college students to be cybersecurity experts for defending our digital resources but also to make them keen of offensive cyber skills. We expect to (1) make students competitive by educating them with the knowledge, skills, and abilities in the field of cybersecurity and (2) train students to be successfully adaptive to the changes in this dynamic cybersecurity field quickly and efficiently.

Since the first time that stepping-stone concept was defined in [3] informally, lots of techniques have been proposed and developed to detect stepping-stone intrusion. The techniques can be categorized as two primary types. One type of the technique focuses on the incoming and outgoing network traffic of a computer

J. Yang (✉)
TSYS School of Computer Science, Columbus State University, Columbus, GA, USA
e-mail: yang_jianhua@ColumbusState.edu

K. Daimi, G. Francia III (eds.), *Innovations in Cybersecurity Education*,
https://doi.org/10.1007/978-3-030-50244-7_13

host to decide if the host is used as a stepping-stone. This includes checking the number of packets, packet payload, and timestamp gaps between captured packets and so on. Generally speaking, the main idea used in this type of stepping-stone intrusion detection technique is to model the behavior of the incoming and outgoing network connections of a host, respectively, and then to decide if the two behaviors are close enough or not. However, this type of detecting approaches suffers from high false-positive error rate since there are many applications using stepping-stones legally. Another type of detection technique is to decide stepping-stone intrusion by estimating the number of hosts used as stepping-stones in a connection chain. Even though stepping-stones can be used in many non-malicious applications, it is still rarely seen that three or more hosts are used as stepping-stones in most of the real-world applications surveyed in [1]. It is easy to understand that the second type of technique can reduce false-positive detection error. In this chapter, we first summarize the typical stepping-stone intrusion detection techniques and then determine the ones integrated into cybersecurity curriculum.

Survey of Stepping-Stone Intrusion Detection Techniques

There have been lots of techniques developed to detect stepping-stone intrusion since it was first observed. In this chapter, we divide the detection techniques into two different categories: host-based and connection chain-based detection techniques. In host-based detection techniques, we survey the typical approaches including content-based thumbprint, time-based thumbprint, using packet amount, using the number of RTTs, and random-walk detection. In connection with chain-based detection category, we survey the following techniques: Yung's approach [2], packet matching, step function, MMD (maximum-minimum distance) data mining approach, and using cross-matching packets.

In the category of host-based stepping-stone intrusion detection, we focus on comparing the incoming and outgoing network traffic of a computer host to obtain the similarity between the traffics of the two network connections. If the similarity is higher than a predefined threshold, it is highly suspicious that the host is used as a stepping-stone. In other words, the connection might be used to launch attacks. The key point in the detection techniques of this category is to model the behavior of incoming/outgoing network traffic, respectively. As shown in Fig. 1, host H1 is used a stepping-stone with C_{in} as its one incoming connection and C_{out} as its one outgoing connection. The network traffic in each connection can be identified as two different streams: request packets stream S_i and response packets stream E_j. Request packets are called send packets, and response packets are called echo packets. Different approaches to model network traffic may lead to different detection algorithms.

Content-based thumbprint approach is the first way ever proposed to detect stepping-stone intrusion by Staniford-Chen and Heberlein in 1995 [3]. If a network session passes through a host and the session is not encrypted, then the payload

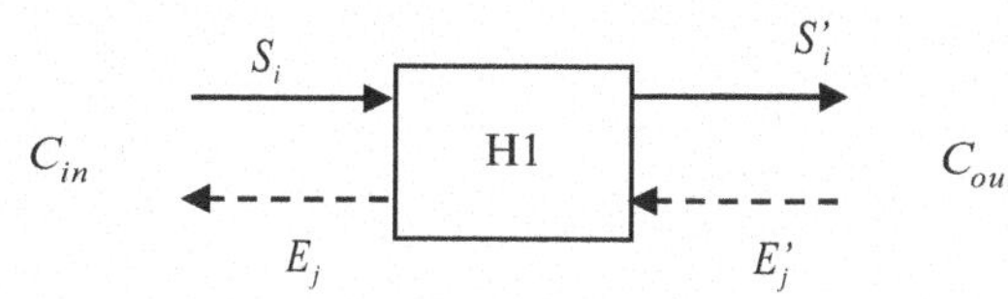

Fig. 1 Modeling a host-based stepping-stone detection

in each packet captured from both the incoming and outgoing connections of the host is observable. Simply comparing the content of each packet can determine if the two connections are relayed and further decide if the host is used as a stepping-stone. In order to make the comparison more efficient and have a quick detection, the researchers proposed to capture a certain number of packets from the two connections, respectively, and apply a hash function to map the contents of a series packets to a 24-byte digest which is called content-based thumbprint. Instead of directly comparing the contents of the packets from the two connections, which is time-consuming, content-based thumbprint approach compares the two 24-byte digests to determine if the host is used as a stepping-stone and further to decide if attacking exists. Unfortunately, this method cannot be applied to encrypted network sessions since its payloads are not observable.

In 2000, Y. Zhang and V. Paxson proposed a time-based thumbprint approach to detect stepping-stone intrusion by using on-off timestamp gaps between the packets collected from the network TCP/IP interactive session of a host [1]. Another similar approach to detect stepping-stone intrusion was proposed by Yoda and Etoh [4] in 2000 as well. Other than the time-based thumbprint, the approach proposed in [4] took advantage of the deviation between the incoming and outgoing connections of a host.

Since the beginning of the twenty-first century, more and more approaches have been developed to detect stepping-stone intrusion. Comparing the amount of the packets from the incoming and outgoing connections of a host, respectively, dominates the techniques to detect stepping-stone intrusion, such as the approaches proposed between 2002 and 2007 [5–7]. Watermark approaches were proposed to detect stepping-stone intrusion by Wang [8–10]. The basic idea of the watermark approaches is to inject a watermark into the incoming connection of a host and then to check if this watermark could be detected at any one of its outgoing connections. The mechanism behind the above approaches is to inspect and compare some features of the incoming and outgoing connections of the same host. If any outgoing connection of a host relays with any incoming connection of the same host, the host can be identified as a stepping-stone. One primary concern is that, sometimes, a user might use a host as a stepping-stone legitimately because some applications may need stepping-stones as well. If so, the approaches summarized in the above to detect stepping-stone intrusion by simply comparing two connections would incur false-positive error.

The category of connection chain-based stepping-stone intrusion detection can reduce false-positive detection error. Different from the host-based detection techniques, the approaches in this category detect stepping-stone intrusion using the

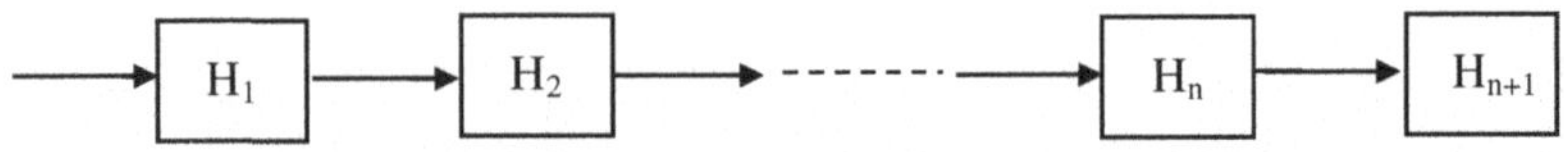

Fig. 2 A connection chain with multiple hosts connected

number of host used as stepping-stones in the same connection chain. Fig. 2 shows a case that multiple hosts are used as stepping-stones. If we detect stepping-stone at host H_1, the connection chain spans the hosts H_2, H_3, ... and H_{n+1}, respectively. In this case, there are n hosts used as stepping-stones. If three or more hosts are used as stepping-stones, it is highly suspicious to be a stepping-stone intrusion. The rationale behind the connection chain-based stepping-stone intrusion detection techniques is based on the observations: (1) it is legitimate that some applications may use one or two hosts as stepping-stones; (2) there are very few applications using three or more than three hosts as stepping-stones. Therefore, estimating the number of hosts used as stepping-stones in a connection chain is the key to reduce the false-positive detection error. But it is still challenging to estimate the total number of stepping-stones in a full connection chain. Most of the approaches proposed so far can only estimate the length (the number of stepping-stones) of the connection chain from a sensor (a stepping-stone host in a connection chain where a detection program resides) to the victim, which is also called downstream connection. As shown in Fig. 2, the host H_1 can be used as a sensor, and the connection chain from H_1 to H_{n+1} is the downstream connection. The part of a connection chain from a sensor to its intruder's host is called an upstream connection.

K. H. Yung first proposed an approach to estimate the length of a downstream connection chain of a host in 2002 [2]. He used the ratio between the ACK round-trip time (RTT) from a sensor to its adjacent host in the same connection and the echo RTT from the sensor to the victim host. His method can approximately tell the length of a downstream connection chain but may introduce false-negative errors. J. Yang and S. Huang proposed a step-function approach based on Yung's approach in 2004 aiming at estimating the length of a downstream connection chain more accurately [11]. Another method through mining computer network traffic to estimate the length of a downstream connection chain was developed in 2007 [12]. Some other approaches have also been proposed in the most recent years to detect stepping-stone intrusion, such as RTT-based random-walk approach [13] and RTT cross-matching [14].

Even though stepping-stone intrusion detection techniques have been developed and widely explored for more than 20 years, as what we know, the techniques have not been integrated systematically into cybersecurity curriculum. It is significant to teach students the knowledge of stepping-stone intrusion detection since more and more intruders have been using stepping-stones to launch their attacks. Most academic educators agree that offering ethical hacking in cybersecurity curriculum is a trend due to two reasons: one is that few of our trained and well-educated students become a malicious attacker; the other reason is that teaching offensive hacking techniques and penetrating skills in information assurance curriculum can yield more and highly qualified security professionals [15]. In this chapter, we develop eight modules based on the detection techniques proposed and integrate them into the cybersecurity curriculum and design eight hands-on labs to help students understand the eight course modules to enhance students' engagement and improve their hands-on learning experience.

The Rationale to Integrate the Techniques into Cybersecurity Curriculum

We are living in a society full of ubiquitous computing. Anywhere using computing involves privacy and security issues especially after the emerging of the Internet which is a double-edged sword. The wide applications of the Internet have changed our world a lot and made our life better and more convenient than before. One serious consequence coming with the Internet is that launching cyberattacks is also much easier than before. It is hard to capture hackers and sometimes even impossible to detect attacks. Another consequence is that the cybercrime grows faster than ever. It is urgent to protect our properties, businesses, and institutions in the digital realm. Cybersecurity becomes more important than ever. Just as Nathan Deal, the GA ex-governor, stressed in an announcement to invest $50 million in funding to establish the Georgia Cyber Innovation and Training Center in Augusta, GA, USA, "Cybersecurity is especially important now that cybercrime is bigger than the global black market for marijuana, cocaine and heroin combined".

Cyberattacks and threats are increasingly faced by businesses, consumers, and all other users of the Internet along with more and more aspects of the national infrastructure that depend on the correct operation of computers, networks, and the Internet. Based on the information from [18], among the 7.7 billion population of the world in the mid-2019, the individuals using the Internet worldwide can be as high as more than 4.5 billion users. The report of 2018 Cyber Incident & Breach Trends from the Internet Society [19] tells us that the financial impact of all types of incidents can be more than $45 billion in 2018, though the results vary widely due to the different methodologies to track data breaches. Today, more individuals and businesses are exposed to cyber threats than ever due to the wide usage of the Internet.

Facing the existing and increasingly potential devastating cyberattacks and threats, in order to protect our society and make the Internet a safe environment to conduct businesses fairly and securely, we need more well-trained and qualified cybersecurity professionals. According to a report on Cyberseek [20], the total number of employed cybersecurity workforce is 997,058 in the USA, and the total number of cybersecurity job openings is 504,316 nationally in 2019. The cybersecurity supply/demand heat map [20] shows that California, Texas, Georgia, New York, Florida, New York, Maryland, Virginia, and North Carolina are among the states having cybersecurity job openings in between 18,290 and 72,123. The national average of cybersecurity workforce supply/demand ratio is 2.0. From the budget for the fiscal year 2019 of the USA, we see that defending the nation and protecting the citizens from cyberattacks and threats have risen to national strategy. The US President's budget for the fiscal year 2019 earmarks $15 billion for cybersecurity-related activities [21]. Therefore, educating and training more qualified cybersecurity professionals is more urgent than ever. It is not only challenging but also an opportunity to the high education.

As 4-year colleges/institutions, we shoulder the heavy responsibility to train and educate cybersecurity workforce. Most colleges/universities offer cybersecurity courses focusing on defensive techniques which could defend computing systems from cyberattacking, such as information security, computer and network security, operating system security, security in programing, computer forensics, cryptography, etc. They all lack of offensive techniques. We all believe an old saying "the best defense is offense." Knowing how to perform ethical hacking is helpful to make defending strategy. Teaching students offensive techniques is more important and necessary than before. Unfortunately, most cybersecurity courses rarely mention how to conduct ethical cyberattacks. One reason is that most parents of students have a concern to possibly educate their children to be hackers. Another one, also the primary one, is that it is hard to teach offensive techniques due to lack of contents, facilities, and hands-on labs. We propose a plan to solve this problem by integrating stepping-stone intrusion detection techniques into cybersecurity curriculum. The course contents and hands-on labs need to be designed carefully to make it affordable to most of the universities/colleges and also alleviate the big concern from students' parents if their children would become hackers or not after learning the offensive techniques.

Course Modules Designed

We develop eight modules based on the typical techniques proposed in the past two decades to detect stepping-stone intrusion. The modules include launch attacks using stepping-stones, network packet matching, stepping-stone intrusion detection using thumbprint, the number of packet's RTTs, random-walk model, the relative length of a connection chain, and the absolute length of a connection chain, respectively. Upon the completion of the eight modules, we expect students to be able to (1) understand the approaches to establish a connection chain to launch cyberattacks through stepping-stones; (2) collect and analyze network traffic to see if there is any malicious traffic; (3) detect and mitigate stepping-stone cyberattacks using various tools and techniques; (4) identify unauthorized, illicit, and anomalous users' behavior based on network traffic; (5) illustrate the trend of stepping-stone intrusion detection techniques; and solve the real-world cybersecurity problems using the techniques [16].

Launch Cyberattacks Using Stepping-Stones

In this module, we introduce the concept of a stepping-stone and how to use stepping-stone to launch cyberattacks. Students learn using OpenSSH to make a connection chain in a local area network (LAN). The connection chain can span at least five hosts including an attacker's computer and a victim host. Each host may install Microsoft (MS) Windows or Linux operating system (OS). Whatever an OS is used, each host must have OpenSSH server installed. To the best of our knowledge, most Linux systems have OpenSSH server and client installed, but most computers with MS Windows OS do not have OpenSSH server installed. Lab system administrator needs to install OpenSSH server separately for the hosts. After a connection chain is established, students can simulate attackers to operate the victims' hosts from the attacker's hosts. Students can also learn using TCPDump/Wireshark to capture TCP/IP packets from the incoming and outgoing connections of the stepping-stones in between the attacker host and the victim host.

The contact hours to complete this module are 2. It is well suited to include in a cybersecurity course for senior cybersecurity/computer science majors. The prerequisite knowledge required for learning this module includes basic MS Windows/Linux OS operations and using OpenSSH. Upon the completion of this module, students are expected to (1) explain stepping-stone intrusions and the benefits of launching attacks using stepping-stones; (2) explain the computer network communication basics; (3) demonstrate how to sniff packets using existing tools such as Wireshark and/or TCPDump; and (4) employ the tools such as Telnet, rlogin, or OpenSSH to make a long interactive connection chain.

Network Packet Matching

Most intruders tend to not only use stepping-stones to launch their attacks but also employ time-jittering and chaff-perturbation manipulation techniques to evade detection. In order to resist intruders' evasion, matching packets and estimating the length of a connection chain become the primary approaches to detect stepping-stone intrusion. Here network packet matching means to match only TCP packets, other than other types of packets, such as UDP, ICMP, and so on. The reason we only match TCP packets is that a TCP request must be acknowledged and/or replied. A matched TCP request/response pair can be used to compute round-trip time which can further be used to measure the length of a connection chain.

Packet matching can bring down stepping-stone intrusion detection false-positive error. It is known that simply deducing an intrusion from a host detected as a stepping-stone is the primary reason to incur false-positive detection error. It is necessary to be clear about the relation between stepping-stone and stepping-stone intrusion. As we mentioned above, some investigations show that even though there exist many applications which take advantage of stepping-stones legitimately, the case using three or more than three stepping-stones was rarely observed. One reason is that the more stepping-stones used to access a server, the less efficiency of the access. If a server can be accessed directly, there is no reason to be accessed indirectly via a connection chain including three or more hosts. If it does, the user must try to hide something which is highly suspicious. So it is reasonable to infer a stepping-stone intrusion if there are three or more hosts used as stepping-stones. If we can estimate the number of hosts used as stepping-stones or estimate the number of the connections in a connection chain, it would not be hard to detect intrusion and lower the false-positive detection error rate.

Packet matching can also resist intruders' evasion, especially the time-jittering and chaff-perturbation manipulation. If we treat regular network traffic as a signal, time-jittering and chaff-perturbation can be considered to generate noisy signal. The noisy signal can defeat the detection approaches we discussed, but they can be filtered out by packet matching.

Time-jittering can hold a request for a while and then release it to the Internet. But this does not affect packet matching results between a request and its response because the request held can also make its response delayed. Can an intruder hold a response for a long time and then release it? It is hard to do so because a response being held for a long time could incur lots of requests resending and further make the network inefficient. Chaffed packets can be easily filtered out by packet matching because each injected meaningless packet must be removed before it arrives at the destination; otherwise, it would affect the command execution inappropriately. So chaffed packets cannot match with any packets. From the analysis, we found that packet matching can not only detect a stepping-stone intrusion but also resist intruders' session manipulation.

Matching computer network packets is challenging, especially in the Internet context. This is because of some inherent design flaws of TCP/IP protocol. TCP/IP protocol was first designed by DARPA in the 1970s for use in ARPANET. Security was not the first consideration then. Lots of designing efforts were put onto its efficiency. Understanding how TCP/IP works would be helpful to learn packet matching and its challenges. Before discussing the challenges facing in packet matching, we review TCP/IP and its working mechanism first.

TCP is a protocol to deliver information in an interactive session reliably from its source host to its destination host. TCP only defines the communication rules between two hosts, other than two routers, or one router and one host. TCP supports the features of pipelining, three-way handshake, cumulative acknowledgement, connection-oriented, flow control, and congestion control. Pipelined protocol can allow a request sent out before the coming of the acknowledgement for its previous requests. This means an on-the-way request may meet the response to its previous requests. This phenomenon is called crossover packets. A protocol without pipelining support can send a request out only after its previous request is acknowledged. Non-pipelined protocol is apparently inefficient, and its utilization is pretty low. But matching requests with responses in a communication using non-pipelined protocol is much easier. Pipelined protocol can keep sending any new requests without caring whether the old requests are acknowledged or not as long as there is enough buffer space at its sender side. If a sender's buffer is full or the receiver notifies its sender that it has not enough space to hold the coming requests, the sender stops the delivery of its new request. This feature makes TCP packet matching not a one-to-one match.

Another feature of TCP protocol to make packet matching complex is "cumulative" acknowledgement. Multiple requests received by a receiver can be acknowledged and echoed in a single response. The initial intent of this design was to improve network communication utilization. However, it makes packet matching much more complex.

Packet resending may also affect the performance of TCP packet matching. As we know, TCP is a reliable communication protocol. The way to implement reliable communication is to check each received packet to see if it is out of order or there exists any errors. If there is any error in a packet received, or out of order is detected, the receiver can then ask the sender to resend the same packet. If after a request is sent out, the sender could not receive the acknowledgement in a predefined time controlled by a timer, the request would be resent automatically. It is also possible that a resent packet was previously correctly received. So resending may make duplications of the same request at a receiver. All of these could definitely complex the situation to match TCP packets.

Packet loss can also make packet matching more complex. A request may be lost; a response (an echo packet) may be lost; and an acknowledgement packet may also be lost. Due to the cumulative feature of TCP protocol, a lost acknowledgement and echo packet can be ignored as long as the timer at the sender side is not expired. However, a lost request must be resent as long as the timer is expired.

We discuss the rationale behind packet matching. Based on the TCP/IP protocol design, every request is acknowledged first and then replied. In most cases, a packet request can be both acknowledged and replied in one packet for networking communication efficiency. Acknowledgement is required for each request due to TCP reliable data transfer. Each request is also replied by a packet, called an echo packet as we mentioned in the above. A request can be defined as a send packet. Correspondingly, an acknowledgement packet is called an Ack packet. TCP protocol defines a point-to-point communication in which a point is a computing host. In between source point and destination point in a TCP session, there may be one or more routers. The RTT between a source point and a destination point consists of the propagation delay along the links from a source host to the destination host, plus the delays occurring in each router in between. It is impossible to estimate the RTT between two hosts via the delays from each router that a TCP session passes through. We use the timestamp gap between each send and Ack or echo packet to estimate the RTT of a TCP session. It is apparent that the RTT may vary upon different send packets. However, the RTTs of send packets in the same TCP session are bounded.

In this module, we introduce four different algorithms to match TCP/IP packets. One is to match TCP send and echo packets by examining the sequence and acknowledgement numbers of the packets. But due to the complexity of computer network traffic, this approach is only applicable to the packets captured in a local area network which has only few crossover packets. On the Internet, due to crossover packets observed frequently, several methods such as First-Match [11], Conservative Match [17], and Greedy Match algorithms [17] were proposed to match TCP packets.

In order to study this module, students are required to take computer network class. The length to complete this module is estimated in between 6 and 8 contact hours. It is well suited to include in a cybersecurity course for seniors in computer science or cybersecurity major. Upon the completion of this module, students are expected to (1) describe the basic concepts of packet matching, (2) explain the significance of packet matching, (3) discuss the challenges of packet matching, (4) discuss the rationale behind packet matching, and (5) explore using packets' sequence numbers to match TCP packets.

Detection Using Thumbprint

Stepping-stone intrusion can be detected using content-based thumbprint and time-based thumbprint. In this module, students first learn how to make a content-based thumbprint based on the occurrence frequency of each packet captured [3]. Using the TCP packets captured at the same time interval from the incoming and outgoing connections of a host can determine the occurrence of the same character at the two connections, respectively. For example, at an incoming connection, if students capture the packets '*l*', '*s*', '-', and '*l*', the thumbprint can be {2'*l*', 1'*s*', 1'-'}. If students can obtain the same or close enough thumbprint at the outgoing connection in the same time interval, then it is not hard to tell the host is used as a stepping-stone. The more packets captured, the higher the probability a stepping-stone is detected. For an encrypted network session, since the packet contents cannot be seen, so the timestamp of each captured packet is used to make a time-based thumbprint. For example, if we monitor the incoming and outgoing connections of a host for 6 min, then we can count the number of packets captured in each minute regardless of their contents. We assume a number sequence, such as {16, 27, 1, 47, 98, 4} from the incoming connection, are obtained, as well as the sequence {15, 27, 1, 46, 97, 4} obtained from the outgoing connection. We can conclude if the host is used as a stepping-stone by comparing the two sequences.

This module is designed for senior undergraduate students in cybersecurity or computer science major. To complete this module, it is suggested 4 to 6 contact hours. In addition to the prerequisite knowledge required from the previous modules, students should have taken operating systems and computer network courses. Upon the completion of this module, students are expected to (1) explain the details of the techniques used to detect stepping-stone intrusion in a host including content-based thumbprint, time-based thumbprint, packet count, random walk, and crossover packet; (2) employ the host-based techniques for detecting stepping-stone intrusion including content-based and time-based thumbprint; (3) describe the time-jittering skill used by hackers to manipulate time-based thumbprint; and (4) demonstrate the chaff-perturbation skill used by hackers to inject some meaningless packets into an existing connection.

Detection Using the Number of Packet's RTTs

Time-based thumbprint can apply for an encrypted network session to detect stepping-stone. However, this detection technique can be easily defeated by intruders' manipulation, such as time-jittering and/or chaff-perturbation. Time-jittering and chaff-perturbation can alter the number of packets per time-unit in the incoming and/or outgoing network connections of a host. We found that the injected packets cannot be matched with other send/echo packets. If the packets captured are matched, chaffed packets can be filtered out. In both incoming and outgoing connections, we use the matched packets to compute the packets' RTTs. Each RTT can be obtained by the difference between the timestamp of a send and its matched echo packet. So we can get the number of RTTs from both the incoming and the outgoing connections of a host. If a computer host is used as a stepping-stone, the two numbers should be very close.

The length of completion of this module is between 6 and 8 contact hours. The prerequisite knowledge required for learning this module is to take computer networking class and packet matching module in this chapter. Upon the completion of this module, students are expected to (1) understand how to apply packet matching for intrusion detection, (2) demonstrate computing RTTs from collected packets, and (3) illustrate using the number of packets' RTTs to detect stepping-stone intrusion.

Detection Using Random-Walk Model

It is observed that the difference between the number of packets (either send or echo) from the incoming and outgoing connections of a host, respectively, can be modeled as a one-dimensional random walk. If we monitor the incoming and outgoing connections of a host and collect all the packets including send and echo packets in each connection, we get the total number of packets from the incoming connection, denoted as $N_{p\text{-in}}$, as well as $N_{p\text{-out}}$ from the outgoing connection. We assume the number of send and echo packets in $N_{p\text{-in}}$ are $N_{s\text{-in}}$ and $N_{e\text{-in}}$, respectively, and similarly for $N_{p\text{-out}}$, they are $N_{s\text{-out}}$ and $N_{e\text{-out}}$, respectively. Due to chaff-perturbation, it is hard to say that $N_{p\text{-in}}$ and $N_{p\text{-out}}$, $N_{s\text{-in}}$ and $N_{s\text{-out}}$, or $N_{e\text{-in}}$ and $N_{e\text{-out}}$ are close enough, respectively. However, the differences between $N_{p\text{-in}}$ and $N_{p\text{-out}}$, $N_{s\text{-in}}$ and $N_{s\text{-out}}$, or $N_{e\text{-in}}$ and $N_{e\text{-out}}$ follow random-walk behavior, respectively. This indicates that if we observe the differences between $N_{p\text{-in}}$ and $N_{p\text{-out}}$, $N_{s\text{-in}}$ and $N_{s\text{-out}}$, or $N_{e\text{-in}}$ and $N_{e\text{-out}}$ present random-walk behavior, the host can be detected as a stepping-stone.

The prerequisite knowledge is to take probability and statistics class. The estimated contact hours to complete this module are between 4 and 6. After the completion of this module, students are able to (1) describe the concept of a random-walk process, (2) discuss the random-walk approach for stepping-stone intrusion detection, (3) employ the routine random-walk algorithm for stepping-stone intrusion detection, and (4) demonstrate using some tools, such as chaff-perturbation and/or time-jittering, to evade detection.

Detection by Estimating the Relative Length of a Connection Chain

In order to detect stepping-stone intrusion with a low false-positive error rate, K. H. Yung [2] proposed an approach to estimate the length of a connection chain relatively. As shown in Fig. 2, instead of estimating the number of hosts used as stepping-stones, Yung estimated the connection length from H_1 to H_{n+1}. To obtain how long the downstream connection chain is, Yung introduced using the length of connection from H_1 to H_2 as a yardstick to measure the length of the connection from H_1 to H_{n+1} approximately.

The timestamp gap between a send packet and its matched echo packet collected at H_1 can be used to indicate the length of the connection chain from H_1 to H_{n+1}, denoted as $\mathrm{RTT}_e = t_e - t_s$. Even though it is still hard to know how long the chain from H_1 to H_{n+1} is, K. H. Yung proposed to use the length of the connection from H_1 to H_2 as a scale to measure the length of the connection chain from H_1 to H_{n+1}. It is apparently impossible to use the timestamp gap between a send and its matched echo to represent the length of the connection from H_1 to H_2. What Yung did was to use the timestamp gap between a send packet and its acknowledgement packet collected at H_1 to represent the length of the connection from H_1 to H_2, denoted as $\mathrm{RTT}_a = t_a - t_s$. Since H_2 is directly connected to H_1 in an interactive TCP connection, so any packet sent from host H_1 must be acknowledged by host H_2. The acknowledgement packet can sooner or later go back to H_1. Upon receiving a packet, an acknowledgement packet can be generated and sent back immediately. From the process of acknowledging a packet, we understand even though it is not accurate to use RTT_a to represent the length of a connection chain, it still makes some sense. It is not accurate because RTT_a tends to be smaller due to lacking of packet processing time at the receiver side. Any acknowledgement packet is generated at transport layer which needs less time than an echo packet which is generated at application layer in an interactive TCP session.

The ratio ρ between RTT_a and RTT_e: $\rho = \mathrm{RTT}_a/\mathrm{RTT}_e$ can approximately estimate how long the connection from H_1 to H_{n+1} is. If the ratio is close to 1, it indicates the downstream connection chain length is not long. However, if this ratio is close to 0, it strongly indicates the length is long.

0.40	ST→e2→zi→e3→cp→e4→ls→sp→e6→cs→e7→xb→df→bs→e8
0.36	ST→e2→zi→e3→cp→e4→ls→sp→e6→cs→e7→xb→df→bs
0.28	ST→e2→zi→e3→cp→e4→ls→sp→e6→cs→e7→xb→df
0.39	ST→e2→zi→e3→cp→e4→ls→sp→e6→cs→e7→xb
0.42	ST→e2→zi→e3→cp→e4→ls→sp→e6→cs→e7
0.44	ST→e2→zi→e3→cp→e4→ls→sp→e6→cs
0.21	ST→e2→zi→e3→cp→e4→ls→sp→e6
0.68	ST→e2→zi→e3→cp→e4→ls→sp
0.57	ST→e2→zi→e3→cp→e4→ls
0.45	ST→e2→zi→e3→cp→e4
0.70	ST→e2→zi→e3→cp
0.62	ST→e2→zi→e3
0.92	ST→e2→zi
0.99	ST→e2

Fig. 3 Sample of a connection chain made by OpenSSH

K. H. Yung verified his idea using the computers located throughout the USA, and some of them were located in Europe. The operating systems running in the remote machines included Linux, FreeBSD, Solaris, VMS, and S390. Fig. 3 shows part of the experimental results done by Yung. In the experiment, Yung's team tested his idea with a connection chain across multiple countries in different lengths from one connection to 14 connections. Their program ran at ST which was used as a sensor in the experiment. All other hosts, such as e2, e3, e4, e6,e7,e8, zi, cp, ls, sp., cs, df, and bs, were used as either stepping-stones or victim host depending on the connection chain established.

The first connection setup was from host ST to e2 which has only one connection. The ratio ρ was 0.99 which is close to 1, but it is still smaller than 1 because RTT_a is smaller than RTT_e just as what we discussed before. When the connection chain had two connections, the ratio dropped slightly from 0.99 to 0.92 but dropped sharply to 0.62 with one more connection extension. When the connection chain extended to four connections, the value climbed a little; obviously, this is not what Yung's team expected. What happened in the experiment can just justify network traffic is fluctuated. But by observing the length of the connection chain extended to length 14, we could see the whole trend is that the ratio dropped gradually. Unfortunately, from their experimental results, we found that the smallest ratio is not when the connection chain has the maximum length; instead, it is when the chain has eight connections. We also found there are four ratios including 0.21, 0.28, 0.36, and 0.39 with shorter connection chains (less than 14) being below 0.4. We do not think this happens to be. It must have something incorrect behind Yung's approach.

We analyzed Yung's approach and found it needs to match send and echo packets to compute RTT_e, as well as matching send and Ack to obtain RTT_a. The approach used to match packet can work well under local area network or under the situation that it does not have lots of network traffic. In other words, if each send can be acknowledged, or echoed quickly, Yung's packet matching method works well. However, if under the Internet, or in the situation that there are lots of crossover packets, false-negative error can be introduced in Yung's approach due to estimating RTT_a and RTT_e incorrectly. We believe this is the primary reason that the ratios of four connections were not the results they expected.

In addition to the above issue, Yung's approach also suffers from yardstick issue. This issue arises from the yardstick used to measure the length of the whole downstream connection chain. If the yardstick is a fare ruler, it would give a reasonable result; otherwise, the measurement would not be accurate. Let's consider a case that if RTT_a is very small due to the connection from host H_1 to H_2 that happens to be very short, even though the downstream connection is not long, the ratio ρ might be very small which still indicates a long connection chain detected. This would bring false-positive detection errors. On the other hand, if RTT_a is very large because the single connection from H_1 to H_2 is very long, even though the downstream connection is pretty long, the ratio might be very large which indicates a short connection chain detected. This may introduce false-negative detection error. Anyway, yardstick is a big issue to make the length estimation inaccurate in Yung's approach.

The prerequisite is the packet matching module. This module is designed for senior undergraduate students in a cybersecurity course for the length of 6 to 8 contact hours. After the completion of this module, students are expected to (1) explain the rationale of using the length of a connection chain for stepping-stone intrusion detection, (2) describe the basic idea of Yung's method, (3) explain the definition of the RTT of a connection chain and the rationale behind a connection chain length estimation, and (4) apply Yung's approach to estimate the length of a connection chain.

Detection by Estimating the Length of a Connection Chain

Yung's approach cannot estimate the actual number of connections in a connection chain. This may incur high false-negative errors. There are some other approaches proposed to estimate the number of connections in a connection chain more precisely. The representatives are the step-function approach [11] and clustering-partitioning data mining approach [12].

Step function is an approach which can detect stepping-stone intrusion by estimating the length of a connection chain. As shown in Fig. 2, the host H_1 is a sensor where we can monitor a connection chain which passes through hosts H_2, H_3, ..., and finally reaches the victim host H_{n+1}. If we can collect all the TCP packets when the chain only connects to H_2 from H_1 and compute the RTT for each send packet, we can get a set of RTTs which are different but are close enough in their values. We denote this set of RTTs as RTT_1. When the chain extends to host H_3 from H_2, we would get RTT_2. The difference between RTT_1 and RTT_2 is that RTT_1 represents the connection chain which has only one connection from the sensor,

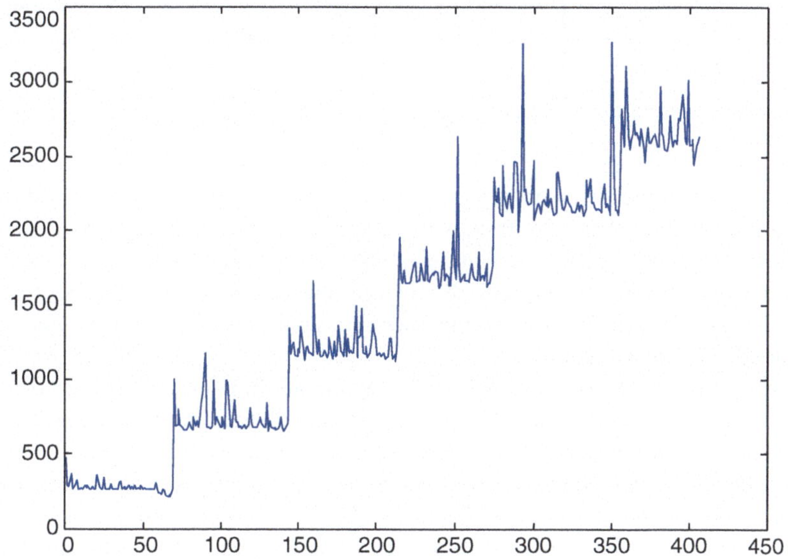

Fig. 4 Verifying step-function stepping-stone detection

but RTT_2 represents two connections in the chain. Most of the values in RTT_2 are larger than those in RTT_1. Similarly, as long as the chain has one more connection extended, we would get a different RTT set. At the end of connection chain, we would get RTT_1, RTT_2, RTT_3, ..., RTT_n. If we put all the RTT sets into a two-dimensional coordinate system, we get different steps with each step representing one connection. The number of steps can tell the number of connections in a chain which is also the number of computer hosts connected by the chain. Fig. 4 shows the detection results of applying step function. The Y-axis represents RTT values in millisecond, and the X-axis represents the number of packets matched. What is shown in Fig. 4 is the situation including six hosts in the connection chain. The core of step function is to match send and echo packets; therefore, each RTT can be computed correctly.

Step-function algorithm works perfect in a local area network but not well in the Internet context. Clustering-partitioning data mining approach can allow us to estimate the length of a connection chain in the environment of the Internet even without matching TCP/IP packets. We mentioned that TCP/IP is a communication protocol between two hosts. This means host h_i can only know it connects to host h_{i+1} as shown in Fig. 5. Host h_i has no idea about the connecting situation after h_{i+1} in the downstream of the session. If we monitor the outgoing connection of the host h_i, what we could know are the TCP/IP packets coming from and going to host h_{i+1}, rather than any other hosts, such as h_{i+2}, ..., h_n, and h_{n+1}. But if there is a session between h_i and h_{n+1}, each packet sent from h_i must be acknowledged by h_{i+1} first and then forwarded to the hosts after h_{i+1} along the chain until the final host h_{n+1} which is also called destination host. Even though the echo of a send

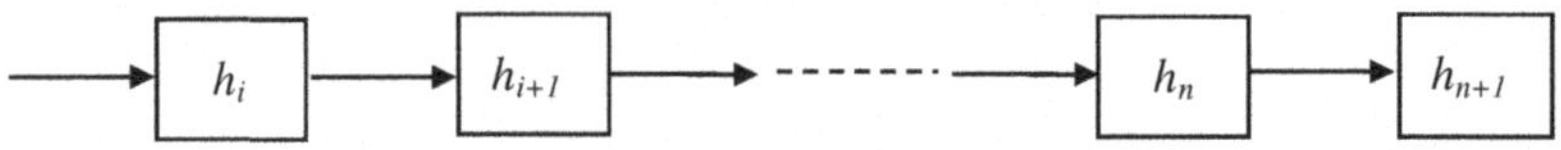

Fig. 5 An interactive connection chain

monitored at h_i comes from host h_{i+1}, the reality is that this packet is echoed by the destination host rather than its directly connected host. This feature motivated us the idea that if we compute the timestamp gap between each send and its corresponding echo coming from h_{i+1}, the gaps should vary but depend on the number of hosts connected after h_{i+1}. The timestamp gaps are increased, while more connections are extended along a session.

As shown in Fig. 5, we monitor host h_i, capture all the sends and echoes from the time that it just connects to host h_{i+1}, and compute all the timestamp gaps between each send and its corresponding echo. We find that those gaps are different but vary slightly. They are bounded within a certain range which is called a "level/step" similar to the one in step function, denoted as L_1. If we monitor this host and capture all the sends and echoes continuously, when one more host is connected, we are supposed to get L_2. In level 2, even though the RTTs are different, on average they are larger than the RTTs in L_1. When more and more hosts are connected one by one, more and more levels are obtained. As long as we monitor this session from the beginning to the end continuously and capture all the sends and echoes, we are able to determine the number of the levels. This number is exactly the same as the downstream length of the connection chain. If we call each level a cluster, this method to detect stepping-stone intrusion is called clustering and partitioning data mining approach [12]. Unlike the policy used in step function to match TCP packets, the algorithm proposed in [12] is to use data clustering and partitioning technique to match TCP/IP packets and to find the RTTs of the packets of a connection chain [12]. The data mining algorithm is a global approach in which it checks all the packets together to determine TCP packet matches. It captures all the send and echo packets of a connection chain in a certain time interval and computes the differences between each send packet and all echo packets received after it. It is true that the correct RTTs are among these differences.

It is obvious that these RTTs can be clustered around different levels. It uses the maximum-minimum distance clustering algorithm (MMD) to find the real RTTs and determine the number of connections in a chain. The experimental result showed that this algorithm can match TCP/IP packets with both high matching rate and high matching accuracy. The mechanism to match packets in the algorithm is to use the distribution of RTTs which can refer to the paper [12]. Figure 6 shows one example of RTT distribution in which Y-axis represents the probability of each RTT value, and X-axis represents different values of RTTs. From this RTT distribution, we learn that it follows Poisson distribution, and more than 95% of the values of the RTTs are bounded around its mean value within the range of one standard deviation.

Suppose we monitor and capture the TCP packets of a connection chain from the time the chain is being established to the time that the chain has four connections.

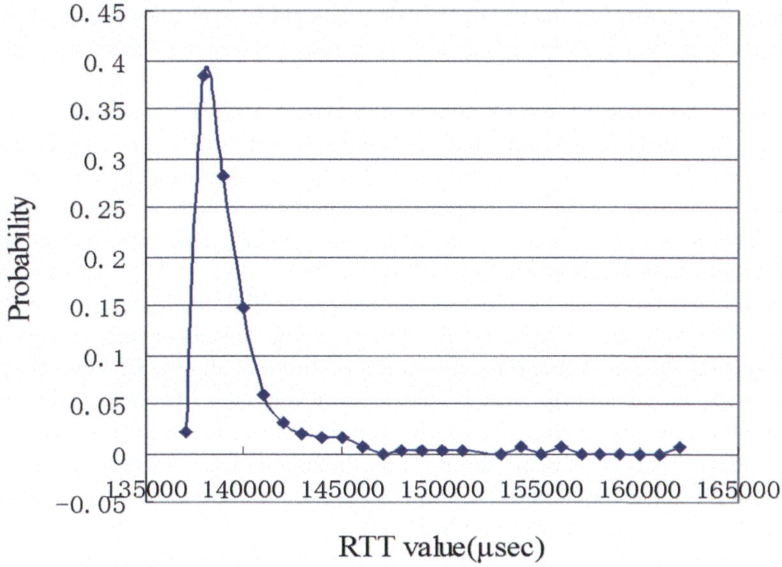

Fig. 6 The distribution of RTTs for a connection chain

At the time when the chain has only one connection, based on the analysis of the distribution of the RTTs of TCP packets in [12], most of its RTTs should be around RTT_1, which is the average value of the RTTs of the chain. Similarly, if the chain is extended incrementally until it has four connections, we have RTTs concentrating around RTT_2, RTT_3, and RTT_4, respectively. This result was first observed in [11]. The clustering-partitioning data mining algorithm can be summarized as the following.

Given n consecutive send packets $S = \{s_1, s_2, \ldots, s_n\}$ and m consecutive echo packets $E = \{e_1, e_2, \ldots, e_m\}$, we assume that these packets are captured in one connection of a host at a certain time interval, and each packet in send set S is echoed by one or more packets in echo set E. We compute the differences between each send packet in S and all the echo packets in E. Since RTT must be positive, we can safely eliminate the negative values. We group these differences in sets according to each send packet in S, forming data sets $S_1, S_2, \ldots, S_n$ where $S_i = \{t(i,1), t(i,2), \ldots, t(i,m)\}$ for $i = 1, \ldots, n$, and $X = \cup S_i = \{t(i,j), 1 \leq i \leq n, 1 \leq j \leq m\}$ where $t(i,j) = e_j - s_i$ represents a potential RTT between the ith send packet and the jth echo packet.

Step 1 of the algorithm is simply a clustering algorithm with a predefined threshold th. The MMD algorithm can result in v clusters. We assume these clusters are sorted in an increasing order of the first index i which is the send packet index. The **range** of a cluster C is defined as the maximum i-index of the elements in C minus the minimum i-index plus one.

In Step 2, we filter out the duplicated elements either with the same send packet or with the same echo packet in each cluster. The preference is given to a smaller RTT.

In Step 3, we measure the likelihood of a cluster truly representing a level in the RTTs. A true RTT cluster should have elements representing consecutive send packets with very few exceptions. So we first define a subset of a cluster containing "connected" elements, i. e., elements having neighbors with a distance of g. A typical value for g is 2 (allowing one missing send packet). "Disconnected" elements are mostly not part of this cluster.

Steps 4 is used to select the clusters that have very high likelihood of true RTTs. Based on the Chebyshev inequality, there are very few clusters outside two standard deviations of the mean. A true RTT cluster is a partition of all send packets satisfying the following two conditions: (1) all clusters are mutually disjointed, and (2) the union of all clusters is equal to the whole send packet set. In reality, condition (1) is easy to satisfy, but it is very difficult to make the union of the clusters exactly equal to the send packet set S. In our algorithm, it turns out to find the clusters, the union of which has the largest distribution in S.

The prerequisite is the packet matching module. This module is designed for senior undergraduate students in a cybersecurity course for the length of 6 to 8 contact hours. After the completion of this module, students are expected to (1) describe the basic ideas of the step-function approach and the clustering-partitioning data mining approach for stepping-stone intrusion detection, (2) employ the step-function approach to estimate the length of a connection chain, and (3) make use of the clustering-partitioning data mining approach to estimate the length of a connection chain.

Techniques to Evade Detection

Monitoring an interactive TCP session and sniffing network traffic can be used to detect stepping-stone intrusion. Host-based and network-based approaches have been developed to detect stepping-stone intrusion. While we develop more advanced methods to detect stepping-stone intrusions, intruders also develop new approaches to evade detection. Time-jittering and chaff-perturbation are the two primary techniques used by most intruders to evade stepping-stone intrusion detection.

Time-jittering is a skill to change the timestamp gap between two consecutive packets. This technique was widely used by intruders to evade most host-based detections. Intruders can use any packet crafting tools, such as packETH, to create and send packet quickly. It can be used to set the count of packets and the delay between packets. Time thumbprint is host-based intrusion detection method primarily depending on the timestamp gaps between consecutive packets. By monitoring an incoming connection of a host, it is trivial to obtain a time thumbprint TT_{in} which basically is a timestamp gap sequence. Similarly by monitoring an outgoing connection of the same host in the same time interval, we can obtain another time thumbprint TT_{out} which is also a timestamp gap sequence. Obviously, *comparing* TT_{in} *with* TT_{out} *can h*elp us to determine if the host is used as a stepping-stone. Intruders can intentionally change TT_{out} by using packETH which can change the delay between packets. Time-jittering technique can easily defeat time thumbprint approach to evade the detection.

Chaff-perturbation is a skill to inject some packets into an existing connection. This skill is also used by intruders to evade most host-based stepping-stone intrusion detection approaches, such as the count of packets, random walk, and crossover packets. Intruders can use any packet crafting tools, such as Fragroute and Snappy, to conduct packet injection. Since packet injection can easily change the count of packets in a connection, so the approaches depending on the packet count to detect stepping-stone can be easily defeated. Intruders can use chaff-perturbation technique to evade host-based stepping-stone intrusion detection.

The prerequisite is computer network class. This module is designed for senior undergraduate students in a cybersecurity course for the length of 4 to 6 contact hours. After the completion of this module, students are expected to (1) describe the time-jittering and chaff-perturbation techniques and (2) demonstrate how to defeat time-based thumbprint by injecting packets into an interactive TCP session.

Hands-on Labs Designed

Making a Long Connection Chain

In this lab, students are expected to learn how to set up a long connection chain using the tool OpenSSH. There are two sections in this lab: the first one is to set up a connection chain in a LAN; the second one is to establish a connection chain in the context of the Internet. In order to build a connection chain in a LAN, students need to have all login confidential for each computer in the LAN. A student can login to any computer of the LAN, such as the host h_1, and then run OpenSSH from the host h_1 to connect to another computer, such as the host h_2. At the host h_2, the student runs OpenSSH again to connect to the third host h_3. So students can establish a connection chain from the host h_1 to the host h_3 via h_2. This connection chain contains two connections: one connection is from h_1 to h_2, and another one is from h_2 to h_3. The length of this connection chain is 2. If $n + 1$ hosts including attacker's host are connected, the length of the connection chain is n.

Each host in a LAN must have both OpenSSH server and client software installed. Students can make a long interactive TCP/IP connection chain across at least five computer hosts in this lab exercise. Such a long interactive session spanning multiple stepping-stones is widely used by hackers to launch their attacks aiming at a victim which hackers are interested in. The learning objectives of this lab are to help students to (1) understand TCP/IP protocols, (2) learn how to establish a long interactive connection chain spanning multiple hosts, (3) understand the concept of stepping-stones, and (4) obtain the knowledge of how an intruder launches attacks through stepping-stones.

Capturing Network Traffic from a Host

Sniffing network packets can not only make students explore different types of packet and their header structure but also understand the behavior of computer network traffic. Network traffic behavior can be used to detect stepping-stone intrusion. There are lots of approaches proposed to detect stepping-stone intrusion. But most of them make use of some characteristics of computer network traffic to detect stepping-stone intrusion, such as packet header, timestamps, packet size, sequence numbers, packet type, round-trip time, and so on. Therefore, capturing network packets is significant in terms of studying computer networks and its security. In this lab, students will be directed to use Wireshark to sniff computer network packets. Students can also use some other popular tools, such as TCPDump or Snort, for packet sniffing. Students can use the connection chain built in Lab A, at h_2, to exercise capturing packets between h_2 and h_3, as well as h_1 and h_2. The learning objectives of this lab are to help students to (1) understand TCP/IP packet header structure and the semantics of each field in the packet header; (2) learn how to save captured packets into a file with different formats; (3) explore the features of packets with different types including but not limited to TCP, UDP, IP, and ICMP; and (4) be familiar with Wireshark to sniff computer network packets.

Time Thumbprint Detection

In this lab, students will learn how to determine if a machine is used as a stepping-stone through a time-based thumbprint. In every captured network packet, it has numerous information including timestamp. It is trivial to convert the timestamp sequence to a set of timestamp gaps by computing the timestamp difference between a packet and its subsequent one. In this way, it is easy to get a set/sequence of timestamp gaps by monitoring each incoming/outgoing connection of a host at the same time interval. Comparing the timestamp gap sequence from the incoming connections with those from the outgoing connections can help us to judge if the host is used as a stepping-stone. Students can make an interactive TCP/IP connection across at least three hosts using OpenSSH and capture the network packets from the one in the middle of the connection chain. After filtering the packets from the incoming/outgoing connection, it can generate the sequences of timestamp gaps and perform the comparison algorithm between two sequences to decide if a stepping-stone host exists.

The learning objectives of this lab are to help students to (1) understand using time-based thumbprint to detect stepping-stone intrusion, (2) illustrate how to

Fig. 7 A connection chain with three hosts

make a time-based thumbprint, (3) demonstrate how to compare two time-based thumbprints, and (4) understand the efficiency of thumbprint comparison algorithm.

Detection Via the Number of Captured Packets

In this lab, students establish a connection chain, as shown in Fig. 7, spanning three hosts which are named attacker, stepping-stone, and victim, respectively. Packets can be captured at the incoming and the outgoing connections of stepping-stone, respectively. The incoming connection of stepping-stone is from attacker to stepping-stone. The connection from stepping-stone to victim is the outgoing connection of stepping-stone. This capturing can last 10 min, while attacker makes packets to victim through stepping-stone. At the incoming connection of stepping-stone, we can capture a certain number of packets in 10 min. We count the number of the packets in the first 1 min, the second 2 min, the third 4 min, the fourth 2 min, and the last 1 min of the 10 min interval and get a number sequence, for example, {10, 17, 69, 21, 18}. Similarly, from the outgoing connection of stepping-stone, we can get a sequence of numbers by counting the captured packets at the same 10-min time interval. We assume it is {9, 18, 71, 21, 18}. Comparing the two sequences, we can know if stepping-stone host is used as a stepping-stone.

Upon the completion of this lab, students are expected to (1) explain how to make packet number sequence-based thumbprint and (2) demonstrate using the number of captured packets to detect stepping-stone intrusion.

Random-Walk Detection

In this lab, students learn how to determine if a host is used as a stepping-stone by using the RTT-based random-walk detection algorithm. In the previous lab, we have discussed the time-based thumbprint, but there are techniques that attackers can manipulate time-based thumbprint, such as time-jittering and chaff-perturbation. The RTT-based random-walk detection algorithm was proposed to detect stepping-stone intrusion and resist intruder's such kind of evasion manipulation. The basic idea of discovering whether a computer is used as a stepping-stone is to compare an incoming connection of the computer with one of its outgoing connections to see if there exist two relayed connections. In this lab, students make a long interactive TCP/IP connection across at least three computer hosts using OpenSSH. Students perform a network packet capture, filter the connections, and save the packets into a file. Once they capture the packets, they match the send and echo packets in

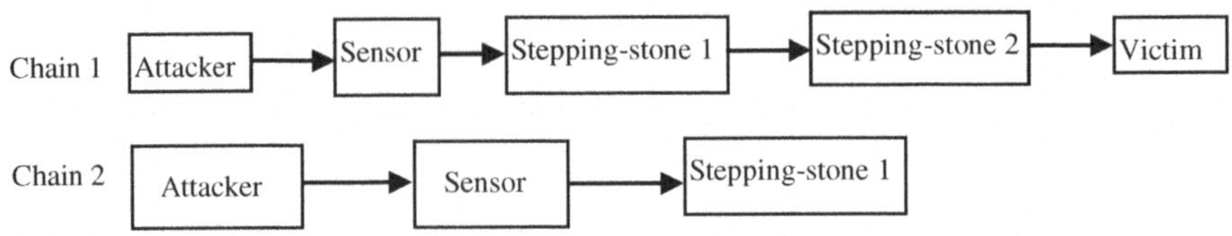

Fig. 8 Two connection chains

each connection and count the number of RTTs in that connection. The difference between the number of RTTs from the incoming connection and the outgoing connection follows random-walk behavior. Students also explore if this difference can be affected by attacker's manipulation.

Upon the completion of this lab, students are expected to (1) understand random-walk model, (2) apply random-walk model to detect stepping-stone intrusion, (3) be familiar with the techniques to evade stepping-stone intrusion detection, and (4) make use of the number of RTTs to resist intruders' evasion.

Detection Via Estimating the Relative Length of a Connection Chain

The objective of this lab is to help students to have a deep understanding of Yung's approach to detect stepping-stone intrusion. Students make a connection chain, Chain 1, as shown in Fig. 8 that goes through attacker, sensor, stepping-stone 1, stepping-stone 2, and victim. At the outgoing connection of sensor, send, echo, and Ack packets are captured. Match each send with its corresponding Ack packet, as well as with the corresponding echo packet. Through the matched send-Ack and send-echo pairs, students can compute the RTT-As, as well as the RTT-Es. Get the average of RTT-As, and denote it as RTTa, as well as the average of RTT-Es which is denoted as RTTe. Check the ratio between RTTa and RTTe. The value should be in between 0 and 1. The closer the ratio to 0, the higher probability sensor is used as a stepping-stone. Make Chain 2 as shown in Fig. 8 to exit from victim and stepping-stone 2 in Chain 1 to include only three hosts: attacker, sensor, and stepping-stone 1. Repeat all the above process, and check if the ratio obtained is larger than the former one.

Step-Function Detection

In the lab of step-function detection, students will use the rises and falls of a step-function graph of RTTs vs. the number of matched packets gathered from a sensor's outgoing connection to estimate the number of stepping-stones in the downstream connection chain. Students need to make a connection chain consisting of six hosts,

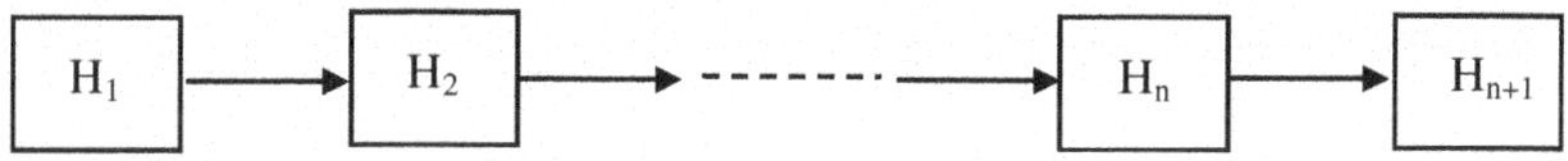

Fig. 9 A connection chain with multiple hosts

namely, intruder's host, victim's host, and four stepping-stones (the first one acts as the sensor), and then collect network traffic packets at the sensor. Send and echo packets are identified from the packets collected and then match the packets using first-match packet matching algorithm [11] to examine the relation between the number of stepping-stones and the RTT levels. In this lab, students can (1) learn how to utilize matched packets to determine the length of a connection chain; (2) understand the packet matching algorithm: first-match; (3) use matched send and echo packets to determine the number of compromised hosts; (4) demonstrate step-function algorithm; and (5) explore the limits of step-function detection approach.

Detection Via Estimating the Length of a Connection Chain

In this lab, students learn how to use clustering-partitioning data mining approach to estimate the length of a connection chain. Students set up a connection chain from the host H_1 to the host H_{n+1}, as shown in Fig. 9, and monitor the network traffic at sensor H_2. Students can only estimate the length of the downstream connection chain from H_2 to H_{n+1}, other than the length of the full connection chain. If students can monitor all the packets passing through the host H_2, applying clustering-partitioning algorithm to the packet set, they can get the number of matched clusters which should be *n* clusters. The objective of this lab is to verify if the number of the clusters is equal to the number of downstream connections which is given to the students.

Conclusion

Integrating stepping-stone intrusion detection techniques into cybersecurity curriculum is necessary and significant since using stepping-stone to launch attacks is popular today. In this chapter, we first introduce the basic techniques of stepping-stone intrusion and then propose eight content modules and eight hands-on labs to help students to study stepping-stone intrusion detection techniques. The detection techniques included in the chapter cannot cover all the approaches developed in the recent three decades, but they are the most typical ones. The goal of this chapter is to open a door to students on stepping-stone intrusion detection. We expect our students can use this door as a starting point to explore and develop more advanced techniques to detect stepping-stone intrusion and prevent such attacks from happening.

References

1. Y. Zhang, and V. Paxson, Detecting Stepping-Stones, in *Proc. the 9th USENIX Security Symposium, Denver, CO*, 2000, pp. 67–81
2. K. H. Yung, Detecting Long Connecting Chains of Interactive Terminal Sessions, in *Proc. the International Symposium on Recent Advance in Intrusion Detection, Zurich, Switzerland*, 2002, pp.1–16
3. S. Staniford-Chen, and L. T. Heberlein, Holding Intruders Accountable on the Internet, in *Proc. IEEE Symposium on Security and Privacy, Oakland, CA*, 1995, pp. 39–49
4. K. Yoda, and H. Etoh, Finding Connection Chain for Tracing Intruders, in *Proc. 6^{th}European Symposium on Research in Computer Security, Toulouse, France, Lecture Notes in Computer Science*, vol. 1985, 2000, pp. 31–42
5. A. Blum, D. Song, and S. Venkataraman, Detection of Interactive Stepping-Stones: Algorithms and Confidence Bounds, in *Proc. of International Symposium on Recent Advance in Intrusion Detection, Sophia Antipolis*, France, 2004, pp. 20–35
6. D. L. Donoho Detecting Pairs of Jittered Interactive Streams by Exploiting Maximum Tolerable Delay, in *Proc. The 5^{th}International Symposium on Recent Advances in Intrusion Detection, Zurich, Switzerland*, 2002, pp. 45–59
7. T. He, L. Tong, Detecting encrypted stepping-stone connections. IEEE Transaction on Signal Processing **55**(5), 1612–1623 (2007)
8. X. Wang, D.S. Reeves, S.F. Wu, and J. Yu, Sleepy Watermark Tracing: An Active Network-Based Intrusion Response Framework, in *Proc. The 16^{th}International Conference on Information security, USA*, 2001, pp. 369–384
9. X. Wang and D.S. Reeves, Robust Correlation of Encrypted Attack Traffic Through Stepping Stones by Manipulation of Inter-packet Delays, in *Proc. The 10^{th}ACM Conference on Computer and Communications Security, USA*, 2003, pp. 20–29
10. X. Wang, The Loop Fallacy and Serialization in Tracing Intrusion Connections through Stepping Stones, in *Proc. the 2004 ACM Symposium on Applied Computing, USA*, 2004, pp. 404–411
11. J. Yang, S.-H. S. Huang, A Real-Time Algorithm to Detect Long Connection Chains of Interactive Terminal Sessions, in *Proc. the 3^{rd}ACM International Conference on Information Security, China*, 2004, pp. 198–203
12. J. Yang, S.S.-H. Huang, Mining TCP/IP packets to detect stepping-stone intrusion. Journal of Computers and Security, Elsevier Ltd **26**, 479–484 (2007)
13. J. Yang, and Y. Zhang, RTT-based Random Walk Approach to Detect Stepping-Stone Intrusion, in *Proc. the 29^{th}IEEE International Conference on Advanced Information Networking and Applications, Gwangju, South Korea*, 2015, pp.558–563
14. J. Yang, Resistance to Chaff Attack through TCP/IP Packet Cross-Matching and RTT-based Random Walk, in *Proc. the 30^{th}IEEE International Conference on Advanced Information Networking and Applications, Crans-Montana, Switzerland*, 2016, pp. 784–789
15. Z. Trabelsi, W. Ibrahim, A hands-on approach for teaching denial of service attacks: A case study. Journal of Information Technology Education: Innovations in Practice **12**, 299–319 (2013)
16. A. McGettrick, *Toward Curricular Guidelines for Cybersecurity, Report of a Workshop on Cybersecurity Education and Training,* (2013)
17. J. Yang, and S.-H.S. Huang, Matching TCP Packets and Its Application to the Detection of Long Connection Chains, in *Proc. the19th IEEE International Conference on Advanced Information Networking and Applications, Taiwan, China*, 2005, pp. 1005–1010
18. Internet World Stats, Internet users distribution in the world-mid-year (2019). www.internetworldstats.com/stats.htm. Accessed 4 Dec 2019

19. Internet Society, Cyber incident & breach trends report (2018). https://www.internetsociety.org/wp-content/uploads/2019/07/OTA-Incident-Breach-Trends-Report_2019.pdf Accessed 4 Dec 2019
20. Cybersecurity Job openings, http://cyberseek.org/heatmap.htm. Accessed 7 Dec 2019
21. Emerging Threats, 10 Cyber security facts and statistics for 2018, Norton by Symantec. www.us.norton.com/internetsecurity-emerging-threats.htm. Accessed 4 Dec 2019

Cybersecurity Scenario Builder and Retrieval Toolkit

Guillermo Francia III, Tirthankar Ghosh, Gregory Hall, and Eman El-Sheikh

Introduction

As our nation's well-being and economy become more dependent on our information infrastructure, the need for a trained cybersecurity workforce has never been so critical. It is imperative that the current and future cyber warriors and workforce be educated and trained on the security of such systems. It is equally important that careful and deliberate considerations must be exercised in designing and implementing educational and training activities to address these issues. Indeed, the call for a disruptive and yet transformative cybersecurity training is upon us.

The innovative cybersecurity education project builds upon two very successful projects on cybersecurity training: an NSF-sponsored capacity building project for faculty development in ICS security [1] and an NSA-sponsored Gamification of Cybersecurity Training project [2]. The faculty development project conducted professional development workshops on ICS security in three successive summers. Overall, the pre- and postworkshop surveys indicate that the topics for the workshop were well chosen and well delivered, and the ICS toolkit was rated as excellent. The results highlight that ICS security is a topic that is not well covered in information assurance/cybersecurity curricula and the workshop, as intended, highlighted the importance of that and other aspects of cybersecurity and provided instructors with tools (the toolkit and the laboratory activities) to integrate ICS security into their courses. The Gamification of Cybersecurity Training project integrated digital games into the learning process. The cybersecurity-based games were tested for efficacy on K–12 students and teachers participating in GenCyber camps, and the results were overwhelmingly positive. The project builds on the success of these two

G. Francia III (✉) · T. Ghosh · G. Hall · E. El-Sheikh
Center for Cybersecurity, University of West Florida, Pensacola, FL, USA
e-mail: gfranciaiii@uwf.edu; tghosh@uwf.edu; ghall@uwf.edu; eelsheikh@uwf.edu

K. Daimi, G. Francia III (eds.), *Innovations in Cybersecurity Education*,
https://doi.org/10.1007/978-3-030-50244-7_14

cybersecurity learning projects to effectively fill a void in cybersecurity education among CAE schools and cybersecurity training for DoD personnel across the nation.

Key Challenges

The key challenges with cybersecurity education and training are the following:

- The state of information technology, both on the software and hardware aspects, is rapidly evolving.
- The human element of cybersecurity is the weakest link.
- The ineffectiveness or lack of availability of hands-on cybersecurity training exercises.
- The lack of laboratories and testbeds for advanced cybersecurity education and training.
- The lack of realistic cybersecurity case scenarios that are delivered interactively in a virtual environment.
- The shortage of cybersecurity educators and trainers.
- The scarcity of cybersecurity professionals.

Objectives

Recognizing these key challenges and the urgent need for cybersecurity training, the project was first envisioned with the following objectives:

- Design, develop, test, and deploy a highly interactive, automated, and intelligent cybersecurity scenario builder and retrieval software toolkit for active cybersecurity learning.
- Develop immersive and realistic scenarios to enable participants the ability to effectively acquire the necessary skills in cybersecurity.
- Build virtual machines (VMs) that will accompany each scenario and provide experiential learning for faculty members, students, and DoD personnel.
- Design, implement, and continuously improve a novel concept of scenario-building: the *Open Virtualization Scenario*.
- Evaluate the efficacy of the scenario-based training modules.
- Devise tools to facilitate the dissemination and sharing of scenario-based training modules for widespread adoption within and outside the Center of Academic Excellence (CAE) community and DoD personnel.

Contributions to the Cybersecurity Education Community

Contributions of this innovative project to the advancement of cybersecurity education include, but not limited to, the following:

Curriculum Materials. The novel scenario-based and competency-focused learning system and materials significantly augment and advance the current cybersecurity curricula and hands-on laboratory exercises for the community.

Development Opportunity. Professional development starts with the faculty members who are deeply involved in developing the case scenarios on cybersecurity. This opportunity will be extended to all faculty members across the nation by enabling free or discounted access using the Clark [3] curriculum repository as part of the project's dissemination efforts.

Practical Training. The scenarios provide off-the-shelf practical training for students and practitioners.

Student Interaction. Students directly contribute to the project as student assistants who helped in designing, developing, and testing the security scenarios. Other students will be the first group to utilize the security scenarios in their courses.

Facility/Lab/Technology Development. The Scenario Builder and Retrieval Toolkit is designed with utmost flexibility for widespread adoption.

Transformation. This innovative learning platform combining virtual systems and scenarios from real-world case studies transforms the methods with which the academic community teaches cybersecurity. This project introduces a novel concept of scenario building: Open Virtualization Scenario, which could advance the standardization of learning scenario development.

To summarize, the project significantly advances the state of cybersecurity education and training, particularly in the area of active learning through real-world case scenarios. As previously mentioned, cybersecurity education and training are in dire need of a transformative means of delivering practical and hands-on exercises to augment classroom learning. The tool and its products provide significant addenda to traditional laboratory exercises to facilitate active learning. The novelty of this approach is the facilitation of achieving the highest level of *Bloom's Taxonomy of Learning Objectives:* **Create**. In addition to the pre-built scenarios, the participants are afforded with the tools, processes, and knowledge to produce new scenarios that meet the demands of a very fluid cybersecurity training field. The project holds excellent potential in transforming and accelerating cybersecurity training for the academic community, the cybersecurity community in general, and the entire world as a whole.

Background and Related Work

Prior and Similar Work

There have been similar efforts to address the need to enhance cybersecurity education and training. Notable-related works by the PI are found in [4–6]. A Critical Infrastructure Protection Training offered by Idaho National Laboratory is intended for the US military and/or the Department of Defense personnel assigned to conduct vulnerability assessment of our nation's critical infrastructures [7]. This 13-day course culminates with a capstone assessment exercise mimicking a live assessment of a critical infrastructure. This project provides a similar scenario augmented with intelligent interactivity to guide the learner. The Cyber Security Education Consortium (CSEC) has created centers of excellence in automation and control systems to provide training on SCADA and control systems security [8]. The courses that were created for this security curriculum are excellent training tools to upgrade the security skills of operators. However, widespread adoption is restricted by the high cost and the lack of hardware resources to support the courses in a portable and affordable setting. The SANS Institute offers a variety of courses on cybersecurity [9] which targets those personnel who are interested in bootstrapping or improving their careers in cybersecurity. The exorbitant registration cost for the course makes it impractical for training workshop adoption. This project offers freely available tools and interactive scenarios using affordable resources that can deliver hands-on and realistic cybersecurity training and education.

Scenario-Based Learning

Scenario-based learning is firmly situated on the learning theory that learning takes place in which the context is applied. Thus, it subscribes to the idea that knowledge is best acquired and fully understood when situated within its context [10]. Using real-life situations, scenario-based learning provides a relatable and highly relevant learning experience through immersive and highly engaging approach [11]. Scenario-based learning works best for training on tasks involving serious consequences such as those in cybersecurity. It offers a simulated environment or situation in which learners can afford to make mistakes without incurring costly repercussions. In [12], Clark proposes the following checklist for determining whether scenario-based learning is the right option:

1. Are the outcomes based on skills development or problem-solving?
2. Does it provide a simulated experience in lieu of a real and dangerous situation?
3. Are the students provided with relevant knowledge for decision-making?
4. Is a scenario-based solution cost- and time-effective?
5. Will the content and skills be relevant to justify the development?

Upon careful consideration, it can be established that scenario-based learning is an excellent option for cybersecurity training.

Problem-Based Learning (PBL)

Problem-based learning encourages learners to apply knowledge to new situations. This learning strategy develops critical thinking and creative skills, increases motivation, improves problem-solving skills, and helps transfer knowledge to new situations. Adults are self-directed, are goal-oriented, are relevancy-oriented, and are practical [13]. Adults take control of their own learning, in particular how they locate appropriate resources [14] and make a decision on which learning strategy to use to facilitate their learning progress. Hung, Jonassen, and Liu [15] state that PBL is self-directed, learner-centered, and self-reflective. The primary goals of PBL include (1) developing scientific understanding through real-world cases, (2) developing reasoning strategies, and (3) developing self-directed learning strategies. In this approach, learning begins with a problem to be solved rather than content to be mastered. Learners actually work on problems in ways that require them to develop expert knowledge, problem-solving proficiency, lifelong learning skills, and team participation skills. This learning technique will be extensively employed by the project through the hands-on scenario-based exercises on the Florida Cyber Range.

Cybersecurity Workforce Development and Competency Strategies

In both scenario-based learning and active learning, measuring learning outcomes and assessing students' knowledge and skills have always been critical. One effective approach is to come up with knowledge, skills, abilities, and competencies (KSACs) and then map learning outcomes to these KSACs. The National Institute of Standards and Technology's (NIST) National Initiative for Cybersecurity Education (NICE) has developed a list of knowledge, skills, abilities, and work roles [16] that has been widely accepted in both public and private sectors. In addition, the US Office of Personnel Management (OPM) has published a comprehensive document with cybersecurity technical and nontechnical competencies [17]. Our scenario-based learning approach is mapped to both these two frameworks, as explained below.

NICE Cybersecurity Workforce Framework

The NICE Cybersecurity Workforce Framework serves as a fundamental reference and resource for describing and sharing information about cybersecurity work and the knowledge, skills, and abilities (KSAs) needed to complete tasks that can strengthen the cybersecurity posture of an organization [16]. The framework categorizes cybersecurity work roles into seven categories, namely, securely provision, operate and maintain, oversee and govern, protect and defend, analyze, collect and operate, and investigate, and identifies the KSAs needed and tasks performed by each work role. The complete list of KSAs for each work role is listed in Appendix B of [16]. The framework serves as a vital resource to bridge the gap between education and industry by providing a common lexicon for organizations to identify and recruit for cybersecurity work roles and for education and training programs to identify the KSAs and prepare professionals for KSAs and tasks required for those work roles. By developing the Scenario Builder and Retrieval Toolkit that allows scenarios to be mapped to the NICE Cybersecurity Workforce Framework, the toolkit can be used to enhance individuals' skills and competencies for cybersecurity work roles and, consequently, strengthen organizations' cybersecurity posture.

Office of Personnel Management (OPM) Cybersecurity Competencies

The US Office of Personnel Management published a report on Attracting, Hiring, and Retaining a Federal ybersecurity Workforce in October 2018 where they discussed their Cybersecurity Competency Model [17]. The model was developed using the following two categories based on OPM's collaboration with the National Security Council Interagency Policy Committee Working Group on cybersecurity education and workforce issues:

- IT infrastructure, operations, computer network defense, and information assurance.
- Domestic law enforcement and counterintelligence.

The set of competencies was created by surveying a select group of employers in various occupations related to cybersecurity. The entire list of competencies can be found in Appendix B of [17]. The scenarios that we have built are mapped to these competencies as described in the next section.

The Solution

The Cybersecurity Scenario Builder and Retrieval Toolkit

In order for the cybersecurity scenarios to facilitate competency-focused learning, they will be required to map to the "NICE Cybersecurity Workforce Development Framework" [16], which is published by the National Institute of Standards and Technology (NIST) and the Cybersecurity Competencies as defined in the "Interpretive Guidance for Cybersecurity Positions" [17], which is published by the US Office of Personnel Management (OPM). Each scenario must be appropriately mapped to one or more category, specialty area, work role, and cybersecurity competency. Each scenario will also be labeled with keywords to enable easy search and retrieval. The scenario description file, which contains the specifics about the scenario, is required. Scenario artifacts such as log files, network capture files, forensic files, etc., may also be uploaded into the system. The main graphical user interface (GUI) is shown in Fig. 1.

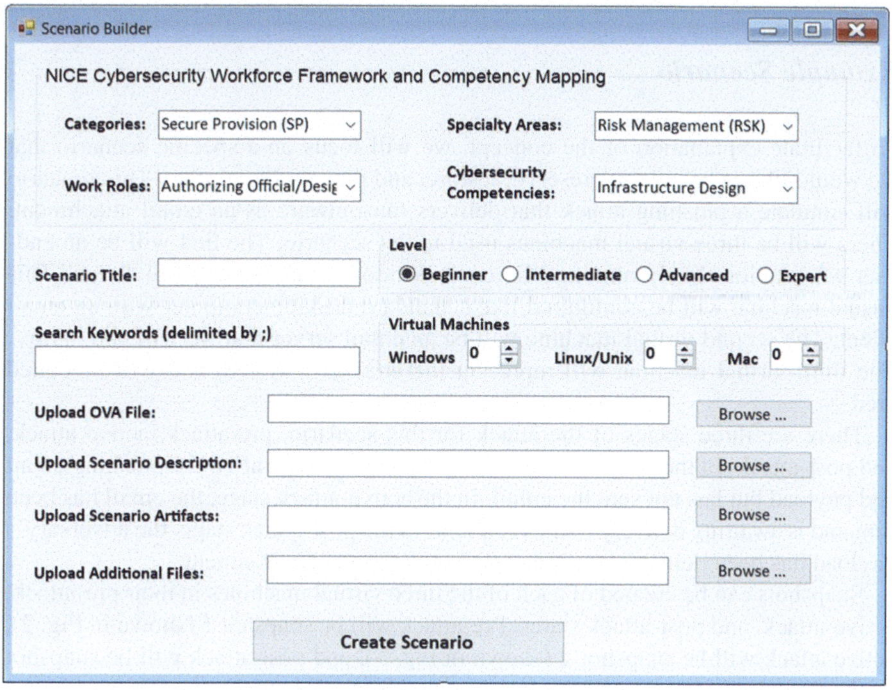

Fig. 1 The main graphical interface (GUI)

Open Virtualization Scenario

Virtualization is an enabling technology that is crucial to the growth of cloud-based storage and service solutions. It is also heavily used in the cybersecurity industry to isolate and protect resources from unknown threats and to provide safe environments to examine potentially dangerous artifacts. The rapid growth and spread of virtualization technology have led to the rise of a variety of companies that offer competing virtualization products. Each product supports a core set of common capabilities, such as the ability to create and restore system snapshots, as well as special capabilities designed to distinguish the product in the marketplace. Each vendor had its own proprietary format for storing virtual machines. The Open Virtualization Format (OVF) arose as a standard for storing and migrating virtual machines that made it possible to migrate machines among products. Open Virtualization Scenario expands upon this notion by combining the abilities of OVF to represent individual virtual machine configurations with additional details of the virtual environment (such as virtual network topology and scheduled events) in order to encapsulate entire education, training, and competition scenarios.

Example Scenario

To facilitate explanation of the concept, we will focus on a specific scenario that we would like to be able to preserve, restore, and potentially migrate. This scenario will simulate a phishing attack that delivers ransomware as an email attachment. There will be three virtual machines used in the scenario. The first will be an end-user workstation (likely running Microsoft Windows) that we will call "blue." This virtual machine will be configured like a standard desktop computer with an email client. The second virtual machine will be an email server that we will call "gray." The third virtual machine will represent the attacker's system and will be called "red."

There are three stages of the attack for this scenario, pre-attack, active attack, and post-attack. In the pre-attack stage, the adversary has crafted the phishing email and payload but has not sent the email. In the active-attack stage, the email has been sent and is awaiting delivery to the recipient. In the post-attack stage, the adversary's payload has been delivered and triggered in the target environment.

Snapshots can be created of each of the three virtual machines in their pre-attack, active-attack, and post-attack states. Pre-attack will be snapshot 1 (shown in Fig. 2), active attack will be snapshot 2 (shown in Fig. 3), and post-attack will be snapshot 3 (shown in Fig. 4). An OVF can be generated from each of these machines in each of the states, resulting in a total of nine possible OVF files.

Fig. 2 Pre-attack snapshot

Fig. 3 Active-attack snapshot

Fig. 4 Post-attack snapshot

Active-Attack Scenario

In this instance, the attack has begun. The phishing email has been sent and resides on the email server. The email server has received the message and likely pushed an alert to inform the recipient that new mail is available. The recipient has not yet downloaded or opened the message. The configuration for this scenario would be red-snapshot 2, gray-snapshot 2, and blue-snapshot 2. The notable changes in the snapshots are that red has the message in its sent folder. Gray has the message in its queue. Blue has a notification of new mail. The scheduled events for the scenario will be delivered from gray to blue and finally accessed by blue.

Post-Attack Scenario

In this instance, the phishing email has been delivered and accessed. The malicious payload has triggered. No further events have to be scheduled. The configuration for this scenario would be red-snapshot 3, gray-snapshot 3, and blue-snapshot 3. The notable changes in the snapshots are that gray would have changed state to indicate the message was delivered and blue would have received and been affected by the malicious email payload. Note that red-snapshot 3 in this case would not have any significant state change and may be omitted.

Scheduled Events

The OVF can already handle representing the snapshot states of the virtual machines. The scenario extension is needed to bundle the snapshot states of each machine as well as the scenario events that still need to occur to advance the scenario. A scenario event describes the machine and snapshot on which it occurs, an action to perform, and a trigger. The trigger can be the passage of a certain amount of time, a significant system event (such as start-up), or manual. The table below captures the scheduled events in the example scenario (Table 1).

Network Topology and Resource Configuration

In addition to virtual machines and scheduled events, there is additional infrastructure information required to configure the scenario environment. The most significant items include the network topology and resource configurations. For network topology, it is important to know which machines are connected to which other machines and how. For this example, we could imagine the email server has two network connections. One connection is to the outside world for incoming messages. The other connection would be to the local area network. The scenario

Table 1 Scheduled events

Pre-attack (snapshot 1)	Active attack (snapshot 2)	Post-attack (snapshot 3)
Send from red to gray	Deliver from gray to blue	Access by blue
Deliver from gray to blue	Access by blue	
Access by blue		

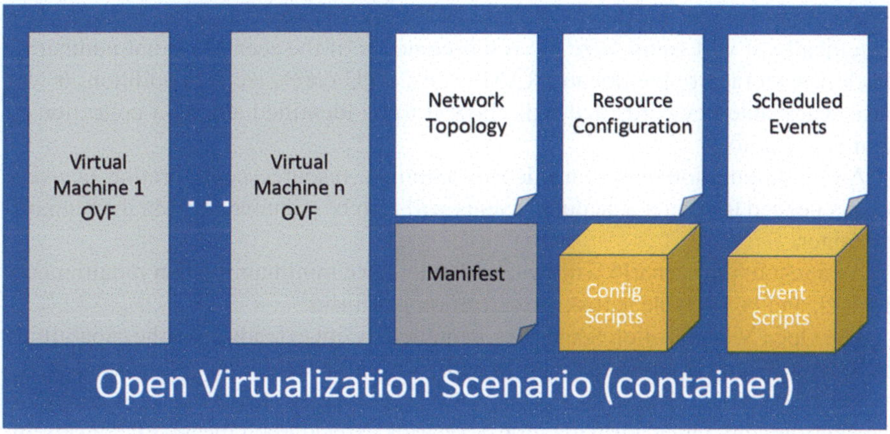

Fig. 5 Open virtualization scenario container scheme

would provide the required details to set up the WAN and LAN virtual networks and configure one virtual network adapter on the email server to have an address on the WAN and the second adapter to have an address on the LAN. The red machine would have an address on the WAN and be able to communicate with the email server (at least as its outgoing mail server). The blue machine would have an address on the LAN and be able to communicate with the email server on that interface. Additionally, the blue machine would have an email client configured to receive mail from the gray machine as its email server. The Open Virtualization Scenario container scheme is depicted in Fig. 5.

Key Benefits

The Open Virtualization Scenario format extends upon the existing portability of virtual machines currently possible via OVF by allowing the packaging of multiple OVF instances representing specific scenario snapshots. It further adds infrastructure details concerning network and service configuration to automate reconstituting the full infrastructure (not just individual machines). Finally, the scheduled events notion allows a scenario to be represented at a certain state of progress, with clearly defined and automated actions to perform to advance the scenario.

The manifest portion of the container will provide meta-data about the scenario. Specifically, it will summarize the resource needs of the scenario (total number of machines, storage requirements, RAM needs, CPU cores, etc.). In addition, it will have searchable keywords that will allow it to be identified among a collection of similar scenarios.

Any configuration scripts needed to automate resource configuration or event scripts needed to trigger scheduled events will also be contained within the scenario container.

As a result, the scenario will be searchable, have minimum system requirements defined, and be portable across virtualization platforms.

The Open Virtualization Scenario container concept extends upon the capabilities present in VMware's Virtual App (VAPP) format. VAPP allows the bundling of multiple OVF files, a network topology, and a "power on" sequence for standing up the environment. Open Virtualization Scenarios expand upon the VAPP concept by being portable (using nonproprietary formats and not being tied to any particular hypervisor) and by incorporating the notion of scenarios and scenario states. At the outset, a middleware component will be needed to translate network topology and resource configuration instructions into the specific API calls needed of each support hypervisor technology. Hopefully, with time and adoption by the cybersecurity education and training community, the various products may expand to directly support OVS as they have done with OVA and OVF.

The Development and Implementation Process

The scenario-based and competency-focused learning exercises will be concurrently designed, implemented, and deployed with the installation and configuration of the Cybersecurity Scenario Toolkit. Student assistants will help the project personnel in the development and testing activities. The goal is to intimately expose the practitioners/students to the scenarios and the toolkit to get a feel on how they would react to such an innovative learning platform.

We believe that this collection of scenario-based and competency-focused exercises is appropriate for the level of expertise that we expect from the students at the CAE-2Y, CAE-CD, CAE-R, and CAE-CO. Further, for each scenario, we will provide multiple problem sets and pathways that will introduce the PBL approach to learning and enable the learners to practice the technique in order to gain a better understanding of the concepts involved. The curriculum modules are embodied as living documents, which will be continuously enhanced and expanded in subsequent years.

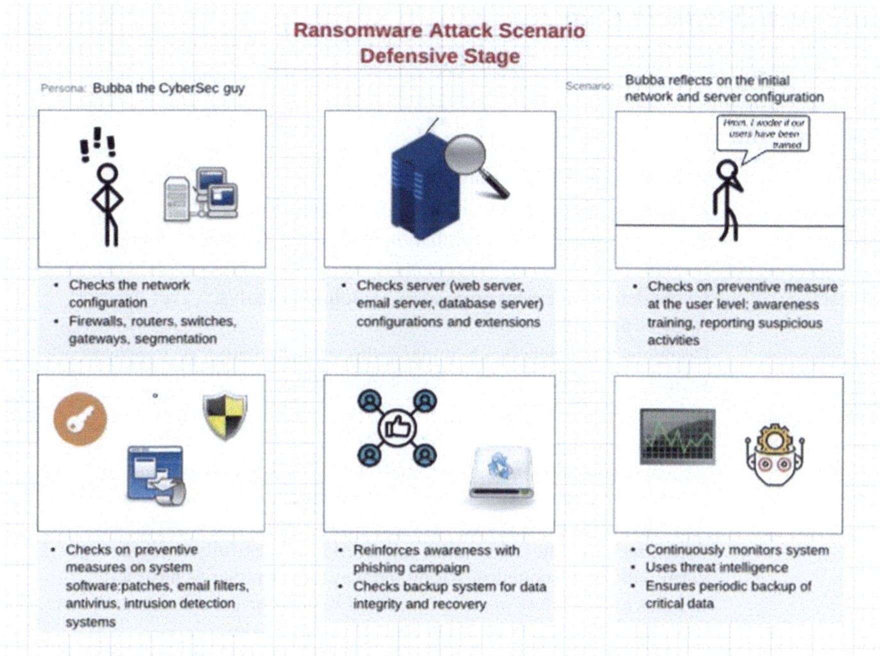

Fig. 6 A sample storyboard created with Lucidchart

The Development Tools

We will strive to make this project as cost-effective as possible. Thus, we will be utilizing open-source tools for development purposes such as the following:

- Twine [18]—an open-source tool for creating adventure style scenarios. This multi-platform tool publishes directly on the web.
- Lucidchart [19]—a visual workspace that combines diagramming, visualization, and collaboration. A free account is available using an academic email address. A sample storyboard for building a cybersecurity scenario is shown in Fig. 6.
- Microsoft Visual Studio Community [20]—this is a fully featured and free interactive development environment for creating multi-platform applications.
- Articulate Storyline 360 [21]—this is an academically discounted tool for developing custom interactive courses that works on any device.

Additional Scenario (Table 2)

Table 2 An exfiltration and lateral movement scenario

Title	Exfiltration and lateral movement
Categories	Protect and defend (PR); analyze (AN); investigate
Specialty areas	Incident response (IR); exploitation analysis (EXP); digital forensics (FOR)
Work roles	Cyber defense incident responder; exploitation analyst; cyber defense forensic analyst
Competencies	Computer forensics; incident management; information systems/network security; internal controls; network management
Level	Advanced
Description	*Case background* ACME corporation, a medium-sized defense contractor, recently suffered a debilitating cyberattack in one of its satellite locations near a naval base. The attack was analyzed by cybersecurity experts. The experts have determined that the cyberattack managed to successfully navigate the entire kill chain sequence. Along the way, the adversaries have left nuggets of indicators of compromise (IOC) that reveal the stages of their campaign. The sequence of events is illustrated below *Day 1:* Early on Saturday, august 10, an IT help desk personnel noticed that network activity appears to be abnormal and the server is running slower than expected. The staff immediately fired up Wireshark with the intent of capturing a snapshot of the network traffic that was inbound to the server. Normal operations persisted *Day 2:* IT management was made aware of the situation. Web traffic monitoring was initiated. No remedial action was taken *Day 3:* A malicious software was flagged by windows defender. FTP activities were also discovered. The IT manager assessed the issues and determined that no action is necessary *Days 4–10:*Nothing substantial happened during those days. Happy days and cruising time! *Days 11–12:* Phishing email document was discovered in the spam folder. FTP activities have resumed. Web server appeared to be very active. The IT staff started to take note of the seriousness of the issues. Normal operations persisted *Day 13:* The staff collected another set of captured packets. Security events revealed security issues. IT manager and staff scramble for solutions. Due to several project due dates, mitigating measures had to be put aside. Daily operations persisted *Day 14:* Lateral movement started to manifest. PowerShell scripts were discovered. Power users were created. Keystroke capture software uncovered. IT manager decided to shut down the server and called in the experts *Day 15:* External entity notified the IT manager that sensitive files, which may have originated from ACME, were found on the internet. The IT manager retrieved those files from the internet and deposited them in C:\users\CIS_Lab1\CheckAuthenticity. Time for post incident review!

(continued)

Table 2 (continued)

<table>
<tr><td>Title</td><td>Exfiltration and lateral movement</td></tr>
<tr><td></td><td>

The analysis process

Acting as a security analyst, you are required to perform a thorough analysis of the security events and artifacts. Your methodology should be guided by the kill chain sequence, and findings should be supported by system artifacts. The kill chain sequence is depicted by the following:

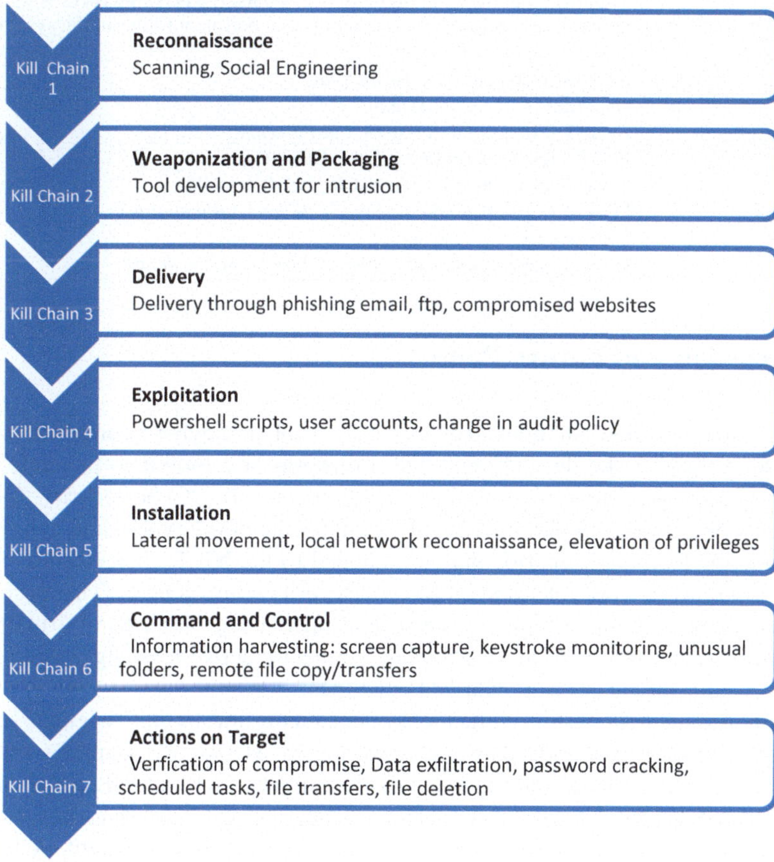

Security analysis requirements

Your analysis of the case study should proceed with guidance from the following requirements:

- Establish a timeline of the entire kill chain.
- For each stage of the kill chain, describe the IOCs that were collected, justify that the IOCs are integral to that stage, describe the consequence of the stage's successful completion, and develop preventive actions to deter its completion. For example, verify that data exfiltration had occurred, describe the IOCs that support the conclusion, and provide preventive methods that could mitigate data exfiltration from recurring.
- Write a report that includes an executive summary, a section for each of the preceding requirements, and a conclusion.

</td></tr>
</table>

(continued)

Table 2 (continued)

Title	Exfiltration and lateral movement
Virtual machines	1-windows 10; 1-Kali Linux
Keywords	Kill chain; indicators of compromise (IOC); lateral movement; network forensics; data exfiltration
Artifacts	The system under investigation has all the tools and the files that you will need to complete a comprehensive analysis. It has no internet connectivity. A brief description of each pertinent folder in the system is provided in the following: Files for web activities: *C:\inetpub* User files: *C:\Users\CIS_Lab1* Event log files: *C:\Users\CIS_Lab1\Events* PCAP files: *C:\Users\CIS_Lab1\PacketCap* PowerShell scripts: *C:\Users\CIS_Lab1\PowerShell* Files retrieved from the internet: *C:\CheckAuthenticity* Sensitive files: *C:\TopSecret* Hash signature of sensitive files: *C:\TopSecret\hash.xml*

Conclusion and Future Plans

This chapter describes an innovative approach to enhance cybersecurity training and education through the development and utilization of a toolkit to facilitate active learning and competency-based skills development. The Cybersecurity Scenario Builder and Retrieval Software Toolkit facilitates the development and deployment of authentic learning scenarios that are mapped to the NICE Cybersecurity Workforce Framework work roles and OPM cybersecurity competencies. These scenarios can be coupled with virtual machines developed using the Open Virtualization Scenario format to create more complete, authentic active learning scenarios that enhance educational development and outcomes.

Opportunities for future work include enhancing the toolkit to guide learners through the scenarios and adapt the scenarios based on each learner's progress. We will utilize artificial intelligence (AI) techniques and methods to develop an intelligent learning environment that can adapt the scenario characteristics and progression based on the user's learning preferences and assessment. A learning model may be tailored and updated for each student in order to maximize the potential of self-directed learning. Another opportunity for future work will focus on developing a curricular framework for the toolkit that can be used to package scenarios with learning outcomes and assessments toward NICE Framework work roles and/or OPM competencies.

Acknowledgments This work is partially supported by the Florida Center for Cybersecurity under Grant Number 3901-1009-00-A (2019 Collaborative SEED Program) and the National Security Agency under Grant Number H98230-19-1-0333. The US Government is authorized to reproduce and distribute reprints notwithstanding any copyright notation herein.

References

1. G. Francia, G. Randall, J. Snellen, Pedagogical resources for industrial control systems security: Design, implementation, conveyance, and evaluation. Journal of Cybersecurity Education Research and Practice **1**, 2017 (2017)
2. G. Francia, D. Thornton, M. Trifas, T. Bowden, Gamification of information security awareness training, in *Emerging Trends in ICT Security*, (Elsevier, Inc., Waltham, 2014), pp. 85–97
3. Towson University. *NSA NCCP National Cybersecurity Curriculum Program* (2019). https://www.clark.center/c/nccp. Accessed 30 Jan 2020
4. G. Francia, N. Bekhouche, T.M. Marbut, C. Neuman, Portable SCADA security toolkit. International Journal of Information and Network Security (IJINS) **1**(4), 265–274 (2012)
5. G. Francia, J. Snellen, G. Richards, Laboratory exercises to accompany industrial control and embedded systems security curriculum modules, in *Cybersecurity and Privacy in Cyber Physical Systems*, (CRC Press, Taylor and Francis Group, 2019)
6. G. Francia, J. Snellen, Embedded and control systems security project. Information Security Education Journal **1**(2), 77–84 (2014)
7. Idaho National Laboratory, *Critical Infrastructure Protection Training* (Idaho National Laboratory, 2020). https://inl.gov/critical-infrastructure-protection-training/. Accessed 30 Jan 2020
8. Cyber Security Education Consortium (CSEC). *Oklahoma Center for Information Assurance and Forensics Education (OCIAFE)* (2017). https://atecentral.net/r3800/. Accessed 30 Jan 2020
9. SANS, "SANS," SANS, 2020. https://www.sans.org/. Accessed 31 Jan 2020
10. R. Kindley, *Scenario-based E-learning: a Step Beyond Traditional E-learning* (2002). http://www.learningcircuits.com/2002/may2002/kindley.html. Accessed 20 Jan 2020
11. A. Pandey. *A 5-step Plan to Create a Captivating Scenario-based Corporate Training* (ELearning Industry, 2019). https://elearningindustry.com/scenario-based-learning-corporate-training-how-create. Accessed 20 Jan 2020
12. R. Clrak, *Accelerating Expertise with Scenario Based Learning* (Learning Blueprint, 2009)
13. S. Lieb. *Principles of Adult Learning* (1991). http://design2learn.ch/downloads/principles_of_adult_learning_lieb.pdf. Accessed 31 Jan 2020
14. S. Brookfield, Adult learning: An overview, in *International Encyclopedia of Education*, (England, Pergamon Press, Oxford, 1995)
15. W. Hung, D.H. Jonassen, R. Liu, Problem based learning, in *Handbook of Research on Educational Communications and Technology*, 3rd edn., (Mahwah, NJ, 2008), pp. 485–506
16. National Institute of Standards and Technology. *National Initiative for Cybersecurity Education (NICE) National Cybersecurity Workforce Framework* (NIST Special Publication 800–181, 2017). https://nvlpubs.nist.gov/nistpubs/SpecialPublications/NIST.SP.800-181.pdf?trackDocs=NIST.SP.800-181.pdf
17. United States Office of Personnel Management. *Interpretive Guidance for Cybersecurity Positions Attracting, Hiring and Retaining a Federal Cybersecurity Workforce* (2019). https://www.opm.gov/policy-data-oversight/classification-qualifications/reference-materials/interpretive-guidance-for-cybersecurity-positions.pdf. Accessed 6 Feb 2020
18. C. Klimas. *Twine* (2009). https://twinery.org/. Accessed 20 Jan 2020
19. Lucidchart. *Lucidchart See more. Know more. Do more* (2020). https://www.lucidchart.com/pages/landing?utm_source=google&utm_medium=cpc&utm_campaign=en_unitedstates_desktop_branded_x_bmm_lucidchart&km_CPC_CampaignId=1458000413&km_CPC_AdGroupID=57044763792&km_CPC_Keyword=%2Blucid%20%2Bchart&km_CPC_MatchType=b&km_CPC_. Accessed 31 Jan 2020
20. Microsoft. *Visual Studio Community*. (Microsoft, 2020). https://visualstudio.microsoft.com/vs/community/. Accessed 31 Jan 2020
21. Articulate. *Articulate 360, Articulate* (2020). https://articulate.com/360. Accessed 31 Jan 2020
22. Massey University. *Massey University of New Zealand*. https://www.massey.ac.nz/massey/fms/AVC%20Academic/Teaching%20and%20Learning%20Cenrtres/Scenario-based-learning.pdf.Accessed 27 Jan 2020

Teaching Cyber Security Through Distance Learning with International Students

Krzysztof Cabaj, Luca Caviglione, Patrick Georgi, Jörg Keller, Wojciech Mazurczyk, and Andreas Schaffhauser

Introduction

In the last years, the importance of cyber security dramatically increased as there have been many incidents with a large influence on real life or economical impact. As paradigmatic examples, we mention the Stuxnet worm, which physically destroyed several industrial machineries [1], a successful attack against a German steel mill [2] and a cyber-attack that deprived almost a quarter of million people of electricity in Ukraine [3]. To face such a set of threats and to be prepared to anticipate new security challenges, various actors in the information technology (IT) world have to be properly trained and educated. Among the others this is the case for software developers, personnel of law enforcement agencies, and system administrators.

K. Cabaj
Institute of Computer Science, Warsaw University of Technology, Warsaw, Poland
e-mail: kcabaj@ii.pw.edu.pl

L. Caviglione
Institute for Applied Mathematics and Information Technologies, National Research Council of Italy, Genova, Italy
e-mail: luca.caviglione@ge.imati.cnr.it

P. Georgi · J. Keller · A. Schaffhauser
Faculty of Mathematics and Computer Science, FernUniversität in Hagen, Hagen, Germany
e-mail: patrick.georgi@studium.fernuni-hagen.de; joerg.keller@fernuni-hagen.de; andreas.schaffhauser@fernuni-hagen.de

W. Mazurczyk (✉)
Institute of Computer Science, Warsaw University of Technology, Warsaw, Poland

Faculty of Mathematics and Computer Science, FernUniversität in Hagen, Hagen, Germany
e-mail: wmazurcz@ii.pw.edu.pl

K. Daimi, G. Francia III (eds.), *Innovations in Cybersecurity Education*,
https://doi.org/10.1007/978-3-030-50244-7_15

Therefore, pursuing a more secure Internet requires proper teaching tools and platforms both for academia and industry. An effective learning path in cyber security should not only focus on computer science as an abstract subject matter, but should also have deep links to everyday life challenges in that space. Hence, apart from the theoretical and conceptual background, students also need hands-on knowledge on how to install, configure, and administrate tools for the protection of networks and computing infrastructures. For instance, an effective information security course must prepare future experts to manage and operate firewalls, traffic sniffers, and analyzers, frameworks to conduct penetration tests as well as intrusion detection systems, just to mention the most popular hardware/software mechanisms employed to build a cyber security posture. Moreover, a comprehensive professional should be also able to perform basic forensics analysis, thus some training on this matter is highly desirable. Unfortunately, the creation of exercises and laboratory trials requires a non-negligible effort and investment in devices and appliances, which need to be properly configured and maintained. Nevertheless, despite the request for cyber security world-wide, the availability of suitable infrastructures or proper professionals could be a critical aspect, especially in rural areas or developing countries [4].

To deal with such issues, a possible approach is to take advantage of distance learning infrastructure and to use virtual learning tools. As a consequence, the laboratory can be offered as a virtual service that is remotely accessed by students. The use of virtual laboratories has been introduced over a decade ago, mainly as a mechanism to reduce the need for physical laboratories to complete many learning paths and curricula. As a possible example, [5] showcases the use of a simulation-based virtual environment for chemical experiments.

In general, early virtual laboratories were based on two technological components: a web interface allowing remote learners to access and control laboratory equipment through a browser or a web-view embedded in a mobile application, and real or virtual devices to complete the learning assessments. With the advent of the Web 2.0 paradigm and the increased availability of computing and storage resources, it is now possible to simulate complex industrial plants, emulate hardware via software in a cycle-exact manner, create device drivers to remotely share real machinery, or provide virtual machines or containerized applications. An important technological challenge to complete the vision of delivering complex and effective virtual learning experiences concerns the development of suitable signaling protocols to exchange data with real or virtual devices and laboratories, as well as provide proper interfaces to guarantee interoperability of software components. A possible solution is to use lightweight and general purpose protocols, similar to the Session Initiation Protocol (SIP), which is now commonly used in Internet Telephony for establishing user calls. The SIP can be used to set up an experiment, initialize the learning environment, manage sessions between the learner and the remote platform without the need of using Hypertext Transfer Protocol (HTTP), as well as to orchestrate different lab components for creating sophisticated learning experiences [6]. Moreover, the SIP can be used to implement part of the multimedia layer responsible of allowing learners and docents to interact via A/V communications.

Even if such requirement is not part of the framework discussed in this paper, such a feature could be of interest to develop platforms enriched with multimedia or real-time services. Additionally, modern virtual laboratories can draw on a large range of information, e.g., gathered from social networks, and offer personalized learning paths and gamification-based assessments, thus making the learning process more pleasant and effective [4]. Thus, a de-facto standard signaling could prevent the need of re-engineering suitable protocols or implement software layers to exchange data or orchestrate events on top of the HTTP.

With regard to the use of distance-learning-enabled paradigms for teaching cyber security, this has been already explored by the Academia and the Industry. For instance, in [7] the authors present an analysis of more than 20 master degree programs completely focused on teaching aspects related to cyber security. Moreover, the analysis of the state of the art in cyber security education shows that online courses or virtual laboratories are increasingly used mainly owing to the flexibility and cost-effectiveness of modern learning platforms. Specifically, in [8] authors describe 35 free online courses covering broad aspects of cyber security. In most cases, courses are available in a Massive Open Online Course (MOOC) environment provided by well-known organizations like Coursera, Cybrary.it, edX, and Udacity. It is notable that not only basic concepts of security are taught, but also deep insights are provided. In fact, from the 35 analyzed courses, 12 are categorized as "intermediate," whereas 7 as "advanced." As today, one of the biggest distributed laboratories for ICT security is co-funded by the Erasmus+ Programme of the European Union. The project is called open Distributed European Virtual CAMPus (DECAMP) [9] and organized by six European countries (Finland, Germany, Italy, Romania, Spain, and United Kingdom). The program takes benefits from European Credit Transfer and accumulation System (ECTS), which allows recognition and transferring credits and grades earned in the other European universities. Each ECTS credit point represents 25–30 h of student workload. Each course in the university is assigned an appropriate number of ECTS points depending on the required amount of work for the average student. The course provided by DECAMP offers the possibility of gaining 6 ECTS points per course each one supervised by a country/university. These six courses are divided into three bachelor and three master courses. The topics cover a wide-range of aspects relevant to understand and enforce cyber security, specifically: Secure Network Management and Computer Networks (Munich University of Applied Sciences), Applied Web Application Security: Attacks and Defense (Metropolia University of Applied Sciences), Cloud Computing Security (University of Cantabria), Security of e-Health Systems (University Politehnica Bucharest), Wireless Network Security (University of Padua), and Applied Computer Forensics and Crime Investigation (University of South Wales). Moreover, the cross-border/cross-institutional organization of DECAMP provides European Union (EU) students the possibility to gain the needed security skills by collaborating in a virtual "green mobility" manner. In this context "green mobility" means, you can study ICT skills abroad while staying at home.

To sum up, past works dealing with distance learning with an emphasis on cyber security mainly focused on technical and architectural aspects as well as performance scalability and network/mobility. Literature offers numerous examples for implementing remote learning infrastructures or for allowing students to access a remote/virtualized physical deployment. The simplest architectural blueprint is built around the client-server paradigm, where a single entity hosts virtual machines used during laboratory activities. In most cases the user interface is based on web technologies. Such a solution simplifies the management and eases the remote access for students. Moreover, it allows providing online materials like detailed descriptions of laboratory activities, useful commands, and a rich set of feedbacks via the user interface. One example is Tele-Lab [10]. For larger student groups, possibly composed of people working simultaneously, the client-server paradigm could cause problems especially in terms of scalability and availability. Due to this, the authors of the Tele-Lab system enhanced their framework to work in a distributed fashion [11], employing many servers in multiple locations: their implementation currently works in two sites, one each in Germany and Lithuania. Another interesting approach for preventing performance and availability issues is to use the services of commercial cloud providers. As an example, in [12] authors describe a virtual laboratory system running within the Amazon Web Services (AWS) infrastructure.

Due to performance problems other solutions that do not use central or cloud infrastructure have been investigated. For instance, in [13] authors present a virtual lab where students use their own computers to run virtual machines to perform experiments both locally and in a distributed way. There is still a central server, but it is mainly used to coordinate and orchestrate actions implemented within local virtual machines. In addition [14] showcases a laboratory that employs a central server running virtual machines that students should protect or attack. The use of virtualization simplifies setting machines back to stable status, e.g., if students with administrative privileges corrupt the installation or if an attack is launched successfully causing a disruption of the guest OS. Another important benefit of adopting virtualization is the cost-efficiency: according to [15], it is estimated that the expenditures to implement a virtual solution are one-third of a fully physically equipped laboratory. Unfortunately, in the presence of a large number of students, centralized virtual laboratories may still need a vast amount of resources such as central processing unit (CPU), memory, and disk storage. To address this, an interesting solution is presented in [16], where authors propose a system in which students perform trials by executing pre-configured virtual machines. Hence, resource-consuming tasks are performed locally, whereas actions requiring some form of cooperation are managed by the centralized component. However, the need arises to protect such a distributed system from attacks, e.g., when results are used to automatically grade students [17]. The work in [18] compares different organizations of virtual labs, yet mostly on a qualitative level, and without a particular focus on cyber security.

Since the design of a variety of architectural blueprints has been already largely addressed, in this chapter we want to investigate the main aspects related to internationalization and how the overall lab experience is perceived by students with different backgrounds and nationality. To this aim, we developed a prototypical virtual laboratory within the FernUniversität in Hagen in order to offer learning materials and five laboratory exercises, each one implementing a software component used to enforce security in real-world scenarios. The setup has been developed within the framework of the International Virtual Lab on Information Security (IVLIS) project granted by FernUniversität in Hagen and external project partners have been invited. Partners are the Warsaw University of Technology and the Institute for Applied Mathematics and Information Technologies of the National Research Council of Italy, and since they possess the relevant expertise in the field of network security and information hiding as well they can contribute to the development of the international virtual lab.

Concerning the design choice of building the IVLIS platform by exploiting a virtualized paradigm, in [19] authors addressed whether virtualization will lead to significant quality losses in teaching. For this they investigated two groups of students: one group worked on the course material in virtualized form, while the other group worked physically in the laboratory. The result of the analysis demonstrated that there was no statistical difference in post course confidence or performance between the virtualized and the physical group.

To implement the laboratory infrastructure, we used virtual machines. This offers some advantages, for instance, virtual machines can be easily reset to their original state in case of misconfiguration and they allow working from home. The latter is a core requirement for FernUniversität's distance learning mission and also promotes and enables (at least in principle) collaboration and use of the lab to a variety of remote parties. Since cyber criminals work beyond national borders, administrators and programmers should also be trained to cooperate internationally. Moreover, as cyber security is a field in constant change, lab exercises, in order to be timely and attractive for students, must be updated frequently and new lab assignments must be offered if needed. This can be very labor-intensive, in particular if expertise in special sub-fields is not at hand.

With the aim of achieving and promoting internationalization, existing lab exercises have been translated into English language. At the same time, they have been updated to the latest versions of standards and tools. In fact, the original vision of developing a virtual lab for security has been presented more than a decade ago [14] and it has been used as the basis for the aforementioned IVLIS project. Moreover, since the Internet evolved in terms of complexity and heterogeneity (e.g., with the advent of the Internet of Things paradigm), new exercises have been designed and integrated. Finally, if the partners want to offer labs of their own on this infrastructure, an English manual for teaching personnel has been created.

To evaluate our design, we compare two passes of the lab: the first with German students only, and the second with students from Germany, Italy, and Poland (countries in alphabetical order). The first run served as a beta-test to remove errors, but also allowed to investigate the influence of internationalization.

Therefore, the contributions of this chapter are: (1) the discussion of a novel platform for the delivery of laboratory trials in the field of cyber security, (2) the evaluation of the "learning experience" of students from different nations, and (3) some hints for the design of courses and the preparation of exercises in a more clear and effective manner.

The remainder of this chapter is structured as follows. Section 15 sketches the platform and Sect. 15 presents the student experiments. Section 15 describes the automated evaluation system for the progress detection, while Sect. 15 describes the Man-in-the-Middle Attack as a representative task of the platform; it also gives a deeper insight in a concrete instance of a course task. Section 15 showcases the role of the mentors in the system and the course and Sect. 15 reports the evaluation of student performance and acceptance of the lab infrastructure. Section 15 discusses the lessons learned from experiments conducted within the IVLIS framework, while Sect. 15 outlines general conclusions and provides an outlook to future work.

Platform

As discussed earlier, the virtual laboratory for teaching cyber security has been built by using a centralized architecture. Originally implemented with virtual machines [14], it was made-over using Solaris Zones [21] to reduce the overhead and drastically increase the number of students able to work concurrently on the platform. Following the technological evolution, it was later converted to a Linux-based system using Docker container technology [22]. Docker containers are an abstraction at the app layer that packages code and dependencies together. They share the machine's operating system kernel and do not require an operating system per application. Furthermore, they do not need a hypervisor and are much lighter than conventional virtual machines. In Fig. 1a and b a schematic illustration of this comparison can be seen.

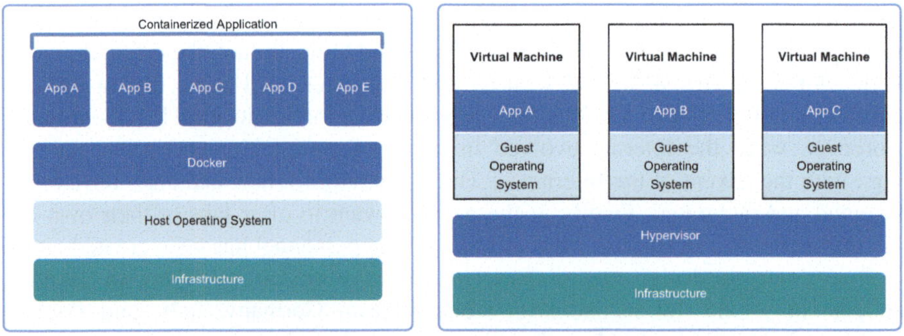

(a) Architecture of Containerized Applications

(b) Architecture of Virtual Machines using Hypervisor and guest Operating System

Fig. 1 Architectural comparison between Docker containers and conventional virtual machines (based on [20]). (**a**) Architecture of containerized applications. (**b**) Architecture of virtual machines using hypervisor and guest operating system

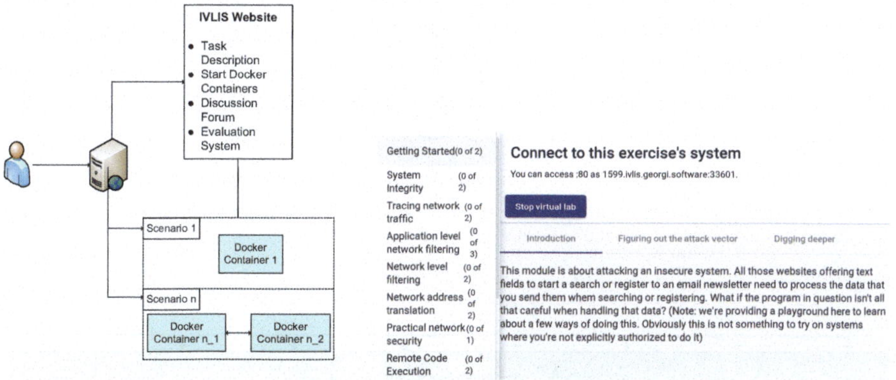

(a) Simplified Architecture of IVLIS (b) Homepage of IVLIS

Fig. 2 Architecture and homepage of IVLIS. (**a**) Simplified architecture of IVLIS. (**b**) Homepage of IVLIS

Figure 2a presents a simplified view on the architecture used by the IVLIS platform as well as the basic components of the system. Figure 3 gives a deeper insight into the technical architecture and start procedures on the server side. The Secure Socket Shell (SSH) network protocol is used so that the students can connect to the respective Docker containers and execute commands on their command line. SSH is a cryptographic network protocol for operating network services securely over an unsecured network. It uses public-key cryptography to authenticate the remote computer and allow it to authenticate the user, if necessary. The standard Transmission Control Protocol (TCP) port for contacting SSH servers is port 22. When the containers are started, the standard SSH port is mapped to the ports of the host, which makes them publicly accessible to students via SSH. The aforementioned Fig. 3 summarizes shortly these important technical details for the core architecture.

To access the platform, students log in on a web page where they are authenticated using username and password. An example for a page used to interact with the IVLIS platform is depicted in Fig. 2b. The web interface is logically divided into four different sections:

- Laboratory modules: they are used to organize the learning experience and each module contains multiple exercises on a well-defined topic;
- Discussion forums: they are used to allow students to interact each other and to have a direct communication path with experts;
- Key management: it allows to manage and upload the SSH keys that students use to log into the virtual machines—those are different from the key to log into the system;
- Evaluation system: it is used to automatically check the progress of the students. It gives feedback to the students, if the task has been successfully completed.

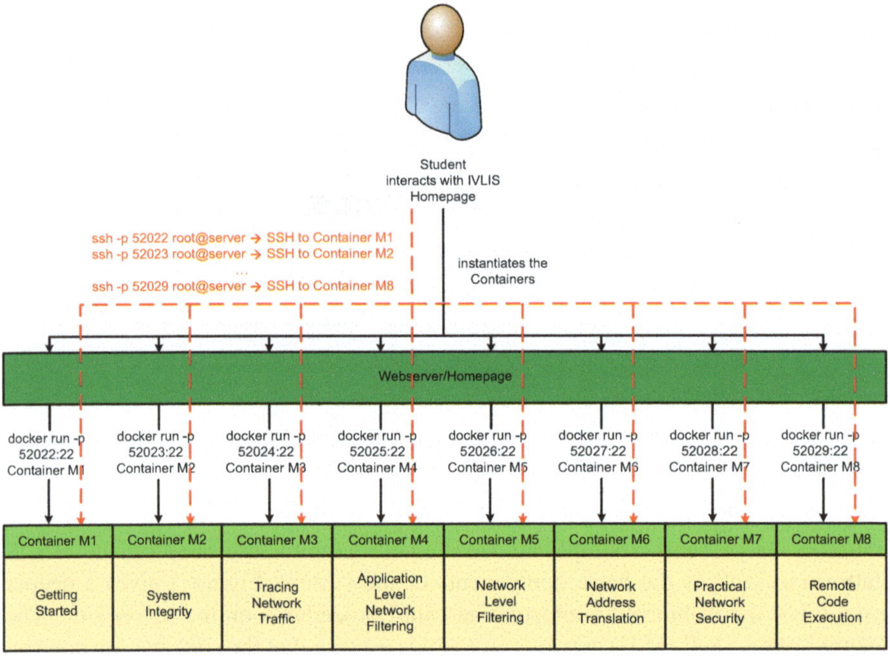

Fig. 3 Core architecture of IVLIS

The network scenarios are implemented using Docker containers that represent virtual servers. For every module, a student has access to one or several servers for their personal use that are configured for the purpose of the module's exercises: the servers are connected with each other through an internal network according to the exercises needs and the required software is also pre-installed. The layout of the internal network and the ports of each virtual system that the system exposes to the outside can be configured per module.

The student can start and stop these instances, and reset them to their initial state in case the configuration has gone astray (e.g., the firewall prevents the student from logging in).

The login is facilitated through SSH, using a server-assigned port per user and module that is presented on the web interface. The connection is authenticated using SSH keys that the user can enter on the web interface. These SSH keys are propagated to all their virtual servers so that they can SSH into all their virtual systems that have SSH enabled.

Once logged in, the student has everything they need to solve the exercise. Once solved, the student can request the system (again through the web interface) to validate the solution. The system will then run an evaluation function that returns either "success" or "failure."

Student Experiments

Figure 2b showcases the navigation bar, which is placed on the left side of the web page. By interacting with the navigation bar, students can select a module covering a certain topic. All the international partners participating in the IVLIS project were involved in developing the modules described in the following.

1. *Module 1—Getting Started:* This module serves as an introduction to SSH and SSH-Tunneling. Specifically, students have to log into the system via SSH, so they can learn how to interact with basic SSH commands and procedures.
2. *Module 2—System Integrity:* The students experiment with the tool `integrit` also by properly editing a config file. The main aim of the module is to do a snapshot of the host system and then verify if a specific file has been altered.
3. *Module 3—Tracing Network Traffic:* This module deals with the basics of Intrusion Detection Systems (IDS). In this case, students use `snort` to setup an alert for Internet control message protocol (ICMP) packets.
4. *Module 4—Application Level Network Filtering:* In this module, students are requested to set up a proxy server by using `squid`. The core of the learning experience is to teach how to deny the access to a specific Uniform Resource Locator (URL).
5. *Module 5—Network Level Filtering:* The basics of port scanning are introduced by experimenting with `nmap` to perform a port scan of a test machine provided by the IVLIS framework.
6. *Module 6—Network Address Translation:* To comprehend the core functioning of firewalls, the students have to set up stateless and stateful firewalls by configuring the tool `iptables`.
7. *Module 7—Practical Network Security:* In this module, students have access to a gateway where two hosts from different subnets communicate via `telnet`. To showcase how a typical Man-in-the-Middle attack operates they have to implement a sort of Man-in-the-Middle threat by using `tcpdump`.
8. *Module 8—Remote Code Execution:* In this case, students have to create a file on a server via remote code execution over an input field on a homepage.

Evaluation System

Once the student has started a task and logged into the container via SSH, they can use the evaluation button to evaluate the task. The evaluation is done, for example, by searching the log file with a task-specific regular expression for artifacts of the desired implementation. The best result and latest results are displayed as "success" or "failure" with a copy of the log file information used for evaluation. In order to get a more concrete idea of this evaluation scheme of a task, a representative problem from Module 1 has been solved. In Fig. 4 the evaluation system can be seen before the task is solved. The aim of this task is a login of the student on a

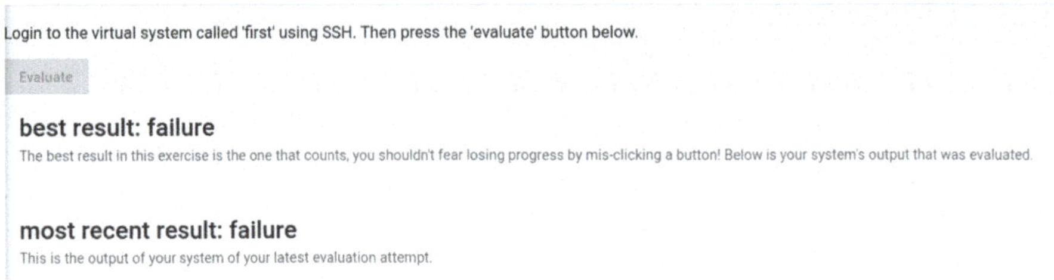

Fig. 4 Evaluation state failure (not solved)

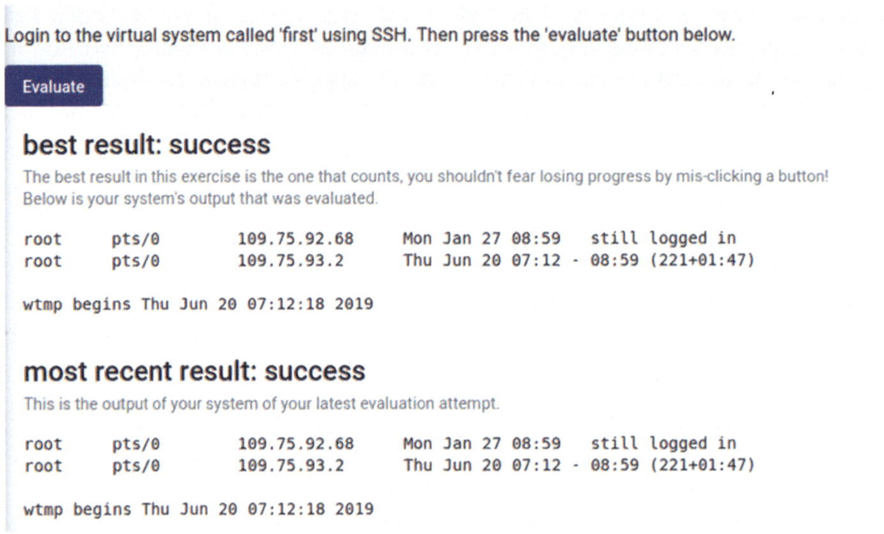

Fig. 5 Evaluation state success (solved)

system called "first" by using the SSH. The best result of this task is marked as failure, which means that no SSH login was detected so far. After the student logs in to the system called "first" and presses the evaluate button, the state of the evaluation result changes: the best result detected for this task now is "success." This means that the student has logged in and thus created the necessary artifact for the task to be completed (see Fig. 5).

Evaluation in IVLIS is done by executing an exercise-specific command for data collection in one of the virtual systems, typically taking the most recent part of a log file or calling a tool that provides some output. In a second step, that output is then analyzed outside the virtual machine using UNIX text processing tools, often using regular expressions to look for certain command/output patterns acting as signals. This second part must print a "1" on successful completion of the exercise. Two tools that are used for this second step are `grep` and `awk`. The first tool (i.e., `grep`) is a command-line utility for searching plain-text data sets for lines that match a regular expression [23]. The second one (i.e., `awk`) is a domain-specific language designed for text processing and typically used as a data-extraction and reporting tool [24]. This separation of data collection and parsing ensures that there is no trivial way for students to cheat (e.g., they cannot just replace some "evaluate" program within the virtual machine with one that always returns success). By moving the parser outside the virtual machine, no analysis of activity within the virtual machine allows a student to figure out the keywords the system is looking for. The only "shortcut" they could take is to replace the executable of the data collection command with one that returns well-formatted output as expected by the analysis command, but for our type of exercise this raises the effort and knowledge required to cheat beyond what is necessary to simply solve the exercise itself.

For the example in Fig. 5, the data collection routine called within the virtual machine is the utility `last`, which prints out the most recent successful logins to the system. Its output is then parsed by using `grep -q '^root\>' && echo 1`, which verifies that there is at least one line beginning with "root," indicating that the student was able to log in.

This method of evaluation is not perfect but it provides a fast and impartial response at any time without the need for staff to do these evaluations manually. More advanced schemes of feedback to students have been devised in [25].

Man-in-the-Middle Attack

The user interface of the system is the same for all modules: for this reason, only the implementation of the Man-in-the-Middle module is reported and it is depicted in Fig. 6. The web site has been developed to have very uniform layout, while the student works on a server terminal via SSH. In this scenario, the student has access to a gateway interconnecting two subnets. Every communication gets routed through the gateway between the two subnets. The latter interact via a telnet session. Telnet is a legacy client-server protocol for bidirectional interactive text-oriented data exchange. By default, telnet does not encrypt any data sent over the connection (including passwords). So, it is often feasible to eavesdrop on the communication and use the password later for malicious purposes. In our scenario, an account name and a password are sent between the client and the server and routed over the gateway. The student shall intercept these packets and get the necessary packets to log in to the client.

After starting the module through the IVLIS website, the student can log in one of their virtual systems named "gateway" via SSH (in this example, we used port 33604). After having obtained the access, the student has to sniff the password from the telnet session between two hosts within the internal virtual network that route their traffic through the gateway system. Since the students have root rights on their Docker container, they have the rights to install various programs that can be used to record network traffic. The larger part of students within the IVLIS project uses `tcpdump`, but also other programs like `tshark` have been made available as to guarantee a suitable degree of freedom. When starting this task, the students

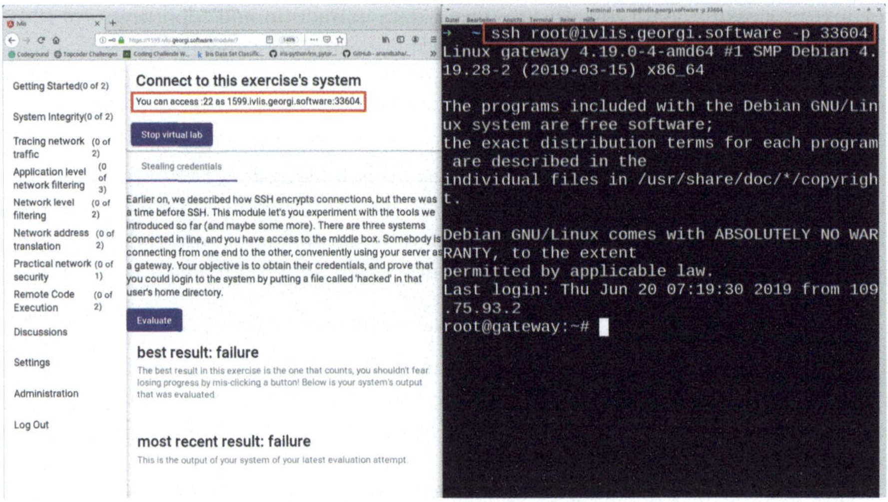

Fig. 6 The Man-in-the-Middle module provided by the IVLIS platform

are in general already familiar with the basic programs as well as basic network knowledge. From a didactic point of view, we would like to give the students the freedom and the opportunity to install and use one of their chosen programs. By knowing the password, the student is then able to log into one of the two hosts and create a local file to prove control over that host. The evaluation system is looking for the presence of that file to grade the student on the exercise. If it can be found, the most recent result will turn success like it can be seen in Fig. 5.

Mentor

The role of the mentor consists of two fundamental tasks. Specifically:

- **the mentor has to supervise their student group(s).** In the current incarnation of the IVLIS platform, the group size has been limited to four students. The mentor can communicate with the student group via a chat forum or individually via e-mail. Moreover, the forum encourages discussion among the students.
- **The mentor has to maintain the system itself.** Like in every human-made IT setting, problems can occur during the use of the IVLIS framework. In our case, there were sometimes problems with the evaluation system and detection of the progress, e.g., if the time between creating two artifacts in a task is so long that one artifact disappears from the log, or if the student has used a non-standard, but nevertheless correct solution approach, which is not properly detected by the automatic evaluation system. This means that a mentor also has to intervene technically in the system if problems arise. In our case, the mentor therefore has

a didactic and technical role. The user interface for mentors is mostly web-based and thus resembles the student interface, with the additional possibility to keep track of student progress. Only the technical intervention needs access via SSH to the virtualization platform itself, to check logs or student code.

IVLIS Evaluation

As a preparation, we first tested the lab with selected students serving as beta testers. They gave valuable feedback which has been used to further improve the IVLIS framework. Then, after polishing the platform and task descriptions based on the collected comments, we ran the virtual lab twice in a regular course setting. The first run was in Spring 2018 with only German master students attending the course. The second run was in Fall 2018 with German, Italian, and Polish master students. During each run, we collected some performance parameters in the platform, e.g., the number of completed experiments until the deadline (which also served as a threshold for passing the lab course). After the courses, we collected feedback of participants by using questionnaires. The questionnaires were anonymized, but the country could still be recognized, to enable interpretation of differences in performance on possibly different previous knowledge of the different student groups. While the total number of responses is small, we would like to point out that a lab course with about 20 participants is already considered large, and response rates have been much higher than usually in course evaluations, which indicates that the responses are representative.

Edition: Spring 2018

The number of participants, which were master students enrolled at FernUniversität in Hagen, was 24. As the lab course is an elective, we can assume that all of them have interest in IT security, especially as the course module on Internet security (10 ECTS credits) was a prerequisite. Of these participants, the 37.5% gave evaluation feedback via a questionnaire. The questionnaire included four closed questions and two open questions where the students could provide a comment. The questionnaire itself can be seen in Appendix 1. In the following, we take a closer look to the results of the four closed questions, as the open comments mainly referred to possible improvements, which we will consider in Sect. 15.

Question 1 aimed to determine whether the topics of the lab tasks have been interesting and insightful for the students. The result of this question was that modules 3 to 7 offered interesting insights for all students (100%) and module 2 offered interesting insights for 88.9%. Module 1 is unlikely to be of interest to every student, as it is an introductory module where students are shown basics of SSH and SSH-tunneling. The summary of obtained results is presented in Fig. 7a.

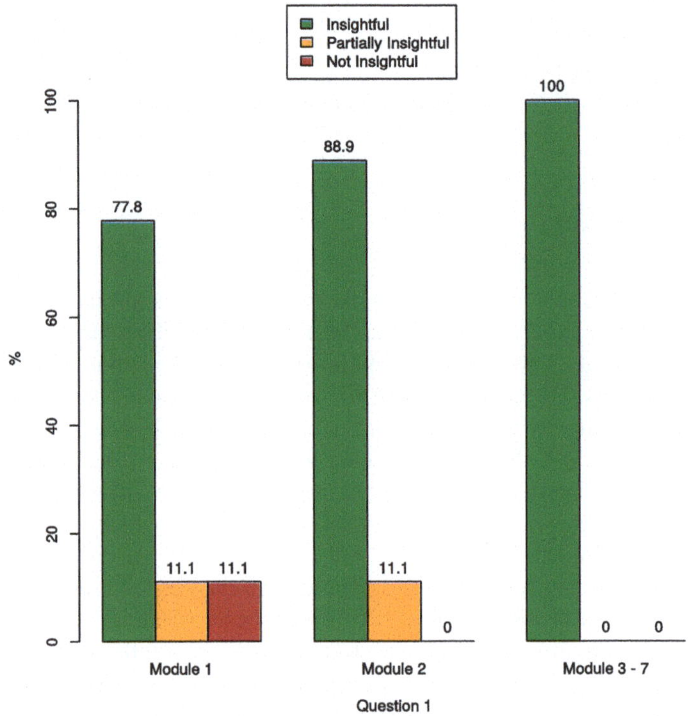

Fig. 7 Results related to insightfulness, skills, materials, and feedback. (**a**) Insightfulness of the offered modules (Note: module 8 was not offered during this edition of IVLIS). (**b**) Skills, materials, and feedback

The second question inquired about the skills improvement perceived by the students. Nearly every student confirmed that their skills have improved during the course (88.9%).

Question 3 asked if the instructions for the lab tasks were sufficient to complete the tasks. Almost half of all participants found some task description insufficient (44.4%). However, none of the students that wished a more detailed description named a task where improvements were necessary. This might be an indication that this critique is more a kind of "not completely satisfied."

Within the fourth question, the usefulness of the traffic light-like evaluation of the lab tasks functionality was assessed. It seems that the function was not sufficient for every student, as 44.4% students found it partially sufficient or insufficient. One reason might be that the automated test of task completion based on scripts is not perfect (see Sect. 15), so the question might better have been worded if the students found this function helpful. The visual summary of questions 2, 3, and 4 is illustrated in Fig. 7b.

Edition: Fall 2018

In the following semester, the virtual lab was run for the first time with an international student population: there were 16 students from FernUniversität, nine students from Warsaw University of Technology, and three students involved in a learning path within the National Research Council of Italy. Evaluation feedback reached us from 50% of the students that participated in the lab evaluated the virtual laboratory, specifically from five students from FernUniversität and nine students from Warsaw University. In the following, when percentages are given in brackets they are referring first to the German participants.

Due to the fact that questions remained unanswered in the first evaluation, e.g., which modules provided insufficient information to complete the tasks, we decided to design a module-oriented questionnaire. The questionnaire was divided into four categories: organizational, content-related, communication-related, and general questions. Through the finer granularity of the questions it should be clearer which modules have an insufficient task description. The altered questionnaire of this semester can be seen in Appendix 2.

In relation to the organizational questions, for the most of the German and Polish students the time frame of the lab was adequate (80% vs. 88.9%). Furthermore, we asked if the lab course corresponded to their ideas and wishes. All of the German students and two-third of the Polish students agreed with that question (100% vs. 66.7%). The last of the organizational questions were related to the contacts established with students from other universities. It turned out that just only 11.1% students made contact with a student from another university. The summary of the organizational questions of the German students is presented in Fig. 8a, while the results for the Polish students are illustrated in Fig. 8b.

The second category contains the content-related questions. First, the students rated the difficulty of each module. The level of difficulty seemed appropriate for the German students, as most of the tasks were felt to be appropriate (ranging from easy to hard). No extreme values (very easy/very hard) have been selected, except for the introductory module 1 and the modules 7 and 8. On the part of the Polish students, the result looks slightly different. Here, almost every task has a certain percentage of students who felt the task very easy, especially modules 2 to 4. The comparison can be seen in Fig. 9a and b. Only module 8 (devoted to learn remote code execution) was perceived by the majority of students of both institutions as very simple (75% vs. 62.5%). This might be an indication that the Polish students had a more advanced previous knowledge about security-related topics.

The task descriptions as well as the provided laboratory materials were understandable to a large extent by the students of both institutions. Thus, the improvements applied after the first lab edition seem to be successful.

The next point of the questionnaire inquired for interest and new insights for the students. It could be seen that for each module in the group of Polish students there existed a subgroup which found the module just partially interesting. That covers with the observation that a fraction of the Polish students found the task very easy.

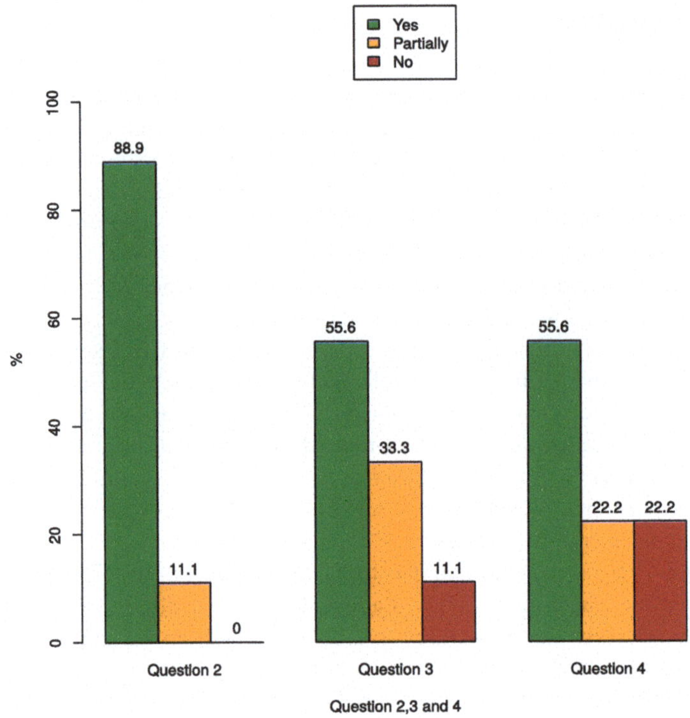

Fig. 8 Statistical results of the organizational questions. (**a**) German students organizational questions. (**b**) Polish students organizational questions

It is conceivable that the Polish students already had more previous knowledge in these areas.

The last content-related question asked about the time investment, it could be also seen that the Polish students invested less time than the German students. Looking at the mean, the time investment of Polish students of modules 1, 2, 3, and 5 were 12, 6, 11, and 8 min less than the investment of the German students, as depicted in Figs. 10 and 11.

Due to the fact that IVLIS is an online course, we also asked participants about the communication habits and how they helped the students to complete the tasks. The intensity of forum usage by German and Polish students was nearly the same (40% vs. 55.6%). Despite smaller group size, the e-mail contact was used more by German than Polish students (40% vs. 11.1%). However, help from the mentors was equally useful for both groups (80% vs. 85.7%).

In the retrospective, the majority of German and Polish students rated the course as an enrichment of their skills (100% vs. 88.9%) and would recommend it to fellow students (80% vs. 77.8%).

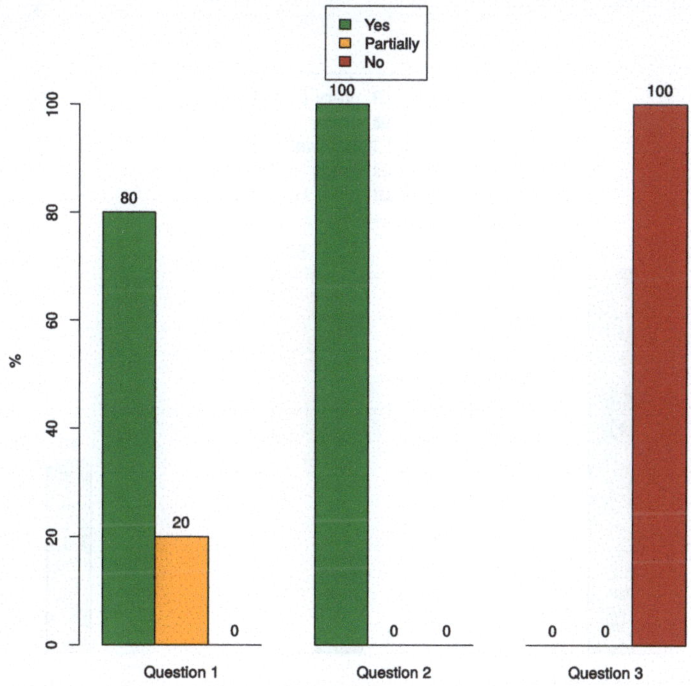

Fig. 9 Difficulty levels for module 2 to 4. (**a**) Difficulty level experienced by German students in modules 2 to 4. (**b**) Difficulty level experienced by Polish students in modules 2 to 4

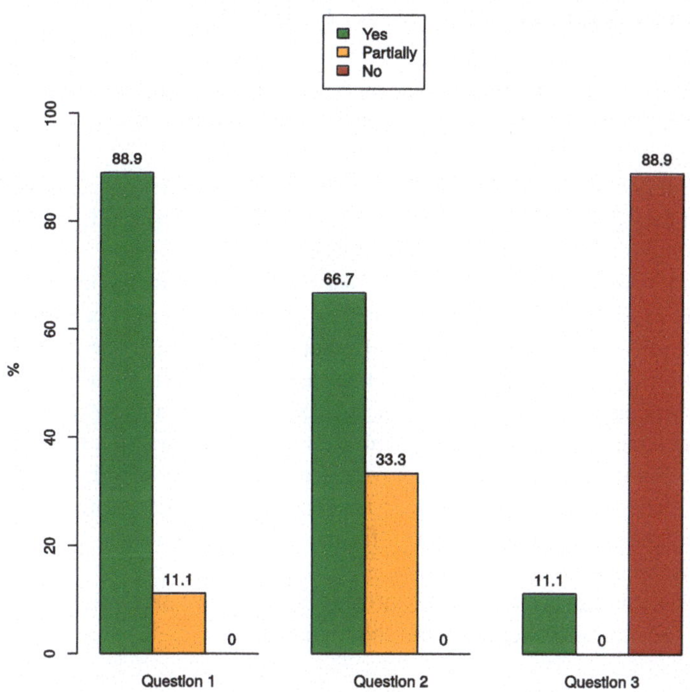

Fig. 10 Time investment of German students in modules 1, 2, 3, and 5

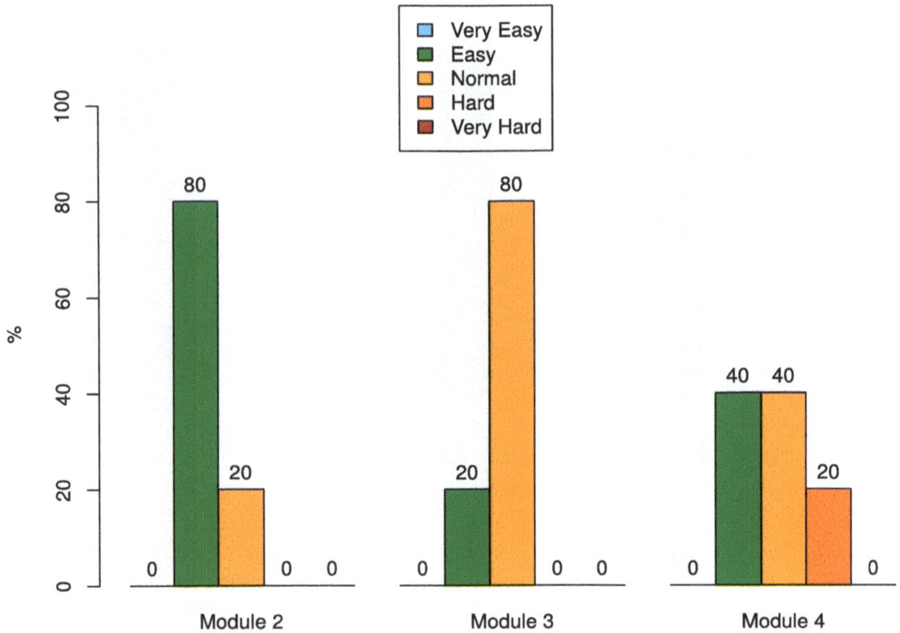

Fig. 11 Time investment of Polish students in modules 1, 2, 3, and 5

Educational Aspects and Lessons Learned

As shown, the way of teaching cyber security by using a distance learning framework has a lot of potential and enables additional benefits for the participants of the course.

First of all, virtual labs have the advantage that the participants can schedule their work when they find it most convenient: this is beneficial, especially for the students combining their studies with professional work.

Second, it provides "virtual mobility" for both distance teaching and on-campus students. It must be also noted that, in the case of typical (not virtual) labs, it would be harder to add an international component, i.e., to engage students from several countries to participate in the same tasks. Moreover, such lab organization potentially encourages communication between the students improving the intercultural and social experience. For example, students can see that different persons and cultures may show different behavior with respect to deadlines. The cultural and social experience can be fostered by special encouragement, e.g. having the students introduce themselves in the forum, and by offering a brief tutorial on cultural differences.

Obviously, the developed IVLIS lab can be extended and improved further by including, for example, the following modifications:

- adding an activity log to find out how much time an average student typically spends on each performed exercise;
- incorporating a communication log in order to observe how many questions/comments about each task were exchanged. In the current setup one explanation for the common lack of communication may be that students communicated locally, in Poland within the campus and in Germany via e-mail;
- extending the number of tasks where students actively work in groups to facilitate cooperation between the participants especially from different countries;
- requesting each student to write a short report about the solution of the given task to be commented by the participant of another nationality.

Finally, the implementation of network scenarios using container technology offers advantages in terms of performance and scalability. Kathara, Netkit's successor framework, also takes advantage of containers. With regard to the startup time and the memory consumption the usage of container within Kathara can decrease these two criteria by the factor 10 and 100 [26]. Older UML-based systems like Netkit are considered to be much heavier and require a hypervisor as well. Therefore, the use of container technology for the realization of the network scenarios seems to us the more advantageous variant.

Conclusions

In this chapter we have discussed the importance of hands-on laboratories, which are core learning tools for teaching cyber security in an effective manner. Since deploying a physical laboratory is not always possible due to both technological and monetary constraints, a possible idea is to rely upon some form of virtualization or remote learning. To this aim, we presented a virtual laboratory framework specifically designed to teach cyber security and developed within the IVLIS Project. Emphasis has been put in providing a simple and scalable environment as well as an easy to administrate and to restore platform. To satisfy such requirements, we showcased the use of virtualization and containers.

An important part of our investigation dealt with efforts needed to provide an international learning environment. In fact, the next generation of cyber security experts are expected to face "borderless" challenges and cooperate despite their physical location or cultural background. Results of our experimental campaign with groups of students belonging to three different nations demonstrate the effectiveness of our proposed architecture. For instance, the use of a forum allows students to aggregate and to discuss the various tasks as to develop collaborative interactions.

Future works aim to include group exercises and develop suitable performance indicators as to allow a detailed evaluation of progress and to provide personalized learning paths.

Acknowledgments This research was supported through the IVLIS project in FernUniversität's Fileh Programme from November 2017 to May 2019.

Appendix 1: Questionnaire Spring 2018

1. Have the topics of the lab tasks been interesting and insightful for you?
2. Do you think that your skills have improved due to the course of the lab?
3. Were the provided labs materials (e.g. instructions, formulation of tasks) sufficient for you?
4. How could the provided lab materials be improved? (free text)
5. Was the feedback that you received on the performed labs tasks sufficient for you?
6. Would you like to comment on issues related to the lab course that have not been considered within the provided questions? (free text)

Appendix 2: Questionnaire Fall 2018

1. The time frame of the laboratory was adequate (start and duration).
2. The organization of the lab course corresponded to my ideas and wishes.
3. I made contact with students from other universities during the lab course.
4. Rate the difficulty of each Module.
5. Evaluate the sufficiency of the provided lab materials of each Module. (free text if no or partially)
6. The targets of the tasks were clearly formulated?
7. Have the topics of the lab tasks been interesting and insightful for you?
8. How many minutes did you invest to solve each module?
9. I used the forum to exchange ideas/approaches for solutions.
10. I had e-mail contact with a mentor to get help on a task.
11. The feedback of the mentors helped me to solve the tasks? (free text if no or partially)
12. Would you recommend the lab to a fellow student? (free text if no or partially)
13. Do you think that your skills have improved due to the course of the lab?
14. Would you like to comment on issues related to the lab course that have not been considered within the provided questions? (free text)

References

1. E. Csanyi, How Stuxnet (PLC virus) spreads. https://electrical-engineering-portal.com/how-stuxnet-plc-virus-spreads-part-1. Accessed 17 May 2019
2. R.M. Lee, M.J. Assante, T. Conway, German steel mill cyber attack. https://ics.sans.org/media/ICS-CPPE-case-Study-2-German-Steelworks_Facility.pdf. Accessed 27 Jan 2020

3. R.M. Lee, M.J. Assante, T. Conway, Analysis of the cyber attack on the Ukrainian power grid. https://ics.sans.org/media/E-ISAC_SANS_Ukraine_DUC_5.pdf. Accessed 17 May 2019
4. S. Manca, L. Caviglione, J. Raffaghelli, Big data for social media learning analytics: potentials and challenges. J. e-Learn. Knowl. Soc. **12**(2), 27 (2016)
5. M.D. Koretsky, D. Amatore, C. Barnes, S. Kimura, Enhancement of student learning in experimental design using a virtual laboratory. IEEE Trans. Educ. **51**(1), 76–85 (2008)
6. L. Caviglione, L. Veltri, A p2p framework for distributed and cooperative laboratories, in *Distributed Cooperative Laboratories: Networking, Instrumentation, and Measurements* (Springer, Boston, 2006), pp. 309–319
7. K. Cabaj, D. Domingos, Z. Kotulski, A. Respício, Cybersecurity education: evolution of the discipline and analysis of master programs. Comput. Secur. **75**, 24–35 (2018) [Online]. https://doi.org/10.1016/j.cose.2018.01.015
8. L. Gonzalez-Manzano, J.M. de Fuentes, Design recommendations for online cybersecurity courses. Comput. Secur. **80**, 238–256 (2019)
9. http://mydecamp.eu. Accessed 6 Jan 2020
10. C. Willems, C. Meinel, Tele-lab it-security: an architecture for an online virtual it security lab. Int. J. Online Biomed. Eng. **4**(2), 31–37 (2008)
11. C. Willems, T. Klingbeil, L. Radvilavicius, A. Cenys, C. Meinel, A distributed virtual laboratory architecture for cybersecurity training, in *2011 International Conference for Internet Technology and Secured Transactions* (2011), pp. 408–415
12. K. Salah, M. Hammoud, S. Zeadally, Teaching cybersecurity using the cloud. IEEE Trans. Learn. Technol. **8**(4), 383–392 (2015)
13. H.P.E. Vranken, J. Haag, T. Horsmann, S. Karsch, A distributed virtual computer security lab, in *CSEDU 2011 – Proceedings of the 3rd International Conference on Computer Supported Education, Volume 1, Noordwijkerhout, 6–8 May* (2011), pp. 110–119
14. J. Keller, R. Naues, Design of a virtual computer security lab, in *Proceedings of the Third IASTED International Conference on Communication, Network, and Information Security, October 9–11, 2006, Cambridge, MA* (2006), pp. 211–215
15. B. Kneale, I. Box, A virtual learning environment for real-world networking, in *InSITE* (2003)
16. J. Haag, T. Horsmann, S. Karsch, H. Vranken, A distributed virtual computer security lab with central authority, in *Computer Science Education Research Conference, ser. CSERC '11* (Open Universiteit, Heerlen, 2011), pp. 89–95 [Online]. Available http://dl.acm.org/citation.cfm?id=2043594.2043602
17. J. Haag, H. Vranken, M. van Eekelen, A virtual classroom for cybersecurity education, in *Transactions on Edutainment XV*, ed. by Z. Pan, A.D. Cheok, W. Müller, M. Zhang, A. El Rhalibi, K. Kifayat (Springer, Berlin, 2019), pp. 173–208 [Online]. https://doi.org/10.1007/978-3-662-59351-6_13
18. P. Li, Centralized and decentralized lab approaches based on different virtualization models. J. Comput. Sci. Coll. **26**(2), 263–269 (2010)
19. J. Chamberlin, J. Hussey, B. Klimkowski, W. Moody, C. Morrell, The impact of virtualized technology on undergraduate computer networking education, in *Proceedings of the 18th Annual Conference on Information Technology, ser. SIGITE '17* (2017), pp. 109–114
20. Containers and virtual machines together. https://www.docker.com/resources/what-container. Accessed 23 Mar 2019
21. P. Georgi, Migration of a virtual computer laboratory to a different migration platform (in German: Migration eines virtuellen Computerlabors auf eine andere Migrationsplattform). Bachelor Thesis, FernUniversität in Hagen, 2009
22. P. Georgi, Container-based virtual laboratory for information science. Master Thesis, FernUniversität in Hagen, forthcoming
23. Manual page grep. https://linux.die.net/man/1/grep. Accessed 23 Mar 2019
24. Manual page awk. https://linux.die.net/man/1/awk. Accessed 23 Mar 2019

25. J. Haag, S. Karsch, H.P.E. Vranken, M.C.J.D. van Eekelen, An exercise assistant for practical networking and IT security courses in higher education, in *Computer Supported Education – 6th International Conference, CSEDU 2014 Barcelona, April 1–3, 2014, Revised Selected Papers* (2014), pp. 84–98
26. G. Bonofiglio, V. Iovinella, G. Lospoto, G. Di Battista, Kathará: a container-based framework for implementing network function virtualization and software defined networks, in *NOMS 2018 – 2018 IEEE/IFIP Network Operations and Management Symposium, April* (2018), pp. 1–9

Cybersecurity in Action

Vickie McLain

Introduction

The most successful students seem to have one factor in common. It's called experience. Employers all want to hire people who have experience. Student workers in our department were very successful for multiple reasons. The two most important reasons were experience and teamwork. It's important to not just have experience at doing homework, but working together in an environment where the decisions that are made will affect the lives of other people. This is commonly known as experiential learning, where the learner is afforded with the facility to perform hands-on learning [1]. Though it would be great to hire every student to be a student worker, that option isn't feasible. The only other option is to give every student the same type of experience on a smaller scale.

Experiential learning has been in practice for a long time with positive results. Experiential learning focuses on learners reflecting on their experience, so they may gain conceptual insight as well as practical expertise [2]. This project used several different types of experiential learning that included project-based, case-based, and problem-based all combined in a realistic business scenario.

Experiential education has value beyond building the social skills, work ethic, and practical expertise that are important in professionally oriented programs (Eyler, 2009). Experiential learning has taken place in the workplace in the past, but with security restrictions it can be quite difficult to provide a workplace experience for each student.

There are very few jobs where you can work independently without requiring some type of regular communication with coworkers. Communication or lack of

V. McLain (✉)
Alexandria Technical and Community College, Alexandria, MN, USA
e-mail: valarie.mclain@alextech.edu

K. Daimi, G. Francia III (eds.), *Innovations in Cybersecurity Education*,
https://doi.org/10.1007/978-3-030-50244-7_16

communication could be the reason a business grows or fails. It was very important that students worked with groups of their classmates to complete their work. Students need to be able to make decisions in a group based on the results of data they determine using their technical skills. The other important concept they need to master is troubleshooting in a team. Solving problems together can also help in skill retention.

The National Initiative for Cybersecurity Education (NICE) Cybersecurity Workforce Framework [3] was used to determine what type of job duties students would be performing in their organization. Students were given a link to NIST Special Publication 800-181, and it was included in their class content. It's important to provide students with resources they can continue to rely on throughout their career.

Students often contact us after graduation describing their experiences. After speaking to a lot of student graduates and employers, it seemed that students were aware of all the concepts and tools of the trade. However, there seemed to be an adjustment period before the new graduates were aware of the appropriate time or situation to use their new set of tools.

It seems many skills are taught independently of each other, but students never get to put them all together in a realistic situation. Students would be much better prepared for the workforce, if they could use their skills the same way that they would be using them in the future. When students enter the real world, they don't have numbered instructions mapping out what they will do at work. They need to use their education to make the right choices for the environment they are working in. That environment often contains policies, some of which are very restrictive. They need to be aware of various policies they may need to work with and know how to keep their organization in business while following the policies. Our students have found out the hard way that having the most secure environment in the world is useless if it prevents the business from running.

Lake Superior College was fortunate to be a recipient of a cybersecurity workforce education grant, which provided the funding for us to create a new lab designed to help students have a better understanding of cybersecurity careers and procedures.

Methodology

Planning

Lake Superior College designed a suite of six student business pods that are utilized by student small groups during courses and extracurricular activities. Students designed and configured a network that they need to protect as if it were a real business. Their business pod conducts regular business-type activities. They utilize the tools they have learned in their networking and cybersecurity courses to keep their business operational. Students will continue working in these "businesses" throughout the semester in multiple classes. The business activities they complete are related to the course content. Businesses are of different types, including healthcare, finance, government, education, manufacturing, and transportation.

The objectives of these activities include the following:

- Increasing the preparedness of cybersecurity students.
- Accelerating their comprehensive cybersecurity skill level.
- Increasing student confidence in their cybersecurity skills.
- Enhancing student communications through working on real problems as a team.
- Providing training opportunities for community.
- Student understanding of cybersecurity work roles.

Providing a more "real-life" model motivated students to learn cybersecurity skills, as well as enhanced their team communication skills. High school students could see demonstrations of a more realistic scenario and are more motivated to enroll in a cybersecurity career when they can see the types of tasks involved in cybersecurity work. There was a lot of collaboration between faculty, student workers, and students before the plans for this project were finalized. Decisions that were required for this project were:

- How many businesses can the project support? This was partially dependent on the room and the class size. A wall had to be torn down to make a bigger space to house six businesses in the project.
- How many students should "work" in each business? It was decided to have four students per group. Our groups pair campus students with online students, so there needs to be at least one campus student in each group.
- How many drops are needed in each business? The number of drops was a little higher than originally anticipated, because it seemed there were always extra students wanting to help troubleshoot the "business" networks.
- What equipment is crucial to each business? The decision was based on the basic equipment most students are required to support in their jobs. Each of our businesses has a switch, wireless access point, servers (some virtual and some physical), firewall, and three desktop workstations. Students also use their laptops in their business.
- How much electrical power will this take? It took quite a while to get all the specifications for equipment calculated to determine electrical requirements for the room. The entire building needed a new piece of electrical equipment installed, because it was almost at capacity before the project started. This installation added about a month to the timeline.
- How will you cool the data center? Air conditioning was installed in the data center room. Temperature monitors were also installed.
- What services are required from a third party and how much will they cost?
  - Carpenters were needed to take down a wall and build an enclosure for our data center. The college has had many handicapped students and some use wheelchairs, so the data center needed to have double doors on each side so

  that they could roll their wheelchairs in to reach equipment. Carpenters also had to construct the individual modules for each business.
  - Electricians were needed to wire the room. The room required a lot more outlets and a 220 line. The company that did the electrical work also installed the Cat6 and all the drops. It's always better to have too many outlets than too little.

  The rest of the work in this project was completed by our faculty or students.

- What options will students have for backups? It was decided to use a storage area network (SAN) drive and that each business would have a physical drive to use as a backup as well.
- What equipment will need to be installed in the data center? A faculty team created the equipment list and installation diagrams. They were assisted by student workers. In our data center, there is a firewall, several servers, a SAN drive, a keyboard, video, mouse (KVM) switch, and some large uninterruptible power system (UPS) units. The specific equipment that was used is listed in Fig. 1.
- What timeline will be used? The grant had a timeline of 1 year. The only delays involved electricity. Some outlets had to be moved, and a standard outlet had to be replaced with a 220 v outlet.
- What are the types/names of the businesses?
  - Finance—Exploit Bank.
  - Healthcare—Hackistan Regional Hospital.
  - Government—Hackistan.
  - Manufacturing—Cyber Flash.
  - Transportation—Aircrack Airlines.
  - Education—Hackster High School and Crypto College.
  - Red team—sophomore hackers and penetration testers.
- What software will be used? After years of dealing with the need for constant costly upgrades, it was decided that to make the project sustainable for the long term, the majority of software must be open source. The other reason to use open source software was that students would still be able to use it when they entered their future jobs or could practice with it on their own. VMWare licenses were purchased because students in our area have a high probability of needing to support VMWare in the future.

  Students surveyed different types of businesses to find out what types of programs were most valuable in their businesses. They were given many names of really expensive programs, so the students looked for open source programs that would perform the same functions. Our students wouldn't really be using these proprietary programs in the future, but they still needed to know what kind of data would be contained in these types of programs and how to protect it.
- Who will install the network equipment? Since our program name is Network Administration and Cybersecurity, there were some very talented students avail-

able who built the entire infrastructure under the direction of faculty. Students had some great ideas of their own about what was needed in each business. It gave the students some very good experience and was almost free labor, as many of them worked on it for their capstone project!

- How will the physical room be configured? Our only requirements were six separate spaces for each business and a data center that could be locked. There is faculty at our school who is an architect that volunteered to design the space, so every inch could be maximized. The room layout is shown in Fig. 2.
- What type of activities will students complete in their business? It was decided to make the experience as realistic as possible, so students started at the beginning. They needed to be interviewed (by other people working at their business), even though they knew their chances of being hired were pretty good! Job roles for businesses were selected from NIST Special Publication 800-181. The jobs were limited to the ones that were possible within the limitations of the student business, so they had a list to pick from. Students needed to prepare resumes to obtain a job at their newly assigned organization. However, students needed to look at the job roles, to determine what types of interview questions to ask the students working in their business, so that they could evaluate the appropriate skills. After getting hired, they were given a limited number of system credentials to begin work.
- Which classes will these businesses be integrated into? The cybersecurity workforce education lab is used for Computer Support, Security, Network Security, Network Forensics, Incident Response, Ethical Hacking, Windows Server, and Intrusion Detection and Prevention. The workplace is integrated into these classes. It is also used to complete skills in most of our other classes, such as testing websites, server installation and configuration, and OS installation.
- How else could the cybersecurity workforce education lab be used? It is used for GenCyber [4] high school cybersecurity camps, Computer Support repair business, and CyberPatriot [5] (high school cybersecurity teams) and as a Cisco networking lab.
- What services will be available in student businesses? (Fig. 3)

Implementation

This is a summary of implementation steps.

Month One

1. Brainstorm ideas for businesses, including the type of business, software, and hardware.
2. Set up project management software that can be shared, so everyone is aware of what tasks are done or in progress.
3. Design a data center to support our six businesses.

Quantity	Item
24	vSphere 6 STD 1 PROC
24	Basic SNS Vsphere 6 Standard I PROC 1 year
1	Vcenter 6 Server Standard Per Instance
1	Basic SNS Vcenter Server Standard 6 Per Instance
12	ProLiant DL160 Gen9 1U RM Xeon 8C E5-2609 v4 1.7GHz / 8GB / 4x3.5" Bays / B140i / 2xGbE / 550W HP Servers
12	Config. Complex Server Configuration - Hardware Install Only Wilmington Config SVC
12	Processor, Xeon 8C E5-2609 v4 1.7GHz / 20MB / 85W for DL160 Gen9 HP Server Accessories
84	8GB PC4-19200 288-pin DDR4 SDRAM RDIMM for Select Models HP Server Accessories
12	Dual 8GB microSD Enterprise Midline USB Kit HP Server Accessories
1	MSA 2042 SAN Dual Controller w / Mainstream Endurance Solid State Drive SFF Storage HP StorageWorks
24	MSA 600GB SAS 12Gb / s Dual Port 10K RPM SFF 2.5" Enterprise Hard Drive HP StorageWorks
1	MSA 2040 1Gb Short Wave iSCSI SFP+ Transceiver (4-Pack) HP StorageWorks
1	24x7 Foundation Care NBD MSA2042 Service HP ESSN/Services
3	**Netshelter SX 42U 600mm Wide x 1200mm Deep Enclosure APC - Data Center**
4	Cat6 UTP Patch Cable, Orange, Snagless, 10ft Belkin-cables
6	**Smart-UPS SRT 6000VA / 6000W 208V RM Online UPS HW Input HW+ (2) L6-20R (3) L6-30R**
6	**Metered PDU 2G, 200 / 208V, 30A, 0U Rackmount, L6-30P Input 3m Cord, (36) C13 (6) C19 Outlets APC - Data Center**
13	Cisco WS-C2960XR-24PS-I Switch IP LITE
7	Cisco ASA 5505 Firewall Edition Bundle
2	PFSense Firewalls
6	Linksys WRT1900ACS - wireless router
2	Hak5 Tool Kits
	Miscellaneous Supplies and Software for Organizations

Fig. 1 List of project equipment

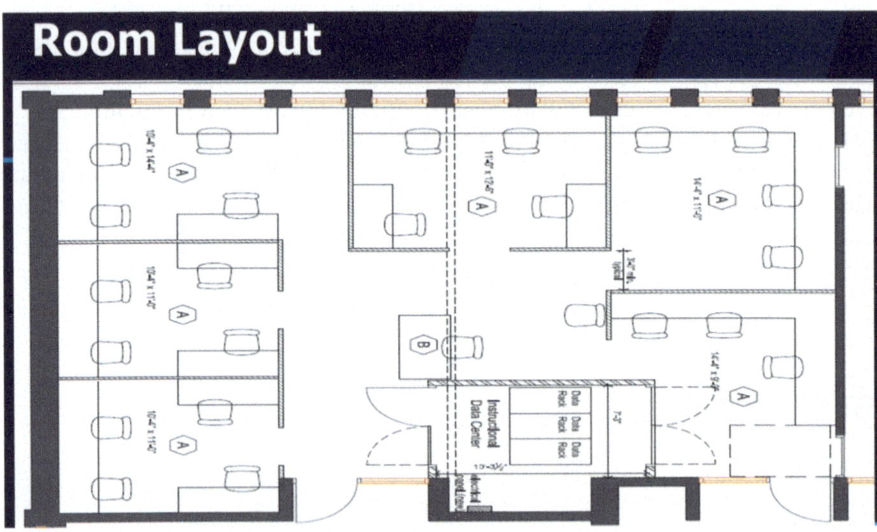

Fig. 2 Cybersecurity workforce education lab room layout

4. Design the infrastructure with associated network map.
5. Survey businesses to find typical software they use and what type of data is most important to them.

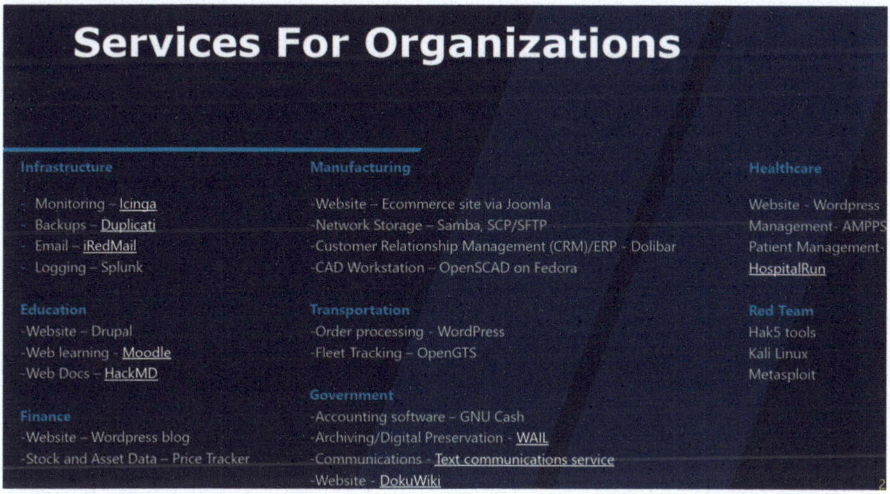

Fig. 3 Student business services

6. Design the location.
7. Order equipment.
8. Create different types of databases for each organization with fictitious names and data.

Months Two to Four

9. Start writing curriculum.
10. Construct the room, including electrical outlets, furniture, and network drops.
11. Install network equipment in data center which includes servers, switches, a SAN, and UPS units.
12. Install equipment in individual businesses.

Months Five to Six

13. Install OS software.
14. Install software proprietary to each type of business.
15. Set up an additional penetration tester business. This used Hak5 tools and open source software such as Kali Linux.
16. Staff test installations.
17. Students test installations.

Months Eight to Ten

18. Write manuals, network documentation, and directions for equipment.
19. Make policy reference manuals. Though students needed to look up business policies themselves, there is a resource list in each business in case they have problems finding them.

20. Make signs and hang typical business posters for each business.

Months Eleven to Twelve

21. Add Scada/IOT equipment to businesses, which includes mini traffic lights, power grids, and a conveyor belt. There was a separate Department of Education grant for this, but it was decided IOT was important for students to gain experience with.

Curriculum Integration

The curriculum was designed for students to complete business tasks in different cybersecurity classes, depending on how it relates to the particular class. For example, they need to configure firewalls in their business for the Intrusion Detection and Prevention class. Some semesters students have stayed in the same business throughout their program, and other semesters they have changed businesses. Students do change to other businesses to make sure all the businesses have approximately the same number of students. It's also a good idea for students to train in more than one type of business.

Example summaries for some class assignments are shown below:

Computer Support

1. Submit a job description that can be used for a business in our cybersecurity workforce education lab (finance, healthcare, transportation, government, education). Use the description of knowledge, skills, and abilities (KSAs) that can be found in the NIST 800-181 documentation in the Content section to write a resume so that it shows all your knowledge and abilities that would make you qualified for the job.
2. Submit the resume clearly showing information about your own knowledge, skills, and abilities as they pertain to the job you are applying for.
3. Be prepared to interview your coworkers based on KSAs that they are expected to have for their job.

Security

Preparation: Students are told they have been hired by a business, and each student is given their assigned "job" by the instructor. The only information they are given about their business is some of the passwords. Students aren't given all the passwords, because students may need to take control of a network where they don't have all or any of the passwords.

Welcome to Your New Workplace.

You have been assigned your job roles. Now you need to have a look at your network.

These are the first tasks students are assigned at their new workplace:

1. Make a network map for your organization.
2. Make a list of regulations and policies you will need to follow in your organization.
3. Log in to your equipment and make a list of your vulnerabilities.
4. Conduct an inventory of your business.
5. Map your network.
6. Develop a risk assessment for your business.
7. Create policies that your business must adhere to.
8. Research all applicable laws/policies that your business must follow and make sure your business complies with the law.

Assets, Threats, Vulnerabilities, and Risk

1. Asset identification: Identify the assets in your organization using a spreadsheet. Include appropriate information about assets such as price, serial number, etc.
2. Threat evaluation: Complete a threat evaluation for your organization. Identify potential threats and their potential impact.
3. Attack tree: Design an attack tree for your organization.
4. Vulnerability appraisal: List the vulnerabilities your organization has currently. Determine how your business will compensate for these vulnerabilities.
5. Risk assessment: Score your risks using a risk scale you create.
6. Risk mitigation plan: Determine how you will handle your risk and write a report describing the actions that will be taken by your business.

Ethical Hacking and Systems Defense

This is a sophomore class, so only students who had completed their freshman year successfully were able to perform the very popular "red team" assignment. This assignment actually motivated a few students to make sure they made it to their sophomore year.

Since there are no doors on our businesses, it made it very easy for the red team to do all types of physical hacking on the unsuspecting freshman businesses. They used several Hak5 tools [6] such as LAN Turtle, Rubber Duckies, Wi-Fi Pineapples, and Bash Bunnies. All these students had access to Kali Linux [7] and Metasploit [8] for software exploits. They were able to implement additional tools of their own. They need to present a detailed report to the freshman business employees when they complete their work.

The sophomore red team is expected to perform the following tasks:

- Use hacking tools to hack an assigned freshman business.
- Write a comprehensive report detailing vulnerabilities, risks, and ways to mitigate the risks.

- Present the report to the freshman business and educate them on how they can make their business more secure.

Network Forensics

The Web Server Investigation aspect entails the following activities:

1. Investigate the web server in your organization:
   (a) What is the IP address?
   (b) What IP addresses has your web server been communicating with?
   (c) What information is available on your website?
   (d) Are there any settings on the web server which don't provide security?
   (e) What information do you think should be available on your website?
   (f) What else do you need to check and why?
2. Discover website vulnerabilities:
   (a) Does your website have any vulnerabilities that can be exploited?
   (b) What could a negative actor use to exploit them with?
   (c) How can you make your website more secure?
   (d) What type of negative actors do you think would target your organization?
3. Investigate and analyze the security of the server:
   (a) Can you find any files on your web server that shouldn't be there?
   (b) How did you determine the files that shouldn't be there?
   (c) Did you find anything on the server that could be dangerous to your organization?
4. Write a report:

   Your team must answer all the Web Server Investigation questions. In addition, the team must write a professional report that will be addressed to your supervisor that summarizes the findings of the team.

   Email Forensics Investigation.

   Each group needs to search for five emails in their business and investigate the five most suspicious looking ones.

- Check the following:
  - Analyze headers.
  - Server investigation (source of the email).
  - Check firewalls for clues.
  - Check for software-embedded identifiers.
  - Check the received header for applications and versions used to send the email.
- Write an email investigation report that includes the following for each email:

Project Peer Evaluation

Rate the members of your team on a scale of 0 – 5. Please rate the team members honestly. I usually know who is contributing and who is not.

0 is the lowest score and 5 is the best.

Member Name	Timely communication	Contributes to equal share to work	Treats team members in professional manner	Displays initiative	Total
EX. Joe Student	3 Took four days to answer email	0 He didn't do anything.	5	3 He came up with lots of new ideas, he just didn't act on any of them!	11

Fig. 4 Project peer review form

- Source and destination.
- IP addresses of sender and receiver.
- Any suspicious text.
- Path the message has traversed.

Programming

Secure coding is the practice of developing computer software that is free of security vulnerabilities [9]. In this project, programming students design "secure" websites for the various student businesses. Students in each business need to find vulnerabilities in the website. Student businesses are supposed to communicate with the programming students who are designing the websites, to make sure it's appropriate for their organization. After cybersecurity students determine vulnerabilities, they need to meet with the programming students who designed the websites so that they can change them to make them more secure.

Peer Review

All students are required to submit a peer review form for every team project. On every team, there are usually more motivated and less motivated students. For that reason, the peer review grade is 25% of their final grade, so that students that don't participate won't automatically get a good grade. An example of a peer review form is in Fig. 4.

Results

Students are very motivated by working in a realistic setting. The first day that students were assigned to the lab at 10 am (the class is over at noon), almost all of them were still there at 5 pm because they were enjoying it so much! High school students use this lab, and it helps them put cybersecurity in a new perspective. A high school student was working in our "Hackistan Regional Hospital," and he decided to shut all the ports on the firewall to secure his system, which ultimately shut down his network. A student worker came up behind him and said: "Now the patients are all dead." That might be an extreme example, but it really makes the students think about the far-reaching consequences of the decisions they make involving their network.

Businesses volunteered to donate some of their old equipment that was connected to the network, so students could be prepared to defend it. The intent is to add more specific industry equipment in the future. It will shorten the training time for their future employees.

Student teams are much better prepared to work with their online coworkers. Students have often expressed the fact that they have no prior professional experience in working with people online. Though initially the perception was that this would be one of the easiest skills for them to master, it turned out to be one of the hardest. It started with simple issues like scheduling meetings when their coworkers weren't available (because they didn't find out when they would be available ahead of time). They also had technical issues choosing which platform to use for communication. It took some teams a long time to figure out how they would divide their work. However, after students finished their first semester working at a "business," their collaboration with their team ran much more smoothly.

It would be beneficial if students were able to log in to their business networks remotely to administer them, but that option is not available at our college. To circumvent that problem, the campus students share all the information they find with their team. In many cases, it is the online students who are directing the campus students in their projects.

Students who went through our program as an employee of one of our "businesses" reported they thought it made them much better prepared to start work. In fact, one student who started an internship working for our state said: "The work is easy here. I just keep doing your homework all the time." It appears that our goal of increasing preparedness and accelerating skill levels has been accomplished.

Student confidence has been increased, because students are much more familiar with the tasks and decisions that will be required in their jobs. Anyone who has spent years coaching knows it's easy to see that the teams that are most confident are the ones that practice the most. The teams that have the most practice usually are more successful as well. People who are less confident may be less willing to contribute in a team environment [10]. Problem-solving can be accelerated by working as part of a team, with members that bring differing skills and perspectives to help determine a solution.

Discussion

The principles that were used in developing this project could be used anywhere in a smaller or larger scale. Colleges could develop virtual business networks to have students run different types of businesses.

However, the importance of students having experience on a physical machine can't be diminished. Though this might not be possible for everyone, students should have an opportunity to see and troubleshoot a physical network. Transitioning from a system administrator to teaching college, it was easy to see some gaps in the virtual experience. For example, in reality RJ-45 plugs were often wired incorrectly or just not plugged in all the way. Connection problems such as this are integrated into the labs in our project. Though virtual machines are very convenient, they need to be installed on a physical machine somewhere, and there will always be a need for employees to be able to manage those machines.

It's difficult for students to know exactly what tasks they will be undertaking in their cybersecurity career. There are several reasons for this, but one of them is the fact that many cybersecurity companies can't let students into certain areas or let them perform some tasks for security reasons. However, there is a still a need for students to understand what they will be doing in the future to determine if it will be the best path for them. Cybersecurity is such a broad field, that without an opportunity to explore some of the specific types of career possibilities, students might not be able to find the career that they can be passionate about.

Students need to think about cybersecurity in a holistic way. Understanding cybersecurity is not just about learning to use tools such as Wireshark; it's about knowing when and why to use that tool. When working with information and technology, students need to think about who and what will be affected by the changes they make. When students work as part of a team, they try harder because they don't want to let down their teammates. Working with others can dramatically increase motivation to complete difficult tasks, even when they may do part of the work alone [11].

Hands-on learning is another important educational tool for students. Many students with 4-year computer science degrees enroll in 2-year colleges to get hands-on experience, because it's more difficult to get hired without experience. Even if students have not been employed before, they are often asked situational questions during interviews that they can't answer without this type of experience. Students who have the educational experience of working with coworkers in a business to solve technical problems can use this experience to advance their careers.

One important factor that can't be overlooked is sustainability of the lab. Documentation and directions are very important, as well as staff to manage the lab. Faculty and student workers may leave, so there needs to be a plan in place that can be easily followed by new staff.

The project at Lake Superior College involves about 100 students per year. Further research could be done on a bigger scale to measure the success of students involved in this type of learning.

Conclusion

The students at Lake Superior College were surveyed during the course of this project. All the students except one reported the experience as beneficial to their education and careers. The one student who reported a negative experience was coincidentally failing the class already.

The students who were part of the design, creation, and installation of this project were exceptionally successful. Many of those students received full scholarships to complete their degree at a 4-year institution and are now very successfully employed in government service.

Those students who worked in the cybersecurity workforce education lab as part of their coursework also reported that they felt more confident about starting their career and felt better prepared to contribute to their organization. Past students had reported it had taken them a little longer to feel confident and skilled enough to contribute. They said they spent more time in observation than the students who had been trained in the cybersecurity workforce education lab.

Students who had been trained in the cybersecurity workforce education lab were more proactive about their careers. Instead of finding out about the specific policies after something happened, they investigated the policies and rules the company had to follow much earlier. They also took the initiative to look at the cybersecurity posture of their organization.

Learning cybersecurity skills by using them in a realistic atmosphere can be valuable to increase student confidence, motivation, and preparedness for their new career. Preparing more students to make a quicker contribution to the cybersecurity workforce is something that could benefit everyone.

Acknowledgments The author would like to thank the National Security Agency (NSA) for offering her the Cybersecurity Workforce Education Grant that allowed the creation of this chapter. The author would also like to thank Thomas Gustafson, Thomas Janicki, Daniel Menze, Brett Karow, Thomas Janicki, Micah Kryzer, Nikolai Mallett, Paul Runnoe, Ronald Williams, Paul Litecky, and Matthew McCullough for their excellent support and hard work throughout this project.

References

1. Just Feeling Like Part of a Team Increases Motivation on Challenging Tasks. (n.d.). https://www.psychologicalscience.org/news/minds-business/just-feeling-like-part-of-a-team-increases-motivation-on-challenging-tasks.html. Accessed 5 Jan 2020
2. A.W.T. Bates. *Teaching in a Digital Age* (2016). https://opentextbc.ca/teachinginadigitalage/chapter/4-4-models-for-teaching-by-doing/. Accessed 25 Mar 2020
3. National Institute of Standards and Technology, *National Initiative for Cybersecurity Education (NICE) National Cybersecurity Workforce Framework* (NIST Special Publication 800-181, 2017). https://nvlpubs.nist.gov/nistpubs/SpecialPublications/NIST.SP.800-181.pdf?trackDocs=NIST.SP.800-181.pdf
4. About GenCyber (n.d.). https://www.gen-cyber.com/about/. Accessed 26 Mar 2020

5. What is CyberPatriot? (n.d.). https://www.uscyberpatriot.org/Pages/About/What-is-CyberPatriot.aspx. Accessed 26 Mar 2020
6. Hak5 Products. (2020), https://shop.hak5.org/. Accessed 20 Mar 2020
7. Leroux, Sylvain. *Kali Linux Review: Not Everyone's Cup of Tea* (2019). https://itsfoss.com/kali-linux-review/. Accessed 26 Mar 2020
8. J. Petters. What Is Metasploit? The Beginners Guide (2019). https://www.varonis.com/blog/what-is-metasploit/. Accessed 26 Mar 2020
9. J. Viega, G. McGraw, *Building Secure Software: How to Avoid Security Problems the Right Way* (MAddison-Wesley Professional, 2001), p. 528. ISBN 978-0201721522
10. L.M. Belcher. *The Impact of Confidence on Work Performance* (2016). https://smallbusiness.chron.com/impact-confidence-work-performance-24235.html. Accessed 5 Jan 2020
11. P.B. Carr, G.M. Walton, Cues of working together fuel intrinsic motivation. J. Exp. Soc. Psychol. **53**, 169–184 (2014). https://doi.org/10.1016/j.jesp.2014.03.015

Part V
Community Outreach

Status of Cybersecurity Awareness Level in Malaysia

M. R. K. Ariffin and M. Letchumanan

Introduction

The threat of cybercrime is expected to increase exponentially as the number of Internet users increases day by day. The development of new technologies such as the Internet of things (IoT), cloud services, big data services and mobile computing further increases the risk of cybercrime. As technology becomes increasingly sophisticated, cyber threats follow the trend of becoming more unique and complex.

Basically, cybercrime involves activities such as theft of personal information, businesses' intellectual property and obtaining knowledge of sensitive information for financial or political gains or other malicious purposes [1]. Meanwhile, types of cyberattacks involve malware, account hijacking, targeted attack, vulnerability, malicious script injection, denial-of-service attack (DDoS), defacement and brute-force or credential stuffing [2]. It is reported that besides individual users, companies, institutions and governments are also facing threats from cybercrime every day [3].

In Malaysia, the Royal Malaysian Police reported that cybercrime is the trendiest crime in the country where it has surpassed drug trafficking as the most rewarding crime. Statistics from the Royal Malaysian Police indicated that 70% of commercial crime cases are now being categorised as cybercrime [1]. On the same note, on February 28, 2018, TREND Micro Incorporated, a cybersecurity solutions provider,

M. R. K. Ariffin
Laboratory for Cryptography, Analysis and Structure, Institute for Mathematical Research, Universiti Putra Malaysia, Seri Kembangan, Malaysia
e-mail: rezal@upm.edu.my

M. Letchumanan (✉)
Laboratory of Ethnomathematics and Didactics, Institute for Mathematical Research, Universiti Putra Malaysia, Seri Kembangan, Malaysia
e-mail: malathi@upm.edu.my

K. Daimi, G. Francia III (eds.), *Innovations in Cybersecurity Education*,
https://doi.org/10.1007/978-3-030-50244-7_17

Table 1 Cybercrime incidents reported in Malaysia (Adapted from MyCERT)

Year	Number of cybercrimes
2019	9805
2018	10,699
2017	7962
2016	8334
2015	9915
2014	11,918
2013	10,636
2012	9976
2011	15,218
2010	8090

reported that Malaysia is ranked first in Southeast Asia (SEA) for the number of malware cases, with 16 million malware threats throughout the year. In addition to that, Malaysia is also categorised as the most heavily hit target in the region, where almost 350,000 malicious URLs are hosted locally which affected 10.5 million victims. Meanwhile, Malaysia is also ranked second behind Singapore for experiencing the most business email compromise (BEC) attacks [4].

According to a senior officer of the Communications and Multimedia Ministry (KKMM), Malaysians reported 2207 cases of cybercrimes involving RM67.6 million in the first quarter of 2019. The three most popular cybercrime activities reported were scamming via telephone calls (773 cases with RM26.8 million in losses), frauds in online purchases (811 cases with RM4.2 million) and the 'African Scam' (371 cases with RM14.9 million) [5]. Table 1 shows the statistics of cybercrime cases reported in Malaysia between the years 2010 and 2019. They comprise incidents such as vulnerability reports, denial of service, cyber harassment, intrusion attempt, content related, malicious codes, intrusion, spam and fraud. The figures show that the number of cybercrime cases reported in Malaysia ranges from 7900 to 16,000 which indicates that cybercrime threats in Malaysia are in a serious condition and need effective solutions. The statistics in Table 1 show that the number of cybercrime cases in Malaysia is not constant every year. For example, the number of cases for the year 2019 decreased when compared to 2018. The same pattern is recorded for the year 2015 where the number of cases is higher when compared with the following consecutive years (2016 and 2017). Although there is no exponential growth in terms of number of cases every year, the total cases recorded is considered high for a small nation such as Malaysia. The statistics were provided by the Malaysian Computer Emergency Response Team (MyCERT), which is the centre of reference for the Internet community in Malaysia to handle computer security cases. Muniandy et al. [6] believed that the actual number of cybercrimes in Malaysia is much higher than what has been reported by MyCERT, as not all victims would come forward and report such incidents to the relevant authorities. Meanwhile, Fig. 1 shows the reported incidents based on the General Incident Classification in the year 2019 in Malaysia.

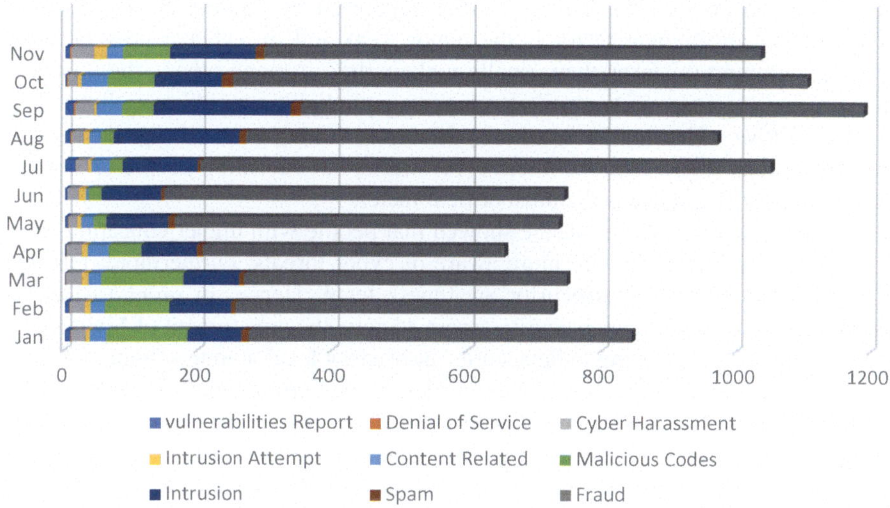

Fig. 1 Reported incidents based on General Incident Classification 2019 (Adapted from MyC-ERT)

The volume of cybercrimes in Malaysia is getting larger every year due to various reasons. One of the main reasons is lack of users' awareness of such threats [26]. As highlighted by Ludwig [7], 'there are numerous controls of IT professionals who can implement strategies to safeguard electronic information from unauthorised users. Practically, the authorised end users that possess the IDs and passwords and have access to data, giving them the ability to print it, share it, alter it or delete it, are the weak link. If they are careless or choose weak passwords, casually discard confidential printed reports in the trash, prop open doors to secured areas, fail to scan new files for viruses, or leave backups of data unsecured, then that information remains at risk'.

It is revealed that the youth group is one of the main target groups of this cyber threat because they may be unaware or too immature to recognise these threats [8]. Besides that, children, adults and senior citizens are also reported to be the victims of various kinds of cybercrimes due to lack of awareness [9]. Interestingly, Kim [10] argued that it is very often that heavy users of Internet and other digital devices that are vulnerable to cybercrimes are the least knowledgeable and unaware of cybersecurity issues and prevention.

Cybersecurity Awareness Status in Malaysia

The chief executive officer of CyberSecurity Malaysia, Datuk Dr. Amirudin Abdul Wahab (2019), pointed out that 99% of successful cyberattacks in Malaysia were due to human errors [11]. This is further supported by Gratian et al. [12] who admitted that the human factor is the major weak link in cybersecurity because of some common mistakes they usually make online. Among others, the mistakes include opening suspicious email attachments and clicking on phishing links. The chief executive officer of Cybersecurity Malaysia further stressed that having the best technology alone cannot combat cyberattacks but educating people on how to be safe in the cyberspace is the most critical factor.

In Malaysia, empirical evidence showed that people with higher education level and better computer skills are more aware of cyber threats. Furthermore, females also recorded higher cybersecurity awareness level. Hence, in order to have a heightened sense about cybersecurity across society, the government of Malaysia is working towards increasing cybersecurity awareness level among school children, senior citizens and people with lower academic background and computer skills. Ong and Chong [13] concluded that individual personal traits such as conscientiousness, extraversion and agreeableness and an individual's behavioural intention towards common information security risk such as email management and malicious software protection determine an individual's self-awareness of information security risks.

Sophos [14] has revealed that in Malaysia executives assumed that their companies 'will never get attacked; will get attacked but there is nothing that they can do; cybersecurity is easy; and cybersecurity professionals over-exaggerate threats and issues'. These statements are alarming and point towards the fact that cybersecurity awareness among highly educated professionals in Malaysia is still poor and needs to be improved to protect their cyberspace. Meanwhile, the executives also stated that organisations are facing difficulties to recruit skilled cybersecurity professionals, there is insufficient budget for cybersecurity and it is challenging to stay up to date with cybersecurity technology. Zainudin and Molok [15] investigated the awareness level of cybersecurity managers on advanced persistent threat (APT) in financial institutions in Malaysia. It was reported that factors that influenced APT awareness among financial institution employees included the emphasis on informal learning on APT, attackers' financial motivation, financial institution's reputational risks and the availability of financial regulatory requirements to protect the financial institution from the risk.

The Ministry of Education Malaysia together with CyberSecurity Malaysia and Digi (a telecommunication provider in Malaysia) [16] conducted a study among primary and secondary school students in Malaysia to determine their cybersecurity awareness level. The findings of the study reported that 2/3 of the students have access to computers with Internet connection and mostly spend 8 h or less in a week on online activities. Most of the students use Internet facilities at home under adult surveillance. They use the Internet for Facebook access, followed by other online activities such as online gaming and downloading music and movies. The students admitted that they do not feel safe when using the Internet but they realise the importance of Internet safety. Moreover, they admitted that they do not know how to protect themselves from becoming cybersecurity victims. It is surprising to note that more than half of the students (56%) admitted that they never shared their passwords with anyone. Those students who shared their passwords normally shared them with parents, close friends and family members. About 32% of the students used one password for all the accounts and never had the habit of changing them. The characteristics of the passwords showed that half of the students used passwords with eight characters long, but they used one type of character such as alphabets. It is noted that 20% of the students had been victims of cyberbullying. From the study, it can be concluded that the cyber awareness level among the selected school students was still at the average level and action is needed to be taken to increase their awareness. The report also suggested that parents need to play an important role in safeguarding their children and exposing their children to cyber safety methods as they mostly use the Internet at home.

In 2017, Zahri et al. [1] conducted a study among primary and secondary school students throughout Malaysia to investigate the cybersecurity situational awareness level in this category. The respondents were divided into three groups based on age, where Group 1 consists of primary school students aged 7–9 years old, Group 2 consists of primary school students aged 10–12 years old and Group 3 consists of secondary school students aged 13–17 years old. Different sets of questionnaires were distributed to each group of students. The aim of this study is to introduce awareness modules in the school syllabus based on the current status of cybersecurity situational awareness level. The authors concluded that there is generally a sense of cybersecurity awareness among the respondents. The study reported that the issues that should be addressed to the students revolve around social media, such as how to use social media responsibly and the type of appropriate posts in social media. Moreover, parents also should be educated in terms of not allowing their children to have social media accounts at a very young age, especially between 7 and 9 years old. Moreover, there is a need to educate teachers and parents on the necessity to teach their children how to use the Internet in a safe way.

In Malaysia, empirical studies on cybersecurity awareness are mostly conducted among higher learning institution students or youths between the ages of 18 and 30. For instance, a study was done among higher education students to understand the cybersecurity behaviour of the youth group [6]. The study reported that the cybersecurity behaviour of the participants was not satisfactory and their behaviours in all aspects of cyber threats were vulnerable. This study investigated higher education students' awareness level on threats such as password usage, phishing, social engineering, online scam and malware. It is important to note that some of the threats faced by the participants could be eliminated if they were aware of these threats. Meanwhile, a study by Khalid et al. [17] showed that university students demonstrated more than 80% of awareness level on certain cybersecurity elements such as cyberbully, personal information and Internet banking. However, their awareness level on cybersex and self-protection is not at the satisfactory level (less than 80%). The authors advised that various role players need to work together to increase the awareness in terms of cybersex and self-protection. Another study conducted among academic staff in the northern part of peninsular Malaysia noted that the majority of academicians have reasonable knowledge on cyber threats such as phishing, computer viruses and trojans [18]. However, many of the academicians are still willing to share their passwords under certain circumstances. This again paves way to increased cyber threat cases.

On another note, research findings also indicate that preuniversity students' awareness level is at an average level and there is no significant difference between male and female students' awareness levels. The findings further pointed out that students with computer skills showed a better cybersecurity awareness level (Suwarna [19]). In addition to that, Hasan et al. [20] indicated that gender, age, knowledge of cyber offences and academic qualification among higher learning institution students determine their level of cybercrime awareness. The authors asserted that female students are more aware and have affirmative insights about cybercrime. Those students who are more than 24 years old showed more awareness level compared with those under the age of 23. Finally, students with knowledge of cyber offences and who possess higher academic qualifications are more aware of cybercrime activities and view the issue seriously. The authors generally investigated the awareness of respondents towards seven security threats such as unauthorised access to computer material, unauthorised access with intent to commit or facilitate commission of further offenses, unauthorised modification of the contents of any computer, wrongful communication, abetments and attempts, cyber terrorism and cyber nuisance.

With regard to how individual differences could mediate higher learning institution students' cybersecurity behaviour and beliefs, another study was conducted to understand the mediation effects of age, gender and education level on factors that influence cybersecurity behaviours and beliefs [21]. The study was guided by the health belief model and protection motivation theory. The health belief model has variables such as perceived severity, perceived benefits, perceived susceptibility, cues to action and perceived barriers. Meanwhile, protection motivation theory comprised constructs such as perceived severity, perceived susceptibility, response efficacy and self-efficacy. Results reported that students' cybersecurity behaviours were mediated by age for factors such as perceived severity, peer behaviour, familiarity with cyber threats, response efficacy and perceived vulnerability. The study also found that security, self-efficacy, computer skills and prior experience were among the cybersecurity scales that were impacted by gender. Finally, educational level differences existed in cues to action and familiarity with cyber threats.

In addition to this, Ishak et al. [22] investigated security awareness among social networking site users. Respondents from various age groups, genders and academic qualifications participated in the study. The majority of the participants fell under the age group of 18–26. The authors assessed the awareness level of respondents based on three categories, namely, basic, technical and advocacy. The result showed that the female group and highly educated respondents were more aware of the cyber threats. Highly educated respondents also took initiatives to educate their children on the proper use of social networking sites.

Meanwhile, Saizan and Singh [23] conducted another research among social networking site users to determine the level of cybersecurity awareness. Furthermore, the study also investigated the factors that contribute to the awareness level of cybersecurity. Respondents are from a private higher learning institution in Malaysia. The authors have validated the influence of knowledge factors that included items to test respondents' knowledge on cyber threats, laws regarding cybersecurity and cybercrime. Besides, the authors also investigated the contribution of environmental factors such as influence of parents, friends and colleagues to increase cybersecurity awareness level. Finally, the influence of attitude factors that comprised measurements such as workshops attended to increase cybersecurity awareness level, knowledge on cybersecurity laws, reference to external resources when confronting cyber threats and the measurements taken to overcome cyber threats was also included. The results showed that the respondents scored medium for knowledge and attitude factors. Sadly, they scored very low for the environmental factor. Thus, it can be concluded that the cybersecurity awareness level of social networking site users is still not at an encouraging state. The authors suggested that the respondents should be exposed to more workshops and training related to cybersecurity awareness. Moreover, organisations are suggested to give initiatives to respondents who had implemented appropriate cybersecurity measurement to combat cyber threats.

A study by Arifin et al. [24] attempted to investigate parental awareness levels on cyber threats, specifically whether they are aware of the risks of the Internet to their children (below 17 years old). The study reported that 80.9% of the parents are aware of cyber threat cases. Parents also admitted that they have a medium level of awareness on their children's online activities, which can lead to Internet addiction among their children.

In 2016, Zahri et al. [25] conducted a focus group study among different kinds of stakeholders such as children, youths, adults and parents and organisations (both public and private sectors). Furthermore, the background of the participants also varied; either they are from management, policy-making, law enforcement and prosecution or research and technical fields. The authors summarised that, in order to achieve successful cybersecurity awareness implementation strategy, three implementation layers such as strategic, programme execution and content development layers are necessary. The strategic layer outlines how the cyber security awareness (CSA) initiatives are planned before implementation. Under this layer, a national CSA forum will be held among experts, target group coordinators and collaborators and partners to establish an accurate plan. The objective of the second layer, the programme execution layer, is to set up a workable process for input gathering and content dissemination to the correct target groups identified. Personnel under this layer will work to collect the relevant input and the content delivery to the targeted audience. This layer will also focus on capacity building to train the trainers responsible in conducting the training programmes or workshops related to cybersecurity awareness. Personnel under the content development layer are responsible to develop the content or programme materials based on the target group. This is because the same awareness topic delivered to different target audiences will most likely require different content and media based on their education background and community status.

Table 2 summarises findings from past research that were conducted in Malaysia regarding cybersecurity. The findings showed that research related to cybersecurity was conducted among primary and secondary school students, higher learning institution students, parents, general public and top management of organisations. It can be observed clearly that research among senior citizens is very low. Hence, it is recommended to conduct further research among senior citizens who can easily become victims of cyber threats. Furthermore, it is recommended to conduct research among all higher learning institution students in Malaysia who are active users of the Internet. This is because the current research only focuses on a few selected higher learning institutions in Malaysia.

Table 2 also shows that interest within past studies in Malaysia in the field of cybersecurity revolved around studying the perception on cyber threat, cybersecurity awareness level, cybersecurity behaviour and belief, factors that contribute to the cybersecurity awareness level (mediating effects of gender, age and education background level) and the strategies that needed to be implemented in order to increase cybersecurity awareness status. In the future, it is recommended to conduct research to evaluate participants' knowledge on the available laws and regulations in Malaysia and the legal consequences that resides within. It will help the participants to realise their rights when they are faced with cyber threats as well as the limitations within the Malaysian context.

Steps Taken to Combat Cyber Threats in Malaysia

The Malaysian government is very much concerned about the increase in cybercrime cases. Accordingly, the government formed an organisation in 1997 known as Malaysian Computer Emergency Response Team or MyCERT. Eventually, in 2007, the MyCERT mandate was restructured under a new entity known as CyberSecurity Malaysia (CSM). CSM was formed to create and sustain a safer environment, to promote national sustainability, to ensure social well-being and to promote wealth creation within Malaysia's cyberspace. Moreover, CSM also provides statistics on cybercrime cases in Malaysia as well as advice and suggestions to the government and nongovernment bodies to combat cybercrime cases. From time to time, they also conduct roadshows to increase the level of cybersecurity awareness among Malaysians.

Meanwhile, in adopting the cyberspace as a new frontier for the Malaysian public to utilise, the Malaysian government has among others taken proactive mechanisms to increase awareness of such facilities, ensure their effectiveness and provide legal framework to mitigate issues that might arise. These initiatives are aimed to propel a seamless migration from the traditional 'brick and mortar' environment towards the digital environment that will in the end increase its potential through global connectivity within a secure environment. Whilst Malaysia has acknowledged the inevitable transition into the digital world, the Malaysian government has taken a unique approach of still playing the role of being the 'guardian' for Malaysians.

Table 2 Summary of past research findings regarding cybersecurity in Malaysia

Authors	Participants	Study focus					Main findings
		Perception on cyber threat	Level of cybersecurity awareness	Cybersecurity behaviour and belief	Factors contribute to cybersecurity awareness level	Strategies to increase cybersecurity awareness level	
Sophos [14]	Executives	X					Executives believed their organisations will face minimum cyber threat risk
Zainudin and Molok [15]	Managers			X	X		Factors that influenced APT awareness among financial institution employees
Fatokun et al. [21]	Higher learning institution students			X	X		Age, gender and education level mediated the effects on factors that influenced cybersecurity behaviours and beliefs
Arifin et al. [24])	Parents		X				80.9% of the parents were aware of cyber threat cases
Khalid et al. [17]	Higher learning institution students		X				Students demonstrated more than 80% of awareness level on certain cybersecurity
Othman et al. [18]	Higher learning institution academic staff		X				Majority of academicians have reasonable knowledge on cyber threats
Saizan and Singh [23]	Public		X				Cybersecurity awareness level of social networking site users is still not at an encouraging state

(continued)

Table 2 (continued)

Authors	Participants	Study focus					Main findings
		Perception on cyber threat	Level of cybersecurity awareness	Cybersecurity behaviour and belief	Factors contribute to cybersecurity awareness level	Strategies to increase cybersecurity awareness level	
Zahri et al. [1]	Primary and secondary school students (aged 7–18)		X				There is generally a sense of cybersecurity awareness among the selected school students
Muniandy et al. [6]	Higher learning institution students			X			Cybersecurity behaviour of the participants was not satisfactory and their behaviours in all aspects of cyber threats were vulnerable
Rani [19]	Preuniversity students		X				Students' awareness level was at an average level, and there was no significant difference between male and female students' awareness levels

Zahri et al. [25]	Children, youths, adults and parents and organisations					X	To achieve successful cybersecurity awareness implementation strategy, three implementation layers, namely, strategic, programme execution and content development layers, are necessary
Hasan et al. [20]	Higher learning institution students		X		X		Female students are more aware and have affirmative insights about cybercrime Students who are more than 24 years old showed more awareness level Students with knowledge of cyber offences and possess higher academic qualifications are more aware of cybercrime activities
Education Ministry of Malaysia, Cyber Security Malaysia and Digi [16]	Primary and secondary school students (aged 7–18)		X				Cyber awareness level among the selected school students was still at the lower level
Ishak et al. [22]	Public		X		X		Female group and highly educated respondents are more aware of the cyber threats

Although this approach has received alternative views with regard that it might hamper the democratisation of Malaysian's cyberspace, in general, it has been well accepted by the Malaysian population. In line with this, the Malaysian government has proposed and implemented several agendas, policies and acts to protect the public against cyberattacks and ensure secure digital privacy. Some of the agendas, policies and acts are:

National Cryptography Policy (NCP) 2013

Cross-border telecommunication networks allow information to be intercepted and accessed illegally. The anonymity issue that exists in cyberspace is now an obstacle that makes it difficult to verify the identity of the user. As such, the widespread use of cryptography in government-to-government electronic affairs, government-to-people, government-to-business and business-to-business is seen as capable of creating a safe and reliable cyber environment. The NCP was formulated to enhance efficiency and achieve independence in the use of cryptography towards economic prosperity, citizen's well-being and national security. It also emphasises the importance of the country to use trusted cryptographic products (which have undergone a process of evaluation and certification by designated government agencies) in all aspects of information security. In fact, this policy is also in line with NCSP and at the same time complements NCSP.

National R&D Roadmap for Self-Reliance in Cybersecurity Technologies 2011

Through this initiative, this roadmap lists the areas of research and development needed by Malaysia for its independence in cybersecurity technology. The plan was formulated by MIMOS, Malaysia's national applied research and development centre, in collaboration with a consortium of 22 organisations representing academia, government, industry and researchers aimed at integrating and managing all cybersecurity research and development (R&D) programmes and projects. This effort is aimed at avoiding duplication and promoting academic, industry and government cooperation. Guided by an integrated R&D framework, this direction will focus on technologies that will protect the Critical National Information Infrastructure (CNII) with the goal of achieving independence in the technology. It is hoped that the above goals will be achieved by increasing efforts to promote cybersecurity research at research institutes to increase the size of the research community in the field of cybersecurity.

MIMOS identified seven prominent fields to conduct research. They are:

1. Secure communications.
2. System that can be used without exception.
3. High availability systems.
4. Network surveillance, response and recovery.
5. Trust relationships.
6. Secure access.
7. System integrity controls.
8. Traceback, identification and forensics.

National Cybersecurity Policy (NCSP) 2006

This policy is aimed towards implementing agendas to ensure the protection of organisations or agencies which are identified as one of Malaysia's Critical National Information Infrastructures (CNII). This is in order to realise Malaysia's vision towards a cyber environment that is resilient and independent. Malaysia's Critical National Information Infrastructure (CNII) is defined as those assets (real and virtual), systems and functions that are vital to the nation whereby their incapacity or destruction would have a devastating impact on:

1. National economic strength; confidence that the nation's key growth area can successfully compete in the global market whilst maintaining favourable standards of living.
2. National image; projection of national image towards enhancing stature and sphere of influence.
3. National defence and security; guarantee sovereignty and independence whilst maintaining internal security.
4. Government capability to function; maintain order to perform and deliver the minimum essential public services.
5. Public health and safety; delivering and managing optimal healthcare to the citizens.

The CNII sectors are:

1. National Defence and Security.
2. Banking and Finance.
3. Information and Communications.
4. Energy.
5. Transportation.
6. Water.
7. Health Services.
8. Government.
9. Emergency Services.
10. Food and Agriculture.

The policy is divided into eight thrusts where each thrust is spearheaded by a relevant Malaysian agency or ministry.

Electronic Commerce Act 2006

The Electronic Transaction Act 2006 was introduced to provide a legal recognition of electronic messages and to fulfil the means and communication of electronic messages. This act will protect any commercial transactions using electronic means and other related matters. The act provides legal effect, validity and enforceability of electronic messages and its contents, creation of content through electronic means, recognition of requirements for writing, signature, seals, witness, originality and copies and allows the service and delivery of electronic documents online. This act is to protect public who are using electronic mode transactions and believed to increase their awareness on measures that can be taken if they counter cyber threats.

Digital Signature Act (DSA) 1997

The Malaysian DSA which was modelled upon the Utah Digital Signature Act (1995) is an initiative to ensure Malaysia will be the preferred investment destination among multinational information technology companies. Specifically, the DSA outlines methodologies to appoint a Certification Authority (CA), the functions and responsibilities of a CA and the needs of digital signature users. The DSA also outlines issues such as the acknowledgement in the eyes of the law on the utilisation of digital signatures, liabilities that CA may incur and the usage of time stamping mechanisms. There are two points to take note: The first is that even though the DSA gives accreditation that a digital signature can be treated in the same manner a traditional signature must be treated, there is no mention on the need for electronic records to be treated in the same manner a hardcopy or contract is treated. This might imply that in the eyes of the law, the need to store records does apply to digital records. The second point to mention is with regard to the amount of liability to be incurred by the CA if the digital signature is forged. Within Sect. 8 of the DSA, it is mentioned that the liability amount is a quantum that the CA will have to discuss with customers. This clause should be read carefully by potential customers in order not to be at the receiving end if forgery does occur.

Computer Crime Act 1997

The Computer Crime Act (1997) which became effective as of June 1, 2000, was created to combat offences related to unauthorised access to computer material, unauthorised access with intent to commit or facilitate commission of further offence, unauthorised modification of the contents of any computer, wrongful communication, abetments and attempts punishable as offences and presumption. The act is applicable to any person regardless of his nationality or citizenship and has an effect outside as well as within Malaysia. If the offense is committed by an individual in any place outside of Malaysia, he may be dealt with in respect of such offence as if it was committed at any place within Malaysia. A person found guilty under this act is liable to fine and imprisonment. In 2019 alone, around 10,772 cases were reported under this act generally [2].

National Information Technology Agenda (NITA) 1996

NITA focuses on the development of human resource, information structure and applications that will increase value and provide fair access for Malaysians to ensure qualitative change towards a society based on knowledge by the year 2020.

NITA identified five critical strategic thrusts that will enable Malaysia's migration into the digital world. The thrusts are:

1. E-Community.
2. E-Public services.
3. E-Learning.
4. E-Economy.
5. E-Sovereignty.

These five thrusts will spearhead Malaysia's ambition to embrace change systematically. Malaysia believes that by addressing these five thrusts in a precise and concise manner, the process of entering into the digital world will be less 'stressful' and can have a high acceptance rate by Malaysians.

It is clear that thrusts 1 to 4 embody the case scenarios of Malaysia's digital space, whilst thrust number 5 reflects Malaysia's awareness of potential security issues that might arise. Through the E-Sovereignty thrust, Malaysia prepares its information technology infrastructure based on current security mechanisms as well as develops local human capital in order to ensure wide acceptance among Malaysians.

Conclusion

This chapter highlights the status of cybersecurity awareness level among Malaysians. Besides, it also discusses the initiatives taken by the Malaysian government to combat cyber threats in Malaysia. It is obvious that cybercrime cases in Malaysia are increasing every year and Malaysians have become the main target of cyberattackers among the Southeast Asian countries. Furthermore, it can be noted that not much initiatives have been taken by the government to understand the level of cybersecurity awareness among the public from a wide range of ages. To date, most of the empirical research had been conducted among higher learning institution students, and not much effort has been taken to educate senior citizens who are mainly ICT illiterate. These senior citizens could easily become the target of cyberattackers, mainly because of their lower level of cybersecurity awareness. Hence, it is suggested for further research to be conducted to assess the cybersecurity awareness level among public from a wide range of age group and construct proposal actions that need to be taken to continuously educate the people.

References

1. Y. Zahri, R.S. Ab Hamid, A. Mustaffa, A, "cyber security situational awareness among students: A case study in Malaysia", World Academy of Science, Engineering and Technology, international journal of social, behavioral, educational, economic. Business and Industrial Engineering **11**, 1654–1660 (2017)
2. MyCERT (2020), https://www.mycert.org.my/, Accessed Jan 2020
3. Y.K. Peker, L. Ray, S. Da Silva, N. Gibson, C. Lamberson, Raising Cybersecurity awareness among college students. Journal of The Colloquium for Information System Security Education **4**, 17–17 (2016)
4. TREND Micro Incorporated, *Trend Micro: Malaysia Encounters the Most Malware Threats In Sea In 2018* (2019). https://www.digitalnewsasia.com/digital-economy/trend-micro-malaysia-encounters-most-malware-threats-sea-2018, Accessed Jan 2020
5. Bernama, RM67.6 Million Lost to Cybercrimes in Q1 (2019). https://www.nst.com.my/news/crime-courts/2019/04/482208/rm676-million-lost-cyber-crimes-q1-2019, Accessed Dec, 2019
6. L. Muniandy, B. Muniandy, Z. Samsudin, Cyber security behaviour among higher education students in Malaysia. J. Inf. Assur. Cyber Secur, 1–13 (2017)
7. K. Ludwig, *Security Awareness: Preventing a Lack in Security Consciousness, GIAC* (2013)
8. E. Kritzinger, M. Bada, J.R. Nurse, A study into the cybersecurity awareness initiatives for school learners in South Africa and the UK, in *IFIP World Conference on Information Security Education*, (Springer, Cham, 2017), pp. 110–120
9. N.V. Kushzhanov, U.Z. Aliyev, Changes in society and security awareness. **ҚАЗАҚСТАН РЕСПУБЛИКАСЫ, 94** (2018)
10. E.B. Kim, Information security awareness status of business college: Undergraduate students. Information Security Journal: A Global Perspective **22**, 171–179 (2013)
11. Othman, *Towards A Safer Cyberspace: Most 'Attacks' Due To Human Error* (2019). https://www.nst.com.my/lifestyle/bots/2018/10/426298/towards-safer-cyberspace-most-attacks-due-human-error, Accessed Jan 2020
12. M. Gratian, S. Bandi, M. Cukier, J. Dykstra, A. Ginther, Correlating human traits and cyber security behavior intentions. Comp Sec **73**, 345–358 (2018)
13. L. Ong, C. Chong, Information security awareness: An application of psychological factors–a study in Malaysia, in *International Conference on Computer, Communications and Information Technology (CCIT 2014)*, (Atlantis Press, 2014)
14. Sophos, *The Future of Cybersecurity in Asia Pacific and Japan – Culture, Efficiency, Awareness* (2019)
15. Z.S. Zainudin, and N.N.A. Molok, Advanced Persistent Threats Awareness and Readiness: A Case Study in Malaysian Financial Institutions," In 2*018 Cyber Resilience Conference (CRC) IEEE 2018* (2019), pp 1–3

16. Education Ministry of Malaysia, Cyber Security Malaysia and Digi (2013). https://digi.cybersafe.my/files/article/DiGi_Survey_Booklet_COMPLETE.pdf. Accessed Mar 2020
17. F. Khalid, M. Yusoff Daud, M.J.A. Rahman, M.K.M. Nasir, An investigation of university students' awareness on cyber security. Int J Eng Technol **7**, 11–14 (2018)
18. M.F.I. Othman, F. Alqahtani, M.A. Bari, A.N.C. Pee, Y.A. Rahim, H.A. Sulaiman, The level of information security awareness among academic staff in IHL. Journal of Telecommunication, Electronic and Computer Engineering (JTEC) **10**, 65–68 (2018)
19. S. Rani, Cyber Security Awareness Among Malaysian Pre-University Students. in *E-Proceeding of the 6th Global Summit on Education* (2017)
20. S. Hasan, R. Abdul Rahman, S.F.H. Tengku Abdillah, N. Omar, Perception and awareness of young internet users towards cybercrime: Evidence from Malaysia. J Soc Sci (2015)
21. F.B. Fatokun, S. Hamid, A. Norman, J.O. Fatokun, The impact of age, gender, and educational level on the Cybersecurity behaviors of tertiary institution students: An empirical investigation on Malaysian universities. J. Phys. Conf. Ser. **1339** (2019)
22. I. Ishak, F. Sidi, M.A. Jabar, N.F.M. Sani, A. Mustapha, S.R. Supian, A survey on security awareness among social networking users in Malaysia. Aust. J. Basic Appl. Sci. **6**, 23–29 (2012)
23. Z. Saizan, D. Singh, Cyber security awareness among social media users: Case study in German-Malaysian institute (GMI). Asia-Pacific Journal of Information Technology and Multimedia **7**, 111–127 (2018)
24. N. Arifin, U.S. Mokhtar, Z. Hood, S. Tiun, D.I. Jambari, Parental awareness on cyber threats using social media. Journal Komunikasi **35**, 485–498 (2019)
25. Y. Zahri, R.S. Ab Hamid, and A. Mustaffa, Development of a cyber security awareness strategy using focus group discussion", in Proc. SAI Computing Conference, London, UK, 2016, pp. 1–5
26. C. Barclay, Sustainable security advantage in a changing environment: The Cybersecurity Capability Maturity Model (CM 2). In Proceedings of the 2014 ITU kaleidoscope academic conference: Living in a converged world-Impossible without standards?, St. Petersburg, Russia, pp. 275–282

Cybersecurity Education: The Skills Gap, Hurdle!

Samuel Ndueso John, Etinosa Noma-Osaghae, Funminiyi Oajide, and Kennedy Okokpujie

Introduction

Cybersecurity is the name given to the goal of protecting information systems (IS) from disruption, data breaches, unauthorised access, exploitation and modification [1]. Cybersecurity is also the amalgamation of concepts, strategies, practices and tools to protect the cyberspace and information assets of organisations. The instrument of cybersecurity ensures the confidentiality, integrity and availability of information. Vulnerabilities in cyber environments are the result of device mobility proliferation and widespread digitisation. Cybercrime like identity theft and phishing is becoming rampant. The dependence of many organisations on third-party enterprise of IT/IS services and outsourcing of either primary or support activities outside the organisations are prevalent nowadays. Some of these essential IT/IS services also increase the risk of cyberattacks. The statistics on data breaches paints a grim situation [2]. Cyberspace's precarious security situation is the direct result of more dependence on mobile devices and the proliferation of online services. The average cost of a data breach is as high as $3.6million [1], and enterprises worry about the next point of action to either mitigate or control the negative effect of the cybercrime.

S. N. John (✉)
Nigerian Defence Academy, Kaduna, Nigeria
e-mail: samuel.john@nda.edu.ng

E. Noma-Osaghae · K. Okokpujie
Covenant University, Ota, Nigeria
e-mail: etinosa.noma-osaghae@covenantuniversity.edu.ng;
kennedy.okokpujie@covenantuniversity.edu.ng

F. Oajide
Nottingham Trent University, Nottingham, UK
e-mail: funminiyi.olajide@ntu.ac.uk

K. Daimi, G. Francia III (eds.), *Innovations in Cybersecurity Education*,
https://doi.org/10.1007/978-3-030-50244-7_18

Cybersecurity is an emerging area of information assurance, and the security of enterprise services is enormous. There is more information system integration in business enterprises. The information system provides the platform for the alignment of mobile interfaces and specialised security, such as blockchain technology. Cybersecurity has garnered more attention in recent times due to the new digital innovation that has brought on the proliferation of iMobile, iRobotics and other digital technologies across all spheres of life. Cybersecurity, as an emerging field, continues to attract more research funding from all parties directly and indirectly affected by its importance.

Also known as information security, it seeks to defeat the waves of cybercrime and cyber threats facing the world today. Securing information today is a process, a continuing process. But there is a fundamental linking gap between the true cybersecurity professional and the needs of the industry. There are also shortages in qualified educators, training facilities, funding and weak policies that slow the progress of cybersecurity education. Job role vacancies, knowledge and skills required, ethics and standards, accreditation and certification are some of the most significant hurdles that cybersecurity education must surmount [3].

As one of the critical concerns of governments throughout the world, it has been discovered that human capital (the skills gap) is the weakest link in the cybersecurity chain. Recent graduates are ill-prepared for the harsh realities of the cybersecurity workplace. Industry leaders bemoan the enormous amount of time in "man-training" hours invested in bringing supposed graduates of cybersecurity up to speed with the capacity to tackle real-world cybersecurity challenges. The supply of cybersecurity professionals is not only inadequate; the majority of the available ones are so unprepared to be of any real value to the industry [4].

A multidisciplinary field of study, cybersecurity has the propensity to become mission-specific to fill identified gaps and solidify institutional social and cultural nuances that makes it possible for governments and organisations to achieve cybersecurity goals effectively. CybEd should prioritise depth over breadth, industry/work experience, practical skill-set development and industry-integrated curricula. Emphasis must be placed on practical relevance and real-world applications to fill the current skills gap in cybersecurity.

Twenty (20) articles bordering on the skills gap in cybersecurity and CybEd were downloaded and intensively reviewed. The materials were downloaded over one week in portable document format and imported using EndNote®. The downloaded documents were scanned for duplicates which were removed. Other documents that did not strictly deal with the issue of the skills gap in cybersecurity were also removed. The process is depicted in Fig. 1. The keywords that were used for getting the reviewed articles were the "cybersecurity skills gap". Databases that were consulted to get the items reviewed include Google Scholar, IEEE Xplore, NIST Standards and ScienceDirect.

The next section of the article presents the identified skills gap in cybersecurity. The following section gives an exposition on how a robust cybersecurity curriculum would address the skills gap issue. The next part elucidates on the holistic cybersecurity education that can help retain cybersecurity talents and encourage

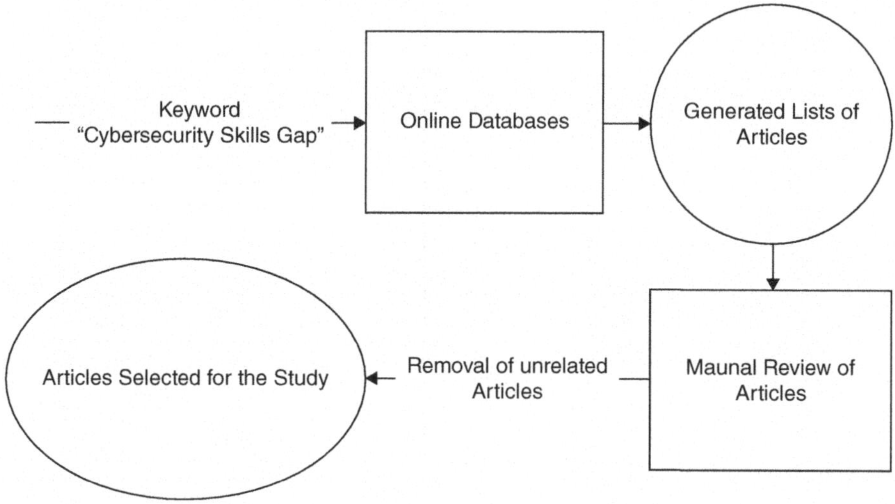

Fig. 1 Research methodology

lifelong learning in developing countries. The study thus provided knowledge to cyber education and then concluded with a list of relevant references.

Cybersecurity Skills Gap

Cybersecurity spans across a wide range of fields, among which are information systems, management (industrial and business), technology and communication. It utilises a host of skills like critical thinking, problem-solving, interpersonal and analytical information security architecture, risk management and compliance and intelligence/threat analysis that are some of the top talents needed in cybersecurity. In the industry, there are lots of cybersecurity roles that are vacant due to the shortage of professionals to fill those roles. As shown in Fig. 2, the primary skill goals of cybersecurity education are to develop professionals who are influential leaders that embrace teamwork as a means of exerting well-developed technical, critical thinking and problem-solving skills to solve cybersecurity challenges with a commitment to lifelong learning due to the rapidly evolving nature of cybersecurity. Many organisations have identified analysts, security assessors, security engineers, consultants and managers as some of the top cybersecurity skills gaps to be filled in the industry [2]. This is shown in Fig. 3, where cybersecurity is seen to be divided into five broad classes of job roles that need to be filled in the industry. The cybersecurity educator gap is not listed because it is quickly supplied by seasoned professionals that have undergone thorough training in any of the other listed areas of cybersecurity.

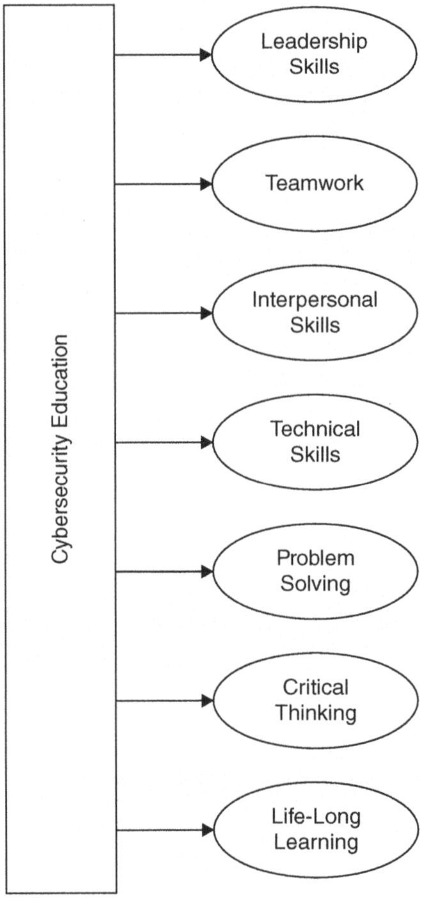

Fig. 2 The seven skill goals of cybersecurity education

The skills gap in cybersecurity seems not to be getting anywhere near close. The difference keeps expanding, and the shortage of adequately skilled cybersecurity personnel has become a cause of concern to all stakeholders in the field. There is a growing demand for cybersecurity professionals to fill entry-, middle- and top-level positions in organisations all around the world. The top gaps can be found in education, research and industry. This is the result of a shortage in the supply of seasoned cybersecurity professional and academic programs to train new crops of cybersecurity practitioners [1].

A lot of small- and medium-scale enterprises (SMEs) do not have the resources to guard against cyberattacks. The skills needed to provide SMEs with adequate information security are expensive. The high cost is due to the gap in skills and the scarcity of cybersecurity professionals. The awareness of SMEs to the dangers of non-existent information security inspections is not encouraging [5]. Digital forensics, which gives the ability to adequately respond to cybersecurity incidents,

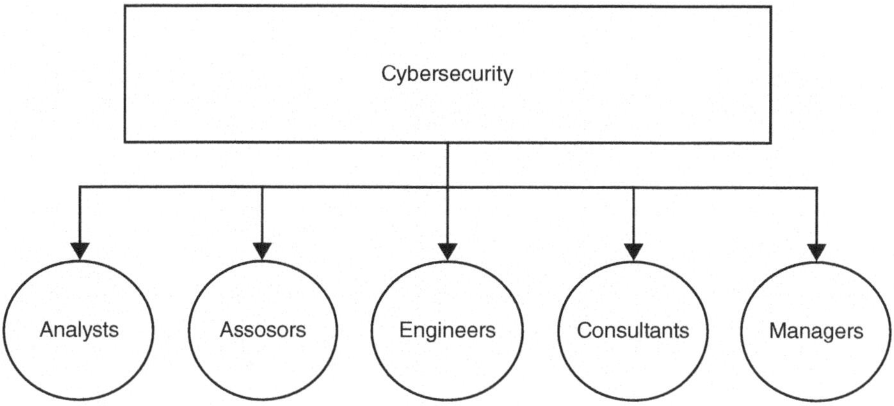

Fig. 3 Cybersecurity skills gap in terms of job roles

also presents a significant skills gap in cybersecurity as many organisations are not able to resolve already accomplished cyberattacks [2].

The skills gap that needs to be filled in cybersecurity is more industry inclined. Some of them are information security risk management and mitigation, security policy development, web application security, security log management, security incidents investigation and resolution, server security, antivirus solutions, access control, enterprise encryption solution and security architecture. The job descriptions are matched to the respective job roles in Fig. 4. The arrows point in the direction of the needed skill. The right-most job role is the cybersecurity manager that provides strong leadership for the other job roles. Cybersecurity professionals work as teams to solve cybersecurity challenges and fill the skills gap [6].

As the number of Internet users continues to grow, there would continue to be new vulnerabilities and new cybersecurity challenges. The skills gap would remain wide in the absence of tangible effort or initiative to proactively cover the gaps. To create keen awareness about the cybersecurity skills gap in the industry and academia, a cybersecurity policy needs to be implemented and sustained. At all levels of education, cybersecurity studies should be offered with great depth to bridge the gap presented by the shortage of cybersecurity educators and professionals. Accrediting bodies and the industry must be ready to cut down on bottlenecks that unnecessarily slow down the progress of cybersecurity education. The workload that comes with developing programs for cybersecurity education needs to be faced headlong if the skills gap must be closed. Rigorous and time-consuming as it may seem, the development of new and effective cybersecurity programs is a very effective way of overcoming the cybersecurity skills gap hurdle [7].

The training to fill the cybersecurity skills gap should allow for enough flexibility and customisation to prevent frequent withdrawals from cybersecurity programs and encourage self-development. Governments should also spearhead security education, training and awareness programs that raise the alertness of communities

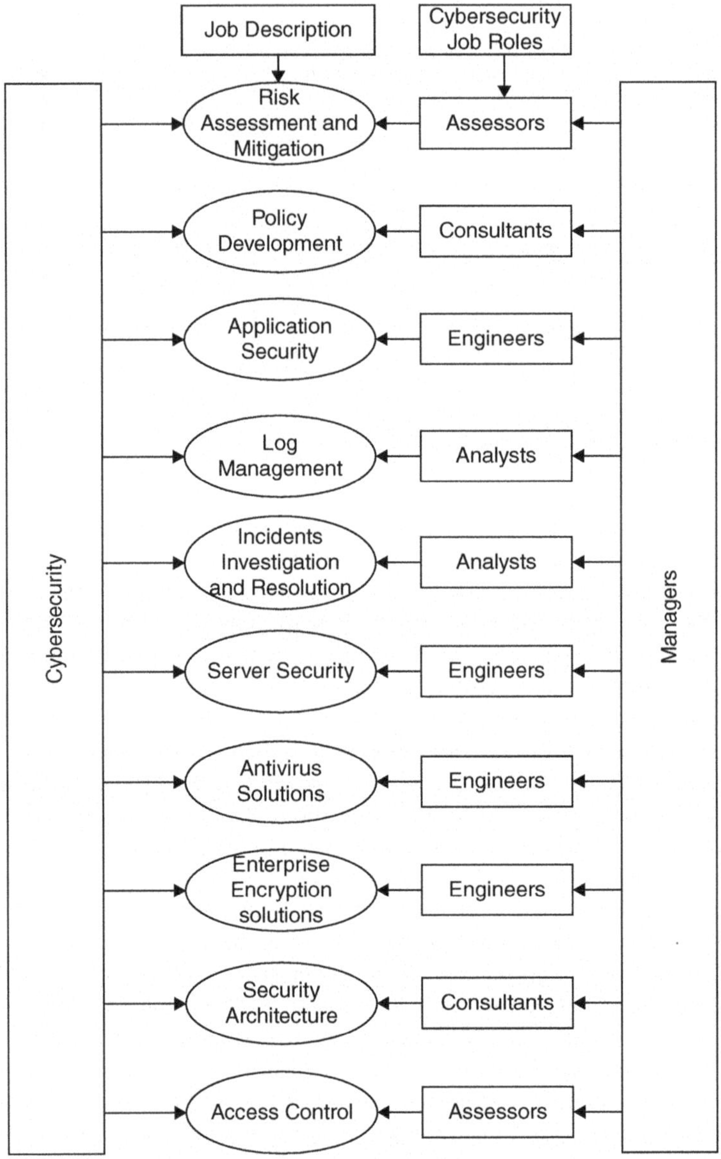

Fig. 4 Cybersecurity skills gap in terms of job descriptions

and citizens to the ever-present danger of cyber threats and cyberattacks, and the skills gap inadvertently created [8].

Time-Blended Problem

The growing dependency on the cyberspace for more interconnectivity creates peculiar challenges of a different kind. The transactions that are done via the Internet are open to a brand new form of threat that is growing in complexity, sophistication and ingenuity. The full benefit of a connected world on timely delivery of services and information makes it impossible to leave the cyberspace for more archaic means of information transfer that are a lot slower with data integrity compliance. In cyberspace with limitless possibilities, there is the ever-burgeoning concern of vulnerabilities that can be exploited by a brand new set of opportunists, criminals and state-sponsored agencies that specialise in making organisations and governments dread the use of the tremendous advantage of the Internet. Just over a century ago, it was almost nonsensical to discuss or suggest the type of vulnerabilities that the modern world has to grapple with presently and in the future. And ironically, some of the critical terms used in ancient warfare and security activities have found their way into the present and ongoing efforts to combat and plug vulnerabilities in cyberspace. Words like Firewall and Trojan Horse borrow underlying connotations from the real-world legends that gave it relevance in the context of the evolution of new information and security paradigms. The uniqueness of possible emergencies, crises and outages in cyberspace is arguably an unequaled problem that governments, organisations and other stakeholders are in a hurry to solve. The amount, in monetary terms, lost to the schemes that exploit vulnerabilities in the cyberspace has reached explosive proportions. Organisations have had full online functionality handicap similar to real-world assassinations and temporary holdups, blackouts and kidnappings. The variety of security issues faced by the cyberspace are so terrifying that stakeholders count the billions in monetary terms that is spent to build secure systems as a fundamental necessity [9].

The peculiar challenges bedevilling the cyberspace pave the way for the niche area of cybersecurity. The integrity of data and its protection from unauthorised access is one of the main goals of cybersecurity. On the surface, it looks classy and well classified, but beneath it is a whole mountain of hidden ice that has given cyberspace stakeholders a considerable concern. The ability to secure data in an evolutionary manner that keeps pace with the ingenuity of hackers and cybercriminals is pure talent. Cybersecurity talents are equipped with fundamental and high levels of technical skills to tackle the issue of safety in cyberspace. These talents ensure that the systems of organisations and governments are secured and free of vulnerabilities that could be exploited by underhand elements in the cyberspace. The goal has always been to gather a sufficient pool of highly skilled cybersecurity professionals with the necessary technical skills to make the cyberspace secure. The discovery, retention and continuing training of cybersecurity talent are driven by the combined efforts of industry and academia. But the demand for cybersecurity talent keeps outpacing the supply of the same. Throughout the world, thousands of cybersecurity-related vacancies remain unfilled because graduates or individuals with the skills needed to take on the cybersecurity job roles advertised are rare to find. The sincere efforts of cybersecurity educators fall short by a vast margin. With each passing year, cybersecurity vacancies continue to increase at an exponential rate, and cybersecurity talents to meet the growing demand for niche cybersecurity skills keep falling short [10].

The gap has since been identified, and while many areas of cybersecurity have professionals in encouraging supplies, an enormous gap remains to be filled. The gaps are thoroughly skills-based and have strong emphasis on practical, hands-on and deep technical skills. The skills gap has its roots in fundamental computer design and architecture. It also has a "blanket covering" that has to do with computer coding and data recovery. The main aim of the cybersecurity push is to blend security into every aspect of information and communication technology. Strong emphasis is laid on building secure systems, writing secure codes and recovering from already suffered cyberattacks. These skills are in very short supply and have given all cyberspace stakeholder cause for great concern. The gap is so much so that despite sincere efforts to fill it, it persists. The gaps are made worse by the "lightning speed" at which changes are occurring in the cyberspace and the amorphous nature of cyber threats that continue to grow in complexity and volume. Cybersecurity experts with the skills needed to hunt down vulnerabilities in secure software systems are so much in short supply that industry players are beginning to regard graduates of cybersecurity as ill-baked and grossly unprepared for the "time-blended" challenge of cyber threats. The design and implementation of secure systems is another skills gap that needs to be filled in cybersecurity. The way communication systems are designed presently and in the future would give security an upper hand in priority for the deployment of information infrastructure. The talents to deliver secure systems also fall short of the demand and have created a vacuum that needs to be filled in earnest; the skill of defending systems against cyberattacks and threats is also falling short of the total worldwide demand and has left many organisations vulnerable to the antics of cybercriminals [11].

The root of the problem is in the type of graduates and professionals churned out by universities and training bodies. Top talent managers and human capital experts all over the world lament the unpreparedness of cybersecurity graduates for the real-world challenges cybersecurity brings. This can be traced to the manner of training students received and the inadequacy of the education process to meet the learning outcomes for dependable and prepared cybersecurity professionals. Apart from the shortfall in core technical skills, there is the shortfall in soft skills required by employers of cybersecurity graduates. The ability to communicate effectively, work as part of a team and write excellent and insightful reports has been ranked as even more important than core technical skills. These soft skills are not possessed by the majority of cybersecurity graduates, and employers are not able to place these recent cybersecurity graduates into job roles that require the delivery of outcomes on the go. Even the metrics to gauge how effectively students may be able to perform in the real cybersecurity workplace are almost non-existent. These metrics make it possible to assess the progress of students in the learning process and evaluate preparedness for the cybersecurity workplace. It remains glaring that filling the existing skills gap in cybersecurity is a significant hurdle that must be completed successfully if the security of information in cyberspace is to be guaranteed [12].

The fundamentals of cybersecurity in individual tracks or career paths also have insufficient depth to deliver job-ready cyber-security graduates. The fundamentals of computer organisation and architecture, data management, secure coding, Linux- and Windows-based systems, low-level programming languages, operating system internals and the core basics of exploitation techniques are not taught to a large number of cybersecurity students, hence the large gap in skill-sets of cybersecurity graduates. These baseline skills have been the complaints of several human capital managers as seen to be inadequate in supply in recent cybersecurity graduates. The world of cybersecurity is enabled by a deep understanding of the fundamentals of networks, hardware and software. Many tasks in cybersecurity are enabled only by a deep technical background and thorough knowledge of the fundamentals of computer systems and networks [13].

The CybEd Curriculum Challenges

Cybersecurity forms a part of information assurance and security research. It is evolving at an exponential rate with a dire need to continually improve, transform and update its body of knowledge. The capacities for advanced research, strong leadership and sound judgment in the area of cybersecurity severely lag the pace at which changes occur in the field. The cybersecurity skills gap can be filled with a proactive and intentional curriculum that addresses the functional requirements of cybersecurity [1].

There are lots of industries that need the services of cybersecurity professionals, and the curriculum for cybersecurity would be more useful if it directly addresses the skills gap and the knowledge required for the industry. A well-drafted cybersecurity curriculum would lead to adequate education and favourable training policies that are appealing to implement. The curriculum must address the institutional, legal, political and social dimensions of cybersecurity. Emphasis should be laid on capability-based approaches that cater to community education, industry partnerships, information sharing, government agencies, international engagements and criminal justice [6].

A curriculum that boosts cyber partnership between government, researchers and industry would go a long way to strengthen leadership and tackle the cybersecurity skills gap. The curriculum must also encourage homegrown solutions that would yield prosperous entrepreneurial startups and indigenous expertise in the area of cybersecurity. Professional bodies like the Institute of Electrical and Electronics Engineers (IEEE) can provide a helping hand in drafting effective curricular for cybersecurity education.

The curriculum must have an implementation plan and inputs from the industry, academia and government, as presented in the cybersecurity education triad shown in Fig. 5. The industry, for the most part, provides industry certifications; the

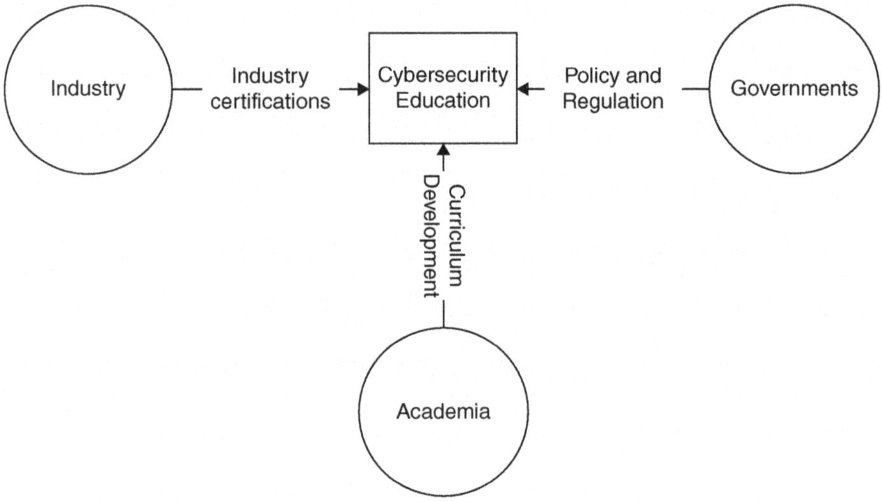

Fig. 5 The cybersecurity education triad

government offers regulation and policy implementation. While academia focuses on curriculum development, the government must continue to supply strong "top-to-bottom" leadership to ensure the standardisation of the cybersecurity education curriculum and its regulation. Funds must be committed and invested in seeing that the cybersecurity curriculum policy is converted into visible outcomes.

The curriculum should be designed in such a way that it is tailored to match students to the appropriate cybersecurity learning track that would engender satisfaction, self-motivation and happiness [9].

Incorporate into the curriculum the dicey issue of ethics. All ethical issues surrounding the teaching of ethical hacking, quality assurance and application design should be addressed by the curriculum. This will give students ethical reasoning skills and foster the use of cybersecurity skills responsibly. In fact, the ever-rising threat of cyberattacks makes it imperative to actually teach students the ethics of cybersecurity [10].

As a multidisciplinary field of study, the cybersecurity education curriculum covers all areas of technical and social knowledge that enable the free flow of reliable data on cybersecurity between the government, industry and academia. The training curriculum for cybersecurity addresses the proactive and reactive as well as the defensive and offensive areas of concern [11]. The product of such a curriculum is learning outcomes-based and effective.

The cybersecurity expertise model shown in Fig. 6 gives a pictorial representation of the level of expertise expected if cybersecurity is proactively integrated into the learning curriculum at all levels of education. The learning outcomes are accurately matched to the level of education, and the cybersecurity expertise expected at the primary level is the "rookie" or one who is just starting. The "enthusiasts"

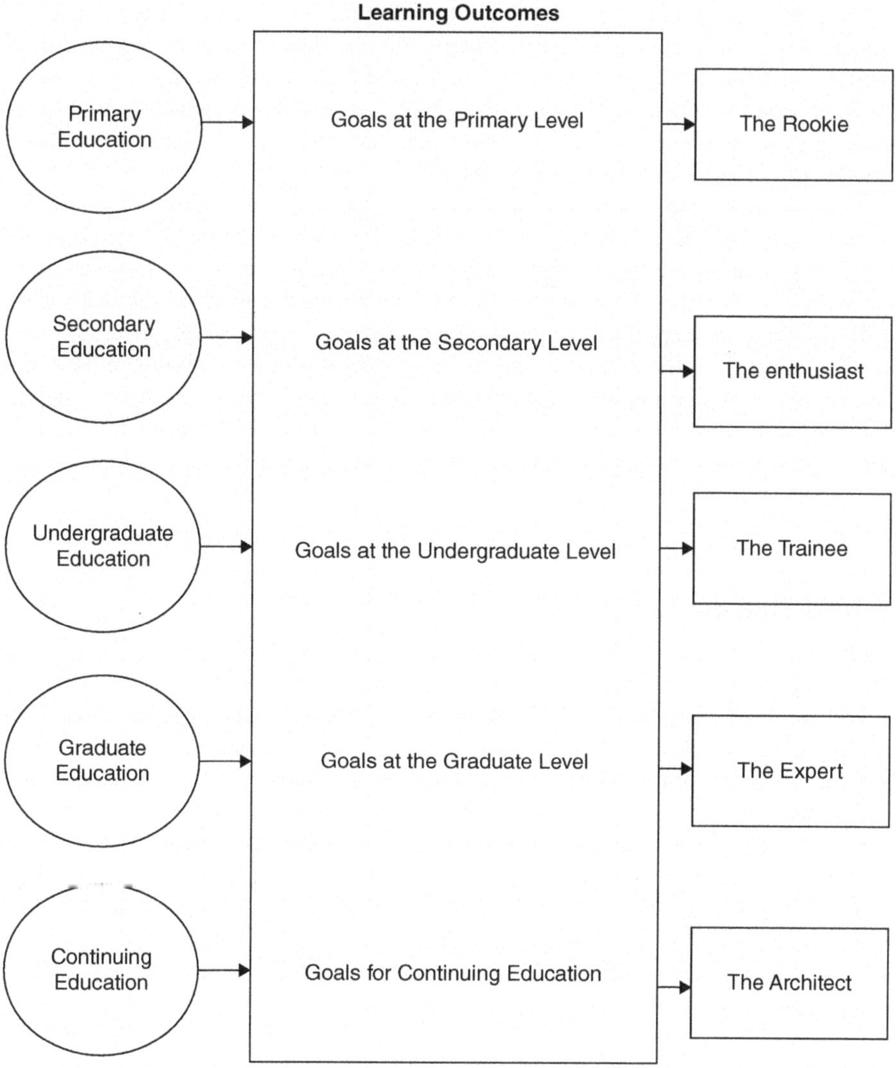

Fig. 6 The learning outcomes-based curriculum

at the secondary level have enough knowledge to partake in competitions and engage actively in simulated cybersecurity scenarios, as seen in virtual classes and games. The "trainees" are fresh graduates who are taken into the entry levels of various organisations and schools. The "experts" are holders of advanced degrees, industry certifications and years of experience tackling cybersecurity challenges. The "architects" are scientists, researchers and educators who are committed to lifelong learning and advanced research in cybersecurity.

As a way of guiding the development of effective cybersecurity education curriculum, policymakers, industry leaders and the academia must satisfactorily answer questions bordering on the scope of the curriculum, learning outcomes, levels at which cybersecurity education can be accessed, acceptable methods of teaching cybersecurity, assessment of students' performance and goals for graduate cybersecurity real-world work preparedness [12].

A curriculum that prioritises prevention, detection and prompt response to cybersecurity issues usually makes provisions for studying human behaviour as it relates to cybersecurity issues. Cybersecurity education at all levels fosters the discovery and retention of talents as well as provides the means to fill the skills gap.

To fill the skills gap, the curriculum must encourage hands-on practices, assessments using real-world situations and industry-rooted problem-solving. Where the creation of a new curriculum is not possible, Bloom's taxonomy and Webb's depth of knowledge may be used to integrate cybersecurity into the already existing curriculum in other related subject areas without increasing students' course credit load [13].

CybEd-Holistic Education

The core goals of cybersecurity education have been identified as innovation, critical thinking, creativity and problem-solving. These goals form the basics of learning outcomes design in cybersecurity education. Cybersecurity education aims to prepare students for academic and professional positions. Collaborative learning that emphasises teamwork, team-based projects and synergy should, as a matter of importance, be incorporated into the goals of cybersecurity education [1].

Higher institutions of learning can provide the platform to cater to emerging issues and challenges in cybersecurity and provide the channel for grooming talents in advanced research. Models for higher education and continuing research in cybersecurity should be designed with the most pressing and future needs of the industry and academia in mind.

Cybersecurity education must address all topics and areas of knowledge to their full depths in varying levels of difficulties [14]. This would pave the way for the classification of CybEd into studies at the primary level to the doctoral research level. The course learning outcome should adequately and sufficiently meet the professional and career demands of cybersecurity. The goals of CybEd must be clearly spelled out and accurately matched to the corresponding learning outcomes.

Provisions should be made for long-distance learning. Online classes on cybersecurity should be encouraged especially for areas where there is a dearth of qualified cybersecurity educators. The ubiquity of wireless communication and increased capacity of telecommunications infrastructure should be harnessed to build a solid online education platform to raise cybersecurity professionals to fill the widening skills gap in developing nations. Although online education may, to some extent, hinder teamwork which is a core requirement in cybersecurity professionals,

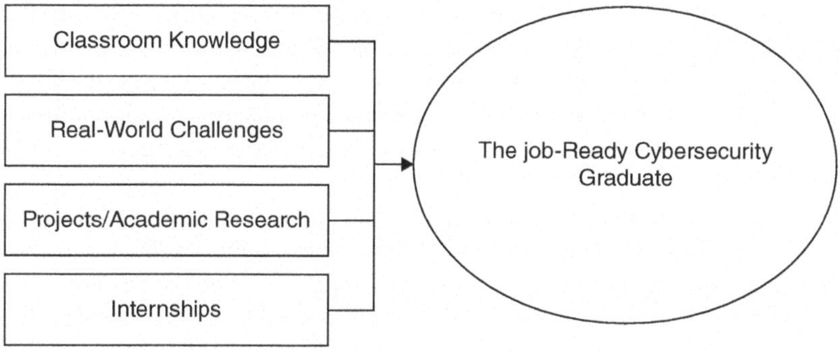

Fig. 7 The job-ready cybersecurity graduate

it can be ameliorated by assignments and mini-projects that can be worked on collaboratively by a team of at least three (3) and at most five (5) persons via online communication channels. Online cybersecurity education must avoid the pitfall of inadequate assessment and lack of depth to achieve the goal of filling the skills gap in cybersecurity education [4].

There is also the need to begin specialised education to address niche areas of cybersecurity such as digital forensics. A lot of challenges are emerging in the field of cybersecurity that requires an investigation that can only be provided by specialists [6].

Centres of excellence can be created to tackle the skills gap in cybersecurity. These centres of excellence can be set up in universities, polytechnics and colleges with the sole aim of training cybersecurity professionals who are instantly ready to fill the skills gap in cybersecurity upon graduation.

Cybersecurity education must be taken as a continuous process that actively seeks students that would be groomed to become job-ready graduates who have an intense personal interest in their chosen cybersecurity career and internal drive for lifelong learning. This can be achieved by recruiting recognised cybersecurity industry experts that can act as instructors, visiting faculty, role models and mentors. Students should tackle real-world cybersecurity challenges from the industry and engage in academic research to develop problem-solving skills. Internships should also be made compulsory to enable the immediate application of theoretical concepts taught in the classroom. The job-ready cybersecurity graduate is described in Fig. 7. The inputs from the class, internship placements, personal studies, projects/academic researches and sample real-world challenges all contribute to making cybersecurity graduates that can hit the cybersecurity workplace "running".

The progress of students must be measured to ascertain the level of competence garnered over the learning period under consideration. Cybersecurity education should be tailored to raise professionals who are adept at the theoretical concepts and the practical implications of acquired knowledge to the real cybersecurity workplace [9].

The innovative use of games, tournaments/competitions, virtual laboratories and rewards makes CybEd fun and downplays its perception as a tough and challenging area of study. Hacking laboratories, both online and offline, can be used to train professionals to cover the digital forensics skills gap. Virtual tutorials and near-real-world challenges can also be used to solidify cybersecurity students' skills and capabilities [15].

CybEd can be classified as live (synchronous) and on-demand (asynchronous). Field trips, physical classroom sessions, virtual classrooms, webinars, mentoring and coaching are contemporary forms of CybEd that can be utilised to train students to fill the cybersecurity skills gap. On-demand of asynchronous CybEd enables collaboration and community efforts to tackle cybersecurity challenges. Portals/websites, blogs, threaded discussions, voice-over Internet protocol, chats and videos (multimedia) are parts of asynchronous CybEd [5].

CybEd for entrepreneurship endeavours or technology-innovation goals is an excellent tool for filling the skills gap in cybersecurity. Students are encouraged to think like entrepreneurs to create solutions to challenging cybersecurity problems. This type of CybEd would create new career directions, sharpen technology skills, create startups and foster partnerships between industry leaders, academia and governments [16].

In developing nations where there are no facilities to groom cybersecurity skills to meet market demands and fill the skills gap, CybEd should leverage on the use of virtual services such as high-fidelity live exercises (HiFliX) [17] to mimic present and future cybersecurity challenges. HiFLiX and others like it depend solely on communication devices and a good Internet connection to deliver its contents to users. This is an excellent pipeline for developing and retaining cybersecurity talents in developing countries. However, industry, government and academia must unite to bring such powerful tools to cybersecurity students in developing nations [18]. The students in developing countries must be proactively engaged from the outset in the real-world application of cybersecurity skills learned in the classroom, and a project-based approach may be adopted when it becomes challenging to get internship placements for all students in the industry [19].

Attention has to be paid to developing human capacity, institutional reformation, robust legal framework, affordability and private-public partnerships to fill the cybersecurity skills gap in developing nations [20]. For the most part, developing countries have partners that assist with the provision of educators, facilities and funding to make CybEd accessible. The success of these partnerships dramatically depends on the free flow of highly reliable information on the state of cybersecurity education in developing nations.

Conclusion

In this study, the existing skills gap in cybersecurity was highlighted. At the core of closing the skills, gap is the role played by cybersecurity education. Suggestions and ideas on the creation of effective curricula for CybEd were given along with solid plans for its implementation in developing countries. CybEd emphasises learning outcomes-driven curricula that prioritise the exposure of students to real-world practical cybersecurity problems. This exposure gives students ample opportunities to apply knowledge gained in the classroom via internships or near-real simulations through software applications. The critical thinking, problem-solving, leadership and collaboration skills required by most employers of cybersecurity graduates can also be honed by learning outcomes-based CybEd that is implemented from the primary-level to the tertiary-level education and driven by the collaborative and sustained effort of industry leaders, governments and the academia.

References

1. P. Wang, Designing a doctoral level cybersecurity course. Issues in Information Systems **19**(1) (2018)
2. A. Parker, I. Brown, Skills requirements for cyber security professionals: A content analysis of job descriptions in South Africa, in *International Information Security Conference*, (2018, Springer, Berlin), pp. 176–192
3. P. Choejey, C. C. Fung, K. W. Wong, D. Murray, and H. Xie, Cybersecurity practices for e-Government: an assessment in Bhutan, in *The 10th International Conference on e-Business, Bangkok, Thailand* (2015)
4. P. Wang, R. Sbeit, A constructive team project model for online cybersecurity education. Issues in Information Systems **18**(3) (2017)
5. R. Creutzburg, Cybersecurity and forensic challenges-a bibliographic review. Electronic Imaging **2018**(6), 100-1-100-16 (2018)
6. J. Slay, G. Austin, Development in training and education for Australian cyber security. Journal of The Colloquium for Information System Security Education **5**(2), 27–27 (2018)
7. S.N. Mogoane, S. Kabanda, Challenges in information and Cybersecurity program offering at higher education institutions, in *Proceedings of 4th International Conference on the, 2019, Vol. 12*, pp. 202–212
8. M. McBride, L. Carter, M. Warkentin, *One Size doesn't Fit all: Cybersecurity Training Should Be Customized* (Institute for Homeland Security Solution, 2012). https://sites.duke.edu/ihss/files/2011/12/CyberSecurity_2page-summary_mcbride-2012.pdf
9. B.E. Endicott-Popovsky, V.M. Popovsky, Searching and developing Cybersecurity talent. Journal of The Colloquium for Information System Security Education **5**(2), 17–17 (2018)
10. N. Radziwill, J. Romano, D. Shorter, and M. Benton, The ethics of hacking: Should it be taught?, arXiv preprint arXiv:1512.02707, 2015
11. G. C. Kessler and J. D. Ramsay, A proposed curriculum in cybersecurity education targeting homeland security students, in *2014 47th Hawaii International Conference on System Sciences*, 2014, pp. 4932–4937: IEEE
12. D. Mouheb, S. Abbas, M. Merabti, Cybersecurity curriculum design: A survey, in *Transactions on Edutainment XV*, (Springer, Berlin, 2019), pp. 93–107
13. M.A. Harris, Using Bloom's and Webb's taxonomies to integrate emerging Cybersecurity topics into a computing curriculum. J. Inf. Syst. Educ. **26**(3), 4 (2019)
14. A. P. Henry, *Mastering the Cyber Security Skills Crisis* (2017)
15. H. Aldawood and G. Skinner, An academic review of current industrial and commercial cyber security social engineering solutions, in *Proceedings of the 3rd International Conference on Cryptography, Security and Privacy*, 2019, pp. 110–115

16. G. Javidi, E. Sheybani, and Z. Pieri, A Holistic Approach to K12 Cybersecurity Education, in *Proceedings of the International Conference on Frontiers in Education: Computer Science and Computer Engineering (FECS)*, 2019, pp. 77–80: The Steering Committee of The World Congress in Computer Science, Computer
17. J. Sigholm, G. Falco, and A. Viswanathan, Enhancing Cybersecurity Education through High-Fidelity Live Exercises (HiFLiX), in *52nd Hawaii International Conference on System Sciences, January 8–11, 2019, Grand Wailea, Maui, USA*, 2019, pp. 7553–7562: IEEE conference proceedings
18. B. Hoanca, B. Craig, Building a K-16-industry partnership to train IT professionals. J. Inf. Syst. Educ. **30**(4), 232 (2019)
19. C. Johnson, University of South Wales national cyber security academy–creating cyber graduates who can 'hit the ground running': An innovative project based approach. Higher Education Pedagogies **4**(1), 300–303 (2019)
20. L. P. Muller, *Cyber Security Capacity Building In Developing Countries: Challenges And Opportunities* (2015)

Index

K. Daimi, G. Francia III (eds.), *Innovations in Cybersecurity Education*,
https://doi.org/10.1007/978-3-030-50244-7

K

The manufacturer's authorised representative in the EU is Springer Nature Customer Service Centre GmbH, Europaplatz 3, 69115 Heidelberg, Germany. If you have any concerns regarding our products, please contact ProductSafety@springernature.com

Printed and bound by CPI Group (UK) Ltd, Croydon, CR0 4YY
30/09/2026
02225508-0015